LEIBNIZ IN HIS WORLD

Leibniz in His World

THE MAKING OF A SAVANT

Audrey Borowski

PRINCETON UNIVERSITY PRESS
PRINCETON & OXFORD

Copyright © 2024 by Princeton University Press

Princeton University Press is committed to the protection of copyright and the intellectual property our authors entrust to us. Copyright promotes the progress and integrity of knowledge created by humans. Thank you for supporting free speech and the global exchange of ideas by purchasing an authorized edition of this book. If you wish to reproduce or distribute any part of it in any form, please obtain permission.

Requests for permission to reproduce material from this work should be sent to permissions@press.princeton.edu

Published by Princeton University Press
41 William Street, Princeton, New Jersey 08540
99 Banbury Road, Oxford OX2 6JX

press.princeton.edu

All Rights Reserved

ISBN 978-0-691-26074-7
ISBN (e-book) 978-0-691-26086-0

British Library Cataloging-in-Publication Data is available

Editorial: Ben Tate and Josh Drake
Production Editorial: Jenny Wolkowicki
Jacket design: Heather Hansen
Production: Danielle Amatucci
Publicity: William Pagdatoon and Charlotte Coyne
Copyeditor: Anita O'Brien

Jacket image: Classic Image / Alamy Stock Photo

This book has been composed in Miller

Printed in the United States of America

10 9 8 7 6 5 4 3 2 1

To Bernard, who works in mysterious ways

Au milieu de l'hiver, j'ai découvert en moi un invincible été.
—ALBERT CAMUS

Qu'est-ce que le bonheur sinon le simple accord entre un être et l'existence qu'il mène?
—ALBERT CAMUS

CONTENTS

Abbreviations · xiii

Introduction   1

PART I   13

CHAPTER 1 Leibniz with Boineburg: Acquiring a New
           Patron, Negotiating Contingency   23

CHAPTER 2 Encountering Paris   33
           A German Publicist and Agent in Paris   37

CHAPTER 3 Learning in Paris   45

CHAPTER 4 Engaging with Cartesianism   53
           'Le grand Arnauld'   54
           Foucher   62

CHAPTER 5 Defending a Particular Conception of the
           Republic of Letters: Leibniz on the *Querelle
           des anciens et des modernes* and the Republic
           of Letters   69
           *Placing Erudition and History in Defence of
           Religion*   76

CHAPTER 6 The Struggle for the Heart of the Republic of
           Letters: Science and the State   83
           'Funny Thought'   89

CHAPTER 7 Infiltrating Colbert's State Republic of Letters   95
           *Louis Ferrand and the Pursuit of Abulfeda's
           Geography*   101

[x] CONTENTS

	Pierre de Carcavy and Leibniz's Calculating Machine	108
CHAPTER 8	Finding a New Patron	119
	Learning from France for Germany	125

PART II   127

CHAPTER 9	Leibniz in His Correspondence with Duke Johann Friedrich (1676–79): The Sincerity of a Projector	129
	Leibniz as Court Librarian	131
	Widening His Sphere of Influence	134
	A 'Walking Encyclopedia' at the Court of Hanover	136
	Leibniz's Own Projects and Specialness	139
	Secrecy and Control	148
	Between Artifice and Sincerity	153
CHAPTER 10	Leibniz with His Fellow Projectors	161
	Johann Daniel Crafft and the Phosphorus Affair	164
	Holding up a Mirror to Himself: Leibniz and Johann Joachim Becher	175
	Leibniz's Double Standards	185
CHAPTER 11	Establishing a European Network from Hanover	191
	Building the Ducal Library: Leibniz with Brosseau	192
	Leibniz's homme à tout faire: *Adolf Hansen*	193
	Leibniz with Christian Philipp	200
	An Impasse in Germany: Leibniz with Hermann Conring	204

CHAPTER 12	Setting up a Scientific and Intellectual Network	211
	Sharing Scientific and Technical Curiosities	214
	Setting Himself up as a Master Informer	218
CHAPTER 13	Facing Dead-ends in France	225
	Instrumentalizing Phosphorus	225
	Conclusion	237

Acknowledgements · 243
Appendix: Translations of Key Texts Discussed · 245
Bibliography · 271
Index · 289

ABBREVIATIONS

A   Leibniz, Gottfried Wilhelm (1923–). *Sämtliche Schriften und Briefe*, ed. Preussische [later Deutsche] Akademie der Wissenschaften. 68 vols. in 8 series to date. Darmstadt [later Berlin]: Reichl [later Akademie Verlag and De Gruyter]. (Cited by series, volume, and page number, e.g., A II, 1:415. Series 2, vol. 1, is cited from the revised second ed., Berlin: Akademie Verlag, 2006.)

LEIBNIZ-ARCHIV   Gottfried-Wilhelm-Leibniz Bibliothek, Niedersächsische Landesbibliothek, Hanover.

MISC. BERL.   Leibniz, Gottfried Wilhelm (1768 [1710]). 'Historia Inventionis Phosphori' [first published in *Miscellanea Berolinensia*, 1710], in *Opera omnia*, ed. Louis Dutens, vol. 2, Geneva: Fratres de Tournes.

Introduction

'HIS TEMPERAMENT APPEARS to be neither purely sanguine, nor choleric, nor nurturing, nor merely melancholic. . . . Ever since he was a boy he led a sedentary lifestyle and did little exercise. Since his youth he began to read a lot and think about a lot of things, so that he was self-taught in most of his knowledge; he also burns with desire to penetrate everything deeper than usual and to invent something new. He does not begin work without fear, but he continues it courageously. . . . Nature gave him an excellent power of invention and judgement and it will not be difficult for him to think up many things, to read (with understanding), to write, to speak extemporaneously, and to develop intellectual concepts, if necessary, through persistent reflection to investigate to the very bottom of what leads me to conclude that he has a dry and spiritual brain.'[1] The Leibniz who described himself in these terms, at the age of around 50, was an established courtier, renowned throughout Europe as a philosopher and mathematician. But his path to this fame was not straightforward.

While Leibniz scholarship has been increasing for the past thirty years, it has focused primarily on his philosophical thought, and relatively few attempts have been made to try to relate Leibniz to his time and social and intellectual milieus.[2] Recent scholarship has contributed to the mythologization of Leibniz as a 'great natural genius' who could apply himself with 'equal vivacity' to any field, as Bernard de Fontenelle put it in his *Eloge* of 1717. According to this type of narrative, the full breadth of Leibniz's work

---

1. Leibniz, 'Imago Leibnitii', in Müller and Kronert, *Leben und Werk* (1969), 1–2.
2. Excellent examples of recent Leibniz scholarship include Antognazza, *Leibniz* (2009); Brown, *The Young Leibniz* (1999); Look, *Continuum Companion to Leibniz* (2010); and Nadler, *The Best of All Possible Worlds* (2008).

and heterogeneous activities can be read in light of an underlying 'unity'.[3] Other approaches have consisted in contrasting his public and private personae, or in interpreting him as esoteric, opportunistic, even Machiavellian.[4] These accounts, while often valid on their own terms, espouse reductive approaches to Leibniz that confine him to a vacuum of ideas without situating him in his social, cultural, or political contexts.

Leibniz is traditionally depicted as a radical rationalist and philosopher who was detached from worldly concerns and impervious to the lot of his fellow men. Although scholarly practice, notably in the history of science, has sought in the past fifty years to break away from such ahistorical interpretations, Leibniz has curiously remained largely immune from this trend. Yet we need to rediscover him within the context of the intellectual effervescence of late seventeenth- and early eighteenth-century European cultural debates, not least because many of the texts on which Leibniz's present reputation are based were not published until long after his death. In fact, in his lifetime and in particular during his so-called Paris sojourn (1672–76), Leibniz was better known as a diplomat, a lawyer, a theologian, and above all a mathematician than as a philosopher.[5]

This book endeavours to rediscover Leibniz as a participant in the learned and courtly communities he frequented in the years 1672–79— that is, the years of his Paris sojourn and those immediately following, when he found employment back in Germany at the court of Duke Johann Friedrich of Hanover. Leibniz arrived in Paris in early March 1672 on a diplomatic mission on behalf of his patron, Johann Christian von Boineburg, aiming with his infamous 'Egyptian plan' to divert Louis XIV from his plans for military expansion in Europe, as will be examined in the first part. Although ultimately unsuccessful, this mission allowed Leibniz to remain in Paris and take full advantage of its intellectual and scientific life at a time when—with the recent creation of several *académies* and the publication of such seminal works as Malebranche's *De la Recherche de la Vérité* and Boileau's *Art Poétique*—the city was the most sophisticated and advanced centre of European culture. I endeavour to clarify how his visit to Paris constituted an important moment in his development, and to what extent it provided him with a new impulse upon his return to Germany. More broadly, this book sets out to examine how those six years affected him and his work, how they were formative years for him,

---

3. Antognazza, *Leibniz* (2009), 5.
4. Russell, *Philosophy of Leibniz* (1900); Stewart, *The Courtier and the Heretic* (2006); Snyder, *Dissimulation* (2009).
5. Garber, 'Leibniz's Reputation' (2009b), 281.

and how he navigated the various worlds he encountered and with what success. I have chosen the years 1672 and 1679 because they demarcate, respectively, Leibniz's departure for Paris from Mainz and the death of Duke Johann Friedrich, one of his more supportive patrons.

'Internalist' readings tend to isolate the evolution of science and portray its pursuit in terms of a disinterested quest for truth, divorced from mundane or petty concerns.[6] Such assessments tend to divorce the production of ideas from their historical contexts and to occlude the forces that have helped shape them. In particular, they obscure the complex dynamics at play in the practice of 'normal science' in Leibniz's time.[7] Far from being a solitary activity, science and more broadly the pursuit of knowledge were fundamentally 'dialogical', their progress lying in a constant stream of exchanges, collaboration, and cooperation between likeminded individuals labouring towards the common ideal of 'citizenship of the mind'.[8] In this manner, the constitution of knowledge was inextricably linked to power structures and established norms, both within and outside the Republic of Letters. Much of Leibniz scholarship has steered clear of delving into the social, cultural, and political circumstances that affected the constitution of knowledge. In early modern societies, one's credibility depended largely on one's social status. Although scholars and patrons alike cultivated the image of pure scientific freedom and the unencumbered pursuit of knowledge, the ability to secure patronage and, increasingly from the 1670s onward, attachment to a research institution played a decisive role in scholars' ability to establish themselves on the intellectual map.[9] Even the most disinterested scholar needed to be a courtier to some extent and to adhere carefully to the norms and prescribed conduct of his intellectual community. Thus, far from being 'pure', the pursuit of science and scholarship were heavily conditioned by the social structures in which they were conducted.[10]

Leibniz was an ambiguous figure whose roles and personae, reflecting his various milieus, particular tasks, and personal aspirations, were far from static. His work cannot be read in isolation from those factors. Exploring how he sought to position himself and navigate the different

---

6. See Shapin, 'The Mind Is Its Own Place' (1991).
7. Kuhn, *The Structure of Scientific Revolutions* (1962).
8. Laerke, *Les Lumières de Leibniz* (2015), 50; see also Ramati, 'Harmony at a Distance' (1996), 451; Blay and Halleux, *La science classique* (1998), 26; Fumaroli, *La République des Lettres* (2015), 73.
9. Stegeman, *Patronage and Services* (2005), examines the patronage of savants.
10. See Shapin, *Never Pure* (2010).

intellectual, scientific, and courtly worlds, this book will reveal him to be a man of multiple identities. In particular, I hope to clarify how, in his early career, Leibniz articulated a niche for himself at the intersection of the sociopolitical and scientific realms. This analysis—based on a more extensive examination than has been attempted previously of his and his peers' correspondence (in French, German, and Latin) during the relevant years—cultivates a sensitivity to the many contextually related dimensions that characterized his personality. My examination will be confined primarily to volumes I, 1; I, 2; II, 1; III, 1, and III, 2, of the ongoing standard *Akademie-Ausgabe* (1923–).

In this work I shall be attending in particular to the following questions. How did Leibniz establish himself as a young scholar? How did his peers perceive him? Which social and professional identities did he endorse at the time? To what extent was he a 'projector' and perceived as such? How much flexibility—social, political, and epistemological—was he afforded within the Republic of Letters and the absolutist political realms he frequented? To what extent did his positions as scholar, courtier, and adviser overlap, and to what extent did they diverge? How did these multiple roles and identities coexist? What kind of rationality did he propound, and did it bring him in conflict with competing intellectual or scholarly models? What was his own view of the Republic of Letters? How did he understand Colbert's enterprise? To what extent did his intellectual aims overlap with his professional duties? And what ultimately drove his various projects?

Considering the many ambiguities and layers at play, this monograph does not pretend to offer a comprehensive exposition of Leibniz's thought or life in the years 1672–79, let alone of the Republic of Letters or of the different milieus in which his thought evolved. It does not purport to be exhaustive but seeks rather to shed light on aspects of Leibniz's life, activities, and modi operandi that have received little or less attention. For this reason, I have excluded or touched only briefly on certain well-trodden topics, including Leibniz's mathematical peregrinations and his formulation of the calculus, his trip to London in 1673, the details of the elaboration of his calculating machine, his encounter with Spinoza, his dispute with Newton, as well as his project towards theological reconciliation, all of which have been covered extensively by remarkable scholars.[11]

11. This includes Mogens Laerke on Leibniz's relationship with Spinoza, Joseph E. Hofmann on Leibniz's mathematical work in Paris, Philip Beeley on his trip to London, Matthew L. Jones on his building of the calculating machine, and Maria Rosa Antognazza on his lifelong work on theological reconciliation.

Any substantial reiteration on my part would be largely superfluous. For reasons of space, I have also passed over elements that feature minimally or not at all in the correspondence, such as the University of Paris, the various Parisian political factions operating at the time, and Leibniz's fledging relationship with Bossuet. Crucially, too, the immensity of the *Nachlass*—which for this period alone spans thousands of pages—the time elapsed, and the frequent gaps in information owing to the number of missing letters, notably in Leibniz's Parisian period, rule out producing categorical statements and definitive conclusions. Accordingly, I have sought to strike a balance between factual certainty and conjecture, and propose to offer specific observations and tentative conclusions which will, I hope, lay the foundations for future research.

Because the relationship between Leibniz and the Republic of Letters was a reciprocal one, I seek by examining Leibniz also to increase our understanding of the nature and workings of that community. The 'Republic of Letters' refers to a transnational community of scholars in the seventeenth and eighteenth centuries who shared the same passion and belief in the exchange and promotion of knowledge through the deployment of rational methods. Transcending local, national, and even confessional difference, this 'universal auditorium' was governed by the principles of reciprocity and obligation and by the ideals of sociability, equality, *politesse*, and *honnêteté*.[12] Each member, at least in theory, had the sense of belonging to a virtual, albeit very real, larger and unified society, especially in times of political turmoil. Such, in any event, is the conventional view of the Republic of Letters, a view that reifies it by interpreting it as a stable institution unified by a set of shared principles and values. But my research has affinities with more recent scholarship that resists the seductive appeal of this beautiful and static ideal. In fact, the Republic of Letters in the second half of the seventeenth century was a highly unstable institution—multifarious, conflicted, diversified, and hierarchical—and it is important to rediscover its complexity and fluidity to understand better both the institution itself and Leibniz.[13]

This reassessment is all the more important considering Leibniz's increasingly central importance in the Republic of Letters. Even before

---

12. Blay and Halleux, *La science classique*, 25. See Borowski, 'Republic of Letters' (2021).

13. Goldgar, *Impolite Learning* (1995), 116. See also Wacquet, *Le Modèle francais* (1989); Wacquet and Bots, *La République des Lettres* (1997); and, on Bayle's attempt to model conduct in the Republic through his *Dictionnaire*, van der Lugt, *Bayle, Jurieu* (2016).

arriving in Paris, Leibniz had established a network of correspondents throughout Europe, many of whom he remained in contact with for the rest of his life. In fact, according to Voltaire, writing decades later in his *Age of Louis XIV*, it was none other than Leibniz, 'perhaps a man of the most universal learning in Europe', who had set in motion the universal network of communication that underpinned this intellectual revolution: 'There was never a more universal correspondence cultivated between philosophers' than in Leibniz's time.[14] An 'homme de réseaux', Leibniz was known to his peers predominantly through his journal articles and letter-writing.[15] His first publication, in the *Journal des Sçavans* of 25 March 1675, was on the topic of portable watches. Perhaps more than any of his contemporaries, Leibniz was a prolific correspondent, and it was above all in letters that he presented himself to his interlocutors.[16] In his letters, which range widely in length and geographical destination and were often circulated beyond their addressees, he presented his views and engaged in discussion on an impressive array of topics spanning the fields of theology, jurisprudence, physics, mathematics, and philosophy. Leibniz's correspondence has rightly been described as an 'integral part of his work',[17] and he continued building his extensive network of correspondents throughout his life.

Through this network he kept abreast of the learned world's *nova literaria* (including new publications, projects, and controversies) and other 'curious' topics, such as inventions and scientific discoveries, as well as information about European politics. This not only satisfied his personal need for information, but in being such an active and willing correspondent, sharing information and assisting his fellow savants, Leibniz so to speak 'performed' his membership of the Republic of Letters.[18]

Crucially, in such an economy of information, letters contained—indeed, constituted—'goods' in themselves, which Leibniz hoped to exchange where they were likely to be valued, and rewarded, the most.[19] By setting up, as we shall see, an effective information network and exploiting the huge value of access to information, Leibniz hoped to build his credit and acquire 'social capital' beyond that of his actual position.[20]

---

14. Voltaire, *Siècle de Louis XIV* (1874), 476.
15. Bertrand, 'Leibnitz et ses réseaux' (1999), 80.
16. Gädeke, 'Gottfried Wilhelm Leibniz' (2005), 263–64, 297; Gädeke, 'Leibniz lässt sich informieren' (2009), 85.
17. Utermöhlen, 'Der Briefwechsel' (1977), 90.
18. Gädeke, 'Gelehrtenkorrespondenz' (2020), 805.
19. Droste, *Im Dienst der Krone* (2006).
20. Gädeke, 'Leibniz lässt sich informieren', 86; Gädeke, 'Gottfried Wilhelm Leibniz', 282.

Consequently, he tended to handle information selectively and strategically, with a view to transcending his often limited official duties and leveraging his growing role within the Republic of Letters into a secure position at court or at the Académie des Sciences.[21] With such an institutional base, combining scholarship with power, Leibniz discerned, he would be better positioned—and have more resources available—to implement his schemes than if he remained merely a savant.[22]

Leibniz is often portrayed as an arch-rationalist: while this is true, it is not the whole story. He inhabited a world in which the intellectual and scientific marketplaces were increasingly saturated with 'curiosities', 'marvellous observations', and novel experiments, all of which challenged the limits of orthodox rationality.[23] Some inventions that were 'foolish, unreasonable and impossible' (*närrisch, irraisonable und ohnmöglich*) succeeded, while others that were well conceived ultimately failed and tested the bounds of possibility.[24] In addition to being a scholar and a courtier, Leibniz, perhaps crucially, operated as an aspiring capitalist and a projector.

The term 'project' (*projet*) first appeared in the Richelet (1680), Furetière (1690), and Académie française (1694) dictionaries with the definition 'design' (*dessein*) or 'plan', and more generally the projection or transfer of a tangible plan to the material world.[25]

Projectors were generally inventors or entrepreneurs who set out to gain the trust and backing of a powerful patron, such as a ruler or aristocratic investor, for what they claimed would be a prestigious and financially profitable venture capable of yielding practical benefits.[26] By instrumentalizing knowledge and technical expertise, early modern states

---

21. Gädeke, 'Gottfried Wilhelm Leibniz', 292–93. Gädeke distinguishes between problem-, strategy-, and communication-oriented letters, but Leibniz's letters often have all three orientations.

22. Gädeke, 'Gelehrtenkorrespondenz', 805.

23. Keller, *Knowledge and the Public Interest* (2015).

24. See Johann J. Becher, *Närrische Weißheit* (Frankfurt am Main, 1682), sig. A2, quoted in Breger, 'Becher, Leibniz' (2016), 37.

25. Smith, *The Business of Alchemy* (1994), 479; Zedler, *Grosses Vollständiges Universal-Lexikon* (1737), vol. 29. On projecting, see Keller and McCormick, 'Towards a History of Projects' (2016); and Ash, 'Expertise and the Early Modern State' (2014).

26. For more on projectors, see Borowski, 'Projectors' (2021); Graber, 'Du faiseur de projet au projet régulier' (2011); Keller, *Knowledge and the Public Interest*; Krajewski, *Projektemacher* (2004); Lazardzig, '"Masque der Possibilität"' (2006); Smith, *The Business of Alchemy*; Stanitzek, 'Projector' (2015); Troitzsch, 'Erfinder, Forscher und Projektemacher' (2004).

hoped to increase their power and become more prosperous.[27] Projectors in turn provided such expertise and positioned themselves as indispensable intermediaries, whose status at court would thus be legitimized.[28]

Projects varied widely in nature and involved almost all types of ventures, including engineering, mining, constructing factories, ameliorating the condition of the poor and infirm, banking, treasure hunting, building projects, and fabricating perpetual motion machines as well as various schemes for turning base metals into precious ones. Some projects, which were rife in the 1670s and 1680s, were of a more political or scholarly nature with a view to reform.[29] This type of activity became so pervasive towards the end of the seventeenth century that Daniel Defoe famously nicknamed his era the 'Projecting Age'.

Most projectors were inclined to present themselves as public advisers and counsellors—rather than as 'projectors' per se—in the context of nascent state absolutism and were parasitic on that office and its rhetoric, overtly placing their ideas or schemes in the service of a state or ruler.[30] Their projects were often highly speculative in nature, and many ended up unsuccessful and money-losing ventures, casting suspicion on the projector as more self-interested than devoted to the public good.[31]

It is worth noting that the idea of such a character, a proposer of projects, in all its ambivalence, antedated the term.[32] In the seventeenth century the idea of the 'projector' could be neutrally descriptive or negatively evaluative, depending on the context. It seems to have acquired an increasingly negative connotation and been deployed as a polemical category as it crystallised over time, becoming synonymous with self-promotion and deceit, and the object of criticism and satire in many quarters.[33] Leibniz's use of

---

27. Keller, *Knowledge and the Public Interest*, 275.
28. See Wakefield, *The Disordered Police State* (2009); and Ash, 'Expertise and the Early Modern State', 13, 21: 'The early modern state, then, was simultaneously a generator, a consumer, and a product of expertise'; 'Operative knowledge thus became the byword of the new natural philosophy.'
29. Borowski, 'Projectors'.
30. See Condren, *Argument and Authority* (2006). In *The Occasion of Scotland's Decay in Trade* (1705), the tremendously successful projector William Paterson did not shy away from launching an attack on projectors and and presenting himself as serving the welfare of his country. My thanks to Ryan Walter for bringing this source to my attention.
31. Graber, 'Du faiseur de projet au projet régulier', 10.
32. Krajewski, *Projektemacher* (2004); Keller and McCormick, 'Towards a History of Projects' (2016).
33. See, for example, Charles-François Lebrun's unforgiving characterization of the '*faiseurs de projets*' in 1790: 'The trader, the manufacturer, the cultivator are not governed . . . by the imaginary calculations of projectors. Projectors' deplorable calculations

the verb (generally the French *projeter*) is consistently neutral during the period under examination, even if his approach to the idea of the projector is much more nuanced.

The projectors' fundamental ambiguity—and the concept's ambiguity, with a valence shifting between ingeniousness and charlatanism—can be ascribed to the grey, ill-defined zone in which they operated, between neatly codified landscapes, disciplines, and emergent roles and personas such as that of the counsellor or philosopher.[34] This need for certainty and mastery did not necessarily match the messiness and unpredictability of reality. The figure of the projector developed at a particular historical juncture, against a background of technological inventions, discoveries, and nascent commercialism.[35]

In this sense, the projector embodies the epoch's contradictions and epistemological ambiguity, attesting to the precariousness of the scientific inventiveness and technological progress that the epoch embraced but often underestimated.[36]

As Pamela Smith has argued, this *homo novus* was a 'liminal individual' whose life embodied the 'fluid cultural moment when "science" had not yet achieved its preeminent modern position as the sole legitimator of truth, but instead had to compete with a number of other intellectual pursuits'.[37]

Projection emerged as the hallmark of an economy of knowledge and action characterised by emulative activity, emerging mercantilism, newly founded natural philosophy, and a crisis of authority. Leibniz's existence as a projector, his grand schemes, and his 'wise foolishness' were made possible by a particular type of fluid epistemology which itself helped 'erect a multivalent vision of reality and knowledge'.[38]

---

are not intended for them but are like nets stretched out to the ignorant multitude that feeds on words and nourishes itself with hopes and chimeras, and which, influenced by the stream of opinions, always remains the plaything of illusions and the victim of credulity.' Lebrun, 'Opinions . . . sur le projet' (1790), 23.

34. Condren, Gaukroger, and Hunter, *The Philosopher in Early Modern Europe* (2006), examines the rise of the figure of the philosopher through forms of self-cultivation and intellectual deportment within particular institutional settings in the early modern period.

35. Borowski, 'Projectors'; Stanitzek, 'Projector'.

36. Graber, 'Du faiseur de projet au projet régulier', 12: 'The problem of the project maker—enthusiasm for innovation and new possibilities of profit, coupled with a deep uncertainty about the morality, the viability, and the interest of these companies which could be merely chimeras or scams—is above all at the heart of the modern development of science and technology.'

37. Smith, *The Business of Alchemy*, 4–5, 10.

38. Smith, 270.

Leibniz was keen to instrumentalize this state of pervasive doubt and uncertainty in order to explore the 'realm of possibilities', to reform knowledge, and to advance his own projects, from his 'Egyptian plan' to the production of phosphorus to his scheme outlined in his essay 'Funny Thought'.

A more experimental and empirical approach to epistemology and reality seemed to prevail over any logical a priorism or static concept of 'truth', and consequently distinctions between rational and irrational thinking, between fact and fiction, could be far from clear-cut: whether a scheme would later be judged illusory was contingent upon whether it could be successfully reduced to practice. 'Speculating' in the attempt to reveal reality in its latent possibilities was one thing; reducing to practice, even though deemed a simple matter of execution, was quite another. The passage from theory to practice—often depending heavily on external and unpredictable conditions—regularly proved impossible. Since the bounds of possibility could not be determined in advance, only experience would establish whether the seemingly impossible was actually achievable.[39]

Part of my task, then, consists in exploring how Leibniz presented himself within the scholarly and courtly circles he came to frequent in Paris and Hanover, and, more generally, how he fashioned various socioprofessional identities for himself according to his interlocutor. This exploration will expose the tensions and discrepancies between Leibniz's self-presentation (including possible contradictions among his different socioprofessional identities) and his perception by others. What will come more clearly into view as a result will be Leibniz as he sought to advance his scientific, philosophical, and theological agendas, to influence intellectual and political debates in his early years—and as he was known to his contemporaries in those years, rather than as he came to be known by later generations (especially after the posthumous publication of some of his most important philosophical and theological writings).

By highlighting the fragmented and precarious nature of his existence in the years 1672–79, full of false steps and setbacks and disappointments, this book seeks to challenge the common presentation of Leibniz as a towering genius. For such an understanding of him bears little resemblance to the complex ways in which he presented himself and his work to his contemporaries or how he was perceived during his lifetime—the latter

---

39. See Keller, *Knowledge and the Public Interest*, 281: 'Leibniz utilized the concept of communally desired objects to advance his particular investigations toward presumed impossibilities.'

being in fact very difficult to gauge. Leibniz did not operate in a vacuum but had to learn to navigate as well as he could a complex historical context and a tense political climate characterized simultaneously by intense plasticity and rigidity on many levels. His various projects were regularly interrupted or deferred, sometimes resumed, often abandoned altogether. His thought and career emerged through, not in isolation from, his social existence: the former could not have emerged as they did without the latter. The present work therefore endeavours to rediscover Leibniz in the complexity, nuances, and contradictions of that social existence in some of his most formative years. Drawing on material that is generally sidelined for being considered superfluous or irrelevant, I have sought to reconstitute, little by little and to the extent possible, the relatively small world in which Leibniz evolved and his place in it. The letters acted as pieces of a puzzle through which I endeavoured to address the following enigma: Why wasn't arguably one of the most brilliant minds of his time revered in his younger years? Why wasn't he better known? And above all, who was he?

Following a broadly chronological as well as thematic structure, this book is divided into two sections. The first section (1672–76) examines Leibniz's rapport with his patron Johann Christian Boineburg and his Egyptian plan. It offers an overview of his political and scientific activities in Paris and the various figures he interacted with there (including the French theologian Antoine Arnauld, the orientalist Louis Ferrand, and the Dutch savant Christiaan Huygens), as well as of Leibniz's involvement in the intellectual and scientific debates of his time (such as Cartesianism, the *Querelle des anciens et des modernes*, and the vision of the Republic of Letters that he sought to promote). The second section (1677–79) examines how, having failed to infiltrate Colbert's state Republic of Letters—either by funnelling books (such as the much coveted *Abulfeda*) to its highly placed members, or by promising new projects to Pierre de Carcavy, the minister's right-hand man—Leibniz found his way to the court of Johann Friedrich in Hanover, from which he set out, despite his relatively lowly official position, to carve out a sphere of influence for himself, not least as the duke's omnicompetent and indispensable commercial and technological adviser, intent on securing beneficial and profitable schemes—such as the production of phosphorus—for his employer. This effort was not without some interesting tensions, including with his fellow projectors Johann Daniel Crafft and Joachim Johann Becher, since Leibniz was himself a projector eager to promote his own schemes. As this book sets out to show, even while Leibniz was busy creating one of most successful epistolary and information networks of the time, he still hoped to insinuate himself into the Parisian Académie des Sciences.

During his frantic activities and networking in his early years, Leibniz encountered little success. He was in a way a misfit, a quasi-public savant obsessed with secrecy, whose mode of interaction could irk as he attempted to develop a new status within the Republic of Letters that exceeded its existing codes. He was a man of contradictions, torn between the ideal of the selfless savant and having to operate in the real world, a political being accountable to his employer, a savant with unclear allegiances seeking to belong to a world and yet, at the same time, incapable of fitting in and, so to speak, staying in place. At a time when the world of scholarship did not exist as a counterweight to absolutist power but, on the contrary, largely depended on it, Leibniz struggled with the politics of scholarship. Above all, the person we have come to perceive as a towering genius was a deeply human—*all too human*—individual.

PART I

HEEDING THE CALL OF BROADER HORIZONS, Leibniz, after a brief stint as secretary of a Nuremberg society devoted to secret alchemical experiments, turned down a professorship at the University of Altdorf in 1666 to enter the service of the Catholic archbishop and elector of Mainz, Johann Philipp von Schönborn, at the time one of the more powerful electors of the Holy Roman Empire.[1] Prior to arriving in Mainz, Leibniz had successfully pursued higher degrees in law and philosophy conjointly, hoping to reconfigure traditional legal jurisprudence as a demonstrative science within a broader logico-philosophical project, in accordance with a dream he had cultivated ever since his early attempt, in the seminal *Dissertatio de arte combinatoria* (1666), to develop the mathematical theory of combinations of simple concepts into a kind of universal logical calculus.

In his 'Specimen certitudinis seu demonstrationum in jure' (A model of certainty or demonstrations in jurisprudence), in fact, he laid out his plans for a rational jurisprudence, weaving together propositional logic, modal logic, and probability with the law of conditions. Both his 'Disputatio arithmetica de complexionibus' and his 'Disputatio de casibus perplexis in jure', the dissertations he wrote to gain his habilitation in philosophy and doctorate in law, respectively, explored how to solve uncertain or difficult cases through combinatorial logic.[2] After dedicating to the Mainz elector his proposal 'Nova Methodus Discendae Docendaeque Jurisprudentiae' (A new method for learning and teaching jurisprudence, 1667), in which he elaborated plans for reform of the legal profession as a whole and reiterated the need for jurisprudence to be grounded on clear universal philosophical principles from which a new order could be deduced according to a geometrical model, Leibniz was appointed to assist the court jurist and counsellor Hermann Andreas Lasser in a project to recodify and rationalize the corpus of Roman law.[3] Here the unexpected opportunity presented itself for him to implement his *jurisprudentia rationalis* and ideally to

1. Arthur (2014), *Leibniz* (2014), 19.
2. Antognazza, *Leibniz* (2009), 62.
3. Antognazza, 83.

help bring about legal and political reform, first in Germany and then across Europe. This collaboration produced in June 1668 a *Ratio corporis juris reconcinnandi*, a rationale for restoring Roman law, which outlined the principles that would guide their reform project, and much of the *Corpus Juris Reconcinnatum* itself, which was to be a revision of Roman law and a new codification of civil law.

Yet Leibniz's interests and hopes for reform extended far beyond jurisprudence.[4] While he disagreed with some of the English philosopher Thomas Hobbes's conclusions, Leibniz was much impressed with Hobbes and indebted to the latter's claim that every operation of the human mind consisted in computation.[5] He also drew substantially on Hobbes's *De corpore* for his own first major works on physics, *Hypothesis physica nova* and *Theoria motus abstracti*, in 1669 and 1670, respectively, appropriating notably the concept of *conatus* to account for the continuous nature of motion. In this manner, Leibniz envisaged bodies in the material world as being composed of an infinite number of unextended and indivisible momentary 'endeavours' or minds, something that Cartesian physics, limiting itself to extension, had been unable to conceive.[6] This early attempt to reconcile modern mechanical natural philosophy with immaterial principles of unity and activity inherited from Aristotelianism prefigured his mature metaphysics.[7]

Significantly, too, Leibniz's first major projects have their beginnings in these years. His impulse for reform found one of its strongest expressions in his early projects to organize science and learning through the creation of societies and scientific academies that would contribute to the intellectual—if not political—unification of the politically fragmented German states. This was to remain a lifelong concern. And although much attention has been paid to Leibniz's later projects—of which his plan for the Berlin Academy of Sciences was one of the most significant and impressive—little has been devoted to his early reflections on scientific and intellectual societies, which he composed either for his private use or (with Boineburg's encouragement) for his patron, Johann Philipp von Schönborn. Founding a learned society was a consuming preoccupation of Leibniz's during his time in Mainz. This endeavour provided a testing

---

4. For more on Mainz's period, see Dingel, Kempe, and Li, *Leibniz in Mainz* (2019).

5. Tönnies, 'Leibniz und Hobbes' (1887).

6. Malcolm, *The Correspondence of Thomas Hobbes* (1998), 2:846. For more on Leibniz's early physics, see Garber, 'Motion and Metaphysics' (1982), 'Leibniz' (1995), and 'Leibniz's Reputation' (2009).

7. Antognazza, *Leibniz*, 52.

ground, so to speak, for the development of many of his later ideas and projects, and it reveals something of the young Leibniz's character, ambition, and vision—which were subsequently challenged and transformed by his stay in Paris. During his early years, Leibniz drafted several proposals for academies of sciences, some more feasible than others.

Leibniz was not the first thinker to conceive of scientific societies. Increasing activity in the natural sciences created a demand for new organizations to support scientific experimentation and encourage scholarly cooperation and exchange. Several models familiar to Leibniz—including Francis Bacon's archetype (in the *New Atlantis*, 1626) of a modern scientific organization, the fictional Salomon's House on the island of Bensalem, with its rational and mathematical arrangement for studying and exploiting the natural world; Johann Amos Comenius's design of a 'house of wisdom'; and the Royal Society of London, founded in 1660 at Gresham College—had been developed during or after the Thirty Years' War (1618–48), when the need to centralize and provide an institutional platform for science had begun to be felt acutely.[8] Leibniz, however, explicitly distanced himself from the more overtly utopian models offered by Bacon, Johannes Andreae, and Thomas More, and he dismissed Girolamo Cardano's and Tommaso Campanella's models as 'extravagant', 'lacking in judgement', and even 'dangerous.'[9] They struck him as particularly pernicious in that they misrepresented progress by exalting unattainable ideals that were not grounded in reality, and by presenting perfectly fulfilled visions while obfuscating the continuous and difficult effort needed to realize them.[10]

In his early effort to design societies or academies of science, Leibniz found more inspiration in the existing established societies, whether in Germany or abroad, than in others' visionary schemes. He regarded the Royal Society highly on account of the great value it placed on experimentation.[11] Leibniz was also familiar with the writings of Johann Heinrich Alsted, whose *Encyclopaedia* emphasized the value of the mechanical arts, as well as with Petrus Ramus's work on practical mathematics, which praised the promotion of mathematics by princes.[12] But given their shaky

   8. Shapin, 'The House of Experiment' (1988), 381; Wahl, 'Naturwissenschaft und Akademiegedanke' (2019), 222–23; and see Lazardig, 'Universality and Territoriality' (2008).
   9. A IV, 1:536, 546.
   10. Rudolph, 'Scientific Organizations' (2014), 546.
   11. Leibniz to Oldenburg, 24 January 1670, A II, 1:38.
   12. Alsted, *Encyclopaedia septem tomis distincta* (1630), and Ramus (1569), 63–66, both quoted in Wahl, 'Naturwissenschaft und Akademiegedanke' (2019), 223. See also Antognazza, *Leibniz*, 39–44.

footing, none of the German societies, such as the Societas Ereunetica (founded by Joachim Jungius in 1622) or the Academia Naturae Curiosorum (founded by a group of peripatetic German physicians in 1652), bore comparison with their English or French counterparts: the Academia obtained imperial patronage only in 1677, when it was renamed the Leopoldina after the emperor, Leopold I. Germany's political fragmentation hindered the formation of an intellectual or scientific centre and impressed on the young Leibniz the need to labour towards the collective advancement of science in the Holy Roman Empire.[13] Hence his resolve to set up an academy on the strongest possible basis, ideally with the backing of committed and financially reliable patrons.

The young Leibniz formulated various schemes for organizing German scientific activity.[14] The first was for a learned society, the Societas Eruditorum Germaniae, to be financed by a tax on paper. In 1669, at age twenty-three, he prepared a more ambitious plan for an international scientific society based in the Netherlands, the Societas Philadelphica, the independence of which would be ensured jointly by the emperor, the pope, and the French king. Although it was intended as an elite organization of intellectually and morally refined researchers, what Leibniz proposed was certainly not a purely contemplative community but rather one firmly entrenched in society and modelled after the ancient Pythagorean school of philosophers.[15] This morally grounded international society was meant to span religious and political divisions, thus laying the groundwork for the reconciliation of the Christian churches in accordance with an idea that Leibniz developed further in a text from the same period, outlining a Societas confessionum conciliatrix.[16] This worldly order would pursue and promote scientific and technological research for the benefit of the public good: the arts and sciences were to be placed in the service of promoting 'the utility of human kind', for 'nothing [was] holier than promoting the usefulness of mankind'. Not unlike the Jesuit order, albeit one excluding

---

13. Leibniz to Oldenburg, 23 July 1670, A II, 1:95: 'Remarkable experiments are not wanting among us, but, such is the current state of politics, on account of the open eagerness of one and the hidden envy of another, that uniting in societies is impossible, and combining so many states into one nation is not easy.'

14. For more on Leibniz and his projects for societies, see Böger, 'Ein Seculum' (1997); Totok, 'Leibniz als Wissenschaftsorganisator' (1966); Wahl, 'Naturwissenschaft und Akademiegedanke'; Schneiders, 'Sozietätspläne und Sozialutopie' (1975); Rudolph, 'Scientific Organizations'; Roinila, 'G. W. Leibniz and Scientific Societies' (2009); and Ramati, 'Harmony at a Distance' (1996).

15. Schneiders, 'Sozietätspläne und Sozialutopie bei Leibniz' (1975), 66.

16. A IV, 1:46.

clerics and motivated solely by a faith in progress, this society would consist of learned men and artisans from all over the world who would collaborate with other institutions to promote scientific and technological progress and would constitute an ideal model for the rest of the world.

Like the Jesuits, too, members would partake in civil society and work as councillors, solicitors, and doctors, in the courts, the universities, the armies, and the navies.[17] By performing their services for free, these agents would win the support of all and infiltrate the highest spheres of power. For care had to be taken in particular that 'our people are promoted to the highest offices; they should hold everything in their hands, the Clerics should become confessors ... so that the youth are educated in the truth by us ... so that little by little it is possible to take on the Orient under the pretext of medicine and learning, and to establish colonies under the pretext of an Oriental Indian society'.

To ensure its economic independence and viability, the society would engage in commerce, sell patents, and obtain privileges. Leibniz intended this society to be the foundation of a new European intellectual elite that would guide and elevate the population morally, intellectually, scientifically, and, ultimately, politically:

> The society can ... easily attain a position in which there is no longer anyone whom it must fear and where it sits at the helm of the state. . . . The military leaders can also be duty bound to the society ... ships and settlers can be sent to America, the whole of the earth can be subjugated not through violence, but through goodness. . . . Finally the whole of mankind will be ennobled, since until that time more than half of them were underdeveloped. The society will even be our referee in wars and easily secure the world against unlawful violence since above all [the members] shall occupy the most important posts, be closely bound with the people, and in charge of regional governments.[18]

Like secretive philosopher kings—an idea appropriated from Plato's political thought—agents of this society would gradually undermine the power of individual states and assume the most important political offices across Europe, including the Dutch state, with the view to spreading wisdom and securing peace, including through gradual colonization—Leibniz's plan includes colonies to be established in America.[19] These agents would

17. Roinila, 'G. W. Leibniz' (2009), 175–76.
18. A IV, 1:556. See Böger, 'Ein Seculum', 80–81.
19. Totok, 'Leibniz als Wissenschaftsorganisator', 304; A IV, 1:556.

not only educate the civil servants but would also subjugate the whole world without violence or danger, gradually erecting a politically effective republic of scholars dedicated to the perfection of the human race.[20] According to this scheme, science would gradually replace conventional politics, especially in the political organization of society, in order to create a metaphysically inspired technocracy that would 'finally help cultivate mankind'.[21]

In 1671 Leibniz wrote two further proposals, the 'Bedenken von Aufrichtung einer Akademie oder Societät in Deutschland zu Aufnehmen der Künste und Wissenschaften' (Reflections on the establishment of an academy or society in Germany for fostering the arts and sciences) and the 'Grundriss eines Bedenkens von Aufrichtung einer Societät in Deutschland zu Aufnehmen der Künste und Wissenschaften' (Outline of a reflection on establishing an academy or society in Germany for fostering the arts and sciences), which, with Boineburg's approval, he dedicated to his employer, Johann Philip von Schönborn, in the hope of winning the latter's support for his national and local schemes. Both read like exhortations to the German nation to unite intellectually—something, Leibniz hoped, that would not fail to strike a chord with the elector, who at the time was engaged in a concerted diplomatic and political effort to unite the German princes against Louis XIV.

Whereas his grandiose plan for the Societas Philadelphica had described an elitist, almost mystical community led by scientist-philosophers, the 'Bedenken' was a more modest endeavour to improve specifically German science. In it Leibniz presented his own version, somewhat glorified and not always strictly accurate, of the history of German contributions to science, especially in chemistry, mechanics, mining, astronomy, and medicine, from Albert the Great to Paracelsus, Copernicus, and Kepler. He particularly emphasized that although Germans had often been the first to invent or discover things, they had generally been the last to benefit from their inventions and discoveries. Germany had excelled in the practical arts and sciences, providing 'all countries with alchemists and laboratory technicians', and had bequeathed to other nations 'military, mechanical and similar arts', only to see their inventions later imported back into Germany with slight modifications, as if they were brand new devices.[22] To revive this tradition of past successes, Leibniz argued for founding a

20. A IV, 1:556.
21. Totok, 'Leibniz als Wissenschaftsorganisator', 304; A IV, 1:556.
22. A IV, 1:547.

PART I [19]

German Academy of Sciences with funding from the aristocracy, on the model of European societies such as the Royal Society. He described how this society would bring together—in what would become, as we shall see, a persistent theme over the next decade—the true promoters of German 'mechanics, artists, and laboratory workers' whose talent too often had been 'buried', either because their judgement was poor or because they had been neglected or ill-treated:

> So many good minds could be cultivated and put to use in the country, so many people preserved from impoverishment, so many families from ruin, so many beautiful designs, inventories, proposals, experiments, rare observations, posthumous works of excellent people from loss and oblivion if in the care of the Republic. Chemists, charlatans, market criers, alchemists, and other busybodies, vagabonds, and 'cricket catchers' [i.e., those given to fantastic or pointless speculations] are generally people of great cleverness, and sometimes experience too; it is just that their disproportion of cleverness to judgement [*disproportio ingenii et judicii*], or sometimes also their great desire to maintain their vain hopes, ruins them and brings them into contempt. Certainly such an individual sometimes knows more from the realities of experience and nature than many a widely and highly respected scholar who adorns the knowledge he has gathered from books with eloquence, exalted language, and other political tricks, and knows how to bring it to the market, whereas another, with his extravagance, makes himself hated and despised. But wise rulers in a well-endowed republic do not pay attention to this, but need such people, and by giving them stable employment, can thus prevent both them and their talents from going to waste.[23]

For Leibniz, societies were not merely repositories of information but should primarily strive towards 'usefulness' and aim to harness science and technology to help better the human condition and achieve the *perfectio generis humani*.[24] Notwithstanding the many achievements of science, its progress was insufficient. As he outlined in his 'Grundriß', great inventions often failed to be communicated or developed owing to a lack of funds and sufficient instruments. This situation could be remedied,

---

23. A IV, 1:549–50. Leibniz in fact would suggest shortly thereafter inviting fifty known researchers to join this task of 'compiling knowledge not from writings, but from nature itself and the treasury of the mind'. Totok, 'Leibniz als Wissenschaftsorganisator', 299.

24. A IV, 1:552–53.

however, through the founding of an academy that would centralize German scientific research and enterprise.[25]

In his plan, Leibniz offered a comprehensive programme that encompassed a wide range of activities (including science, medicine, art, history, and commerce) and affirmed the necessity of drawing practical applications from scientific discoveries and inventions within a broader mercantile system.[26] The benefits of manufacture as well as the treatment of artisans—as opposed to only gentleman and courtier scientists, as in the Royal Society or French Académie—received particular emphasis. For Leibniz there was no doubt that the pursuit of the common good and societal happiness depended on auspicious social conditions and the good health of the populace, and it was therefore incumbent on rulers to promote the advancement of medicine by fostering collaboration among physicians and the collection of data.[27] Education too played a crucial role in Leibniz's scheme: through the founding of a national academy and the adoption of rational teaching and learning methods, a new social, political, religious, and learned order would emerge that could one day hope to rival the achievements of the Royal Society and the French Académie des Sciences. Such a society would yield 'many fine and useful ideas, inventions and experiments . . . means . . . of keeping food production and creating factories within the country . . . attracting commerce, in time workhouses and prisons to put idlers and wrongdoers to work, to facilitate craftsmanship with benefits and instruments . . . to test and produce everything in chemistry and mechanics . . . to obtain [privileges] abroad for new inventions . . . to collect manuscripts and experiments'.[28]

While the extent to which he thought these early projects were realistic or feasible remains unclear, especially since they generally entailed highly independent and mobile functionaries—a prospect that would not necessarily have recommended itself to rulers—Leibniz lent a philosophical justification to his societal projects by relating them back to metaphysical and theological foundations. He defined the role of science as service to the state, and indeed metaphysically, by arguing that scientific discoveries and their practical applications honoured God as contributions to the common good of humanity.[29]

25. Roinila, 'G. W. Leibniz' (2009), 177.
26. Roinila, 174; Totok, 'Leibniz als Wissenschaftsorganisator', 306.
27. Leibniz to Konrad Burchard Vogther, 1712: 'no art is more difficult and more in need of support than the art of restoring health', quoted in Antognazza, *Leibniz*, 100.
28. A IV, 1:536-37.
29. Totok, 'Leibniz als Wissenschaftsorganisator', 312; A IV, 1:535.

Leibniz's 'Societät und Wirtschaft' (1671) articulated this metaphysical defence of applied science and technology most explicitly in a proposal for a rationally and scientifically redesigned society that would be economically fair and just. Citing Holland as an example, Leibniz condemned the economic disparities by which merchants profited at the expense of artisans: Why indeed, he demanded, should 'so many people find themselves in poverty and misery for the benefit of such a small handful of people'? A scientifically conceived society and economic policy that banned economic monopolies and favoured local production would guarantee a fair distribution of reasonably priced goods, thus preventing excessive inequalities. It would help establish a society that would 'keep [its citizen] clean and disease free', as well as provide free education for children, who would be raised in public institutions. What Leibniz envisaged was not a classless society but an ideal harmony of the estates, in which the masters would oversee the more sophisticated tasks and would alleviate the burden of those whose preoccupation with their own survival had hitherto entrapped them in melancholy and even despair. This society would form a collaborative whole that would provide full employment, including to the scholars who were necessary for their 'all-important conferences and amusing inventions', since 'no one will work for themselves, but rather in cooperation'.[30] Each individual—and, by extension, nation—would attend to the particular tasks for which it was most suited:

> Each country should equip itself with the capacities enabling it to produce these necessary goods and manufactured products which previously had to be imported so that it does not have to rely on others for what it can produce itself; each country will be instructed on how to exploit its own resources appropriately. In a country that possesses sufficient wool, factories will have to be built for the preparation of the fabric; another country with an abundance of flax will occupy its population with the production of clothing, and so on. Thus, no country among those which offer the adequate degree of freedom to the society, will prevail over another; rather, each will see itself prospering in the areas where God and nature intended it to excel.

These enlightened princes, in fostering and overseeing the technocracies that Leibniz envisaged, would not only reap financial benefits but achieve eternal fame as benefactors of humankind and reflectors of God's glory.

30. A IV, 1:560.

That nothing substantial came from these early schemes for scientific, economic, and social reform—these were after all highly ambitious, predicated on a radical overhaul of existing institutions—did not deter Leibniz from formulating a range of such projects later, some more successful than others.[31] His next major proposal, in fact, was to be intended for none other than the most powerful ruler in Europe at the time, Louis XIV of France.

31. See Rudolph, 'Scientific Organizations', 548–61.

CHAPTER ONE

Leibniz with Boineburg

ACQUIRING A NEW PATRON,
NEGOTIATING CONTINGENCY

LEIBNIZ'S SITUATION WAS perhaps more precarious than most since, not being the son of a nobleman and lacking both private wealth and a fixed position, his influence was limited. This precariousness made him all the more dependent on his patrons, to whom he remained remarkably faithful. Without the guidance and support of his patrons, especially in his younger days, Leibniz would probably never have become the towering figure he later became. Leibniz had limited personal resources—such as reputation, wealth, rank, family connections, or high office—but he possessed a wide array of skills to draw on. He existed in a relationship of subordination and reciprocal exchange with his patrons, one that was marked by an informal and ongoing process of bargaining and negotiation, whereby, in exchange of material benefits and protection, he provided them with loyalty and service.[1]

It is difficult to overstate the importance of Leibniz's early years at the religiously tolerant court of Catholic Mainz. From 1668 to 1672 he was close to Baron Johann Christian von Boineburg (1622–72), the former prime minister, who arguably became the most important person in the young man's life—a mentor and patron who, like Leibniz himself, moved between the political and scholarly realms. One of Boineburg's strengths was precisely his ability to gather information and books, as well as to establish connections between politicians and scholars, recruiting scholars for political purposes. Eager to participate in the *respublica literaria*,

---

1. See Kettering, *Patrons, Brokers and Clients* (1986), 13.

Boineburg cultivated a learned epistolary network which spanned 'all corners of learned Europe [*ex omnibus Europae doctae angulis*],' helping to circulate the writings of scholars, making his books available to them, and eventually earning him the moniker of 'Employer [of Scholars]' for his patronage of scholars and their activities.[2]

From 1664 to early 1668 Boineburg had been estranged from the archbishop-elector of Mainz and had even been imprisoned briefly on unsubstantiated charges arising from his sympathies with France.[3] Now reconciled with Johann Philipp von Schönborn, Boineburg was seeking scholars who would help him restore some of his former political dignity. He seems to have taken an instant liking to Leibniz, prizing him for his independence of thought and the novelty of his positions, considering him a 'happy mediator between old and new philosophy' and aligned with many of his own positions.[4] On account of Leibniz's considerable talents—he was a 'man of great scholarship, accurate judgement and great manpower'[5]—Boineburg tasked him with numerous projects, including committing to paper some of his 'old desires for which only the right pen was missing'.[6] Boineburg must have felt that he had found in Leibniz the perfect vehicle back into his various employers' good graces and the promotion of his own, especially ecumenical, designs.

Boineburg's assignments often concerned the relationship between religion and philosophy: these included the preparation of a new edition of the Italian humanist Mario Nizolio's *Antibarbarus*, the composition of the magisterial *Demonstrationes Catholicae* dedicated to theological reconciliation, and Leibniz's defence of the mystery of the Trinity. But Boineburg seems to have regarded Leibniz not merely as an employee but as a friend, supporting and encouraging him and his projects as much as he could, including his attempts to secure from the imperial librarian, Peter Lambeck, a privilege for a planned semi-annual journal reviewing new publications, the *Nucleus Librarius Semestralis*.[7] Boineburg enlisted

---

2. Paasch, *Die Bibliothek des Johann Christian von Boineburg* (2003), 25–31 (quote on 26).

3. See Wiedeburg, *Der Junge Leibniz* (1970), 96.

4. Boineburg to Conring, 22 April 1670, in Gruber, *Commercii Epistolici Leibnitiani* (1745), 1287–88.

5. Boineburg to Conring, 26 April 1668, 1208–9.

6. Ueberweg, Frischeisen-Köhler, and Moog, *Die Philosophie der Neuzeit* (1924), 309. Cf. Leibniz to Johann Friedrich, 29 March / 8 April 1679, A I, 2:155.

7. See Boineburg's letter to Lambeck of 18 November 1669, A I, 1:28–33. Boineburg, for instance, discussed his health problems, such as 'asthmatic attacks', with Leibniz, pressing the latter to 'inquire whether there is any method of helping those who are seized by the

Leibniz's skills and talents not only for theological and scientific but also for political and legal purposes, commissioning from him documents in support of the candidacy of Philipp Wilhelm von Neuburg to the Polish throne, and charging him with the legal representation of his family. He also entrusted Leibniz with drafting various diplomatic and political plans—often, it must be said, without the knowledge or approval of the elector. It was a mutually beneficial relationship in which both men's interests and goals often coincided and converged towards their dream of universal improvement.

From the beginning of their relationship, Boineburg—subject of course to the pursuit of his own interests and that of his own patrons— assiduously promoted his protégé to influential potential patrons in Mainz and beyond, and acted as an intermediary for him.

Boineburg was ultimately responsible for Leibniz's trip to Paris on what was initially meant to be a diplomatic mission to persuade Louis XIV to adopt Leibniz's so-called Egyptian plan, which laid out how the French could usurp control of sea and commerce more easily from the Dutch by seizing Cairo and Constantinople than by attacking neighbouring Holland.[8] This project, which according to its deviser would also help unify Christians against the Turks, marked the culmination of several years of unsuccessful attempts to contain France's bellicose foreign policy under Louis XIV—attempts that included a series of diplomatic negotiations and strategic alliances among European nations, such as the Triple Alliance between Holland, England, and Sweden, and an alliance between the duke of Lorraine and the archbishops of Mainz and Trier.[9] Convinced that the Rhineland principalities' interests lay rather in cultivating good relations with France and strengthening their political and military ties with one another, Boineburg had Leibniz write in August 1670 a proposal for a 'Securitas publica' based on a consolidated German constitutional structure and standing imperial army. Leibniz would maintain a lifelong commitment to the idea of a religiously pacified Reich under the joint guidance of an emperor and the pope as temporal and spiritual leaders, respectively.[10] The allied German states would serve as a neutral buffer

---

first stages of this aggravating and difficult disease' (Boineburg to Leibniz, 30 April 1672, A I, 1:269).

8. On the Egyptian plan, see Youssef, *La fascination de l'Egypte* (1998); Beiderbeck, 'Leibniz's Political Vision for Europe' (2014); Cook, 'Leibniz and "Orientalism"' (2008); and Strickland, 'Leibniz's Egyptian Plan' (2016).

9. See Badalo-Dulong, *Trente ans de diplomatie française* (1956).

10. Beiderbeck, 'Leibniz's Political Vision for Europe'.

zone between the two dominant Continental powers, France and Austria.[11] But that same month France invaded Lorraine, thus jeopardizing European peace only a few years after the end of the devastating Thirty Years' War—much to Schönborn's dismay.

Keen to promote his young protégé and the Egyptian plan, Boineburg had written to Louis XIV in the hope of placing Leibniz at the French court in the early days of 1672, for Leibniz's station was too low for him to petition the king directly.[12] Doubtless mindful of the small window of opportunity for relaying Leibniz's Egyptian plan, Boineburg drafted two versions of his letter to Louis XIV: each consists of a short paragraph (penned by Boineburg) introducing a lengthy proposal (penned by Leibniz). By conquering Egypt (rather than the Netherlands), the proposal stipulated, France would successfully cut off the latter from trade with East Asia and gain control of the trade routes between Europe, Asia, and Africa.[13] In addition to accruing political and economic benefits, it would emancipate populations oppressed by barbarians and help unify Christians everywhere through a new 'holy war'.[14]

In both versions, Boineburg puts forward a project he characterizes variably as 'seeming rather flimsy at first', 'somewhat chimerical', and finally as appearing 'a tad extravagant', but which was brought to him by 'a man of considerable qualities', whose 'ability warrants much approval'.[15] In the final draft, presumably sent only once assurance of it receiving the king's attention had been provided, Boineburg shares Leibniz's special project, which, he seeks to reassure Europe's most powerful ruler, 'could be executed without fail within a year', would 'ruin the Dutch' through economic warfare and command of the seas, would help cement an alliance with Austria, and would 'be in conformity with the

---

11. Beiderbeck, 665–65; Antognazza, *Leibniz* (2009), 116–17

12. Johann Christian von Boineburg to Louis XIV, 20 January 1672, A I, 1:249–52.

13. This was part of Leibniz's broader geopolitical thinking, which particularly prized 'strategically important junctions' and 'effective network controls'. For more on this, see Kempe, 'In 80 Texten um die Welt' (2015).

14. In fact, as Claire Gantet points out in 'Leibniz' Sicht von Krieg' (2015), Leibniz relies on the Machiavellian term 'necessitas' to legitimize the French offensive. While he would develop a more nuanced take on Islam in his later years, praising it notably as a form of deism, Leibniz in his *Consilium Aegyptiacum* delivers an unredeemingly negative portrayal of Islam and the Turks, whom he describes as 'barbarous', 'backward', 'lazy', and 'lascivious'. Genuine engagement with Islamic theology within the text—and more generally, over the course of those first few years—remains scant, and Islam and the Turks are invoked purely in political terms.

15. Boineburg to Louis XIV, 20 January 1672, A I, 1:249.

wishes of the Germans, Italians, and the Portuguese'. Crucially, it would confirm France's position as a European military, cultural, and civic power, and Louis himself as a universal leader.[16] The Egyptian plan, incidentally, constitutes a striking example of Leibniz's political duties and personal inclinations dovetailing.

With this proposal, Leibniz presented—one might say, 'projected'—himself as someone who could help channel the forces of contingency and historical uncertainty towards some kind of order. Had the Egyptian plan been implemented, he might well have been able to find more powerful patrons and secure a more favourable position, in the French court or elsewhere. As it was, all of Leibniz's principal correspondents seem to have been aware that they were contending, rather powerlessly, against forces far greater than themselves. And even though it was only rarely explicitly acknowledged, this constant and acute awareness forms a recurring theme throughout the correspondence and the mental backdrop against which all of them were writing. Leibniz and his peers inhabited a universe full of mysteries and uncertainty: the whereabouts of certain individuals remained unknown.[17] The postal services were unreliable, and various 'accidents at Court' resulted in letters being disposed of either because they were deemed unimportant or precisely because their content had already been divined.[18] 'Fortune' was always looming on the horizon and, by its fickle nature, was ready to close an opportunity or favourable historical juncture at any time. Difficulties could arise unexpectedly at any moment and take on any shape.

Novel ideas or projects were of course hardly immune from the vagaries of fortune. Indeed, Boineburg closed his letter by urging the king not to dismiss this 'novelty' and risk 'giving up this auspicious looking juncture',

---

16. Boineburg to Louis XIV, 20 January 1672, A I, 1:251: 'And will turn France into the military school of Europe, and the theatre, where the greatest and most illustrious geniuses of the century in all kinds of professions and civil and military arts can and will want to play a role. . . . And will confer to him . . . the universal direction of affairs and the arbitration between all the princes and all the republics and will attach all famous families to his interests.' We do not have the response to Boineburg's letter, if there was one. The Franco-Dutch War would begin in April 1672 and last until September 1678.

17. Leibniz to Christian Habbeus, January 1676, A I, 1:442: 'For letters from my friends have never taught me anything reliable concerning you; some say that you are in Brunswick, while others in Bavaria: and for my part I believed you in Vienna, or at least I wished you there.'

18. Boineburg to Louis XIV, 20 January 1672, A I, 1:250: 'He adds that he feared the accidents to which the letters are exposed both at the post office and at court . . . and they could be intercepted if their contents were sniffed out.'

but instead to seize this opportunity.[19] This passage anticipates comments made by Leibniz in his correspondence with Duke Johann Friedrich later in spring 1679.[20] Both passages emphasize that a new idea's possible adoption was predicated less on its content than on the conditions and circumstances of its initial reception. Novelties, even when advantageous and at no cost to a ruler's subjects, as Leibniz was careful to point out, were generally treated with suspicion in a context of prevalent fraud, projection, and intellectual hustling.[21] In fact, irrespective of their validity, new projects, even 'good and enforceable' ones, were often discarded for their failure to make a first good impression or because of unfavourable initial circumstances.[22] 'The grace of novelty' having worn off, the potential benefits, and especially the truths they contained, ended up never being placed in the service of the public good.[23] To initiate, complete, or salvage projects that at first glance seemed 'chimerical' or 'extravagant', it was necessary to identify the optimal historical moment in which to propose them.[24]

This pervasive sentiment of precariousness was compounded by regular reminders throughout the correspondence that time was running out. Like historical contingency, time emerged as an omnipotent force in itself, and in particular as a fleeting and precious commodity that should not be wasted, especially since the window of opportunity was a narrow one

19. Boineburg to Louis XIV, 20 January 1672, 1:250.
20. Leibniz to Johann Friedrich, Spring 1679, A I, 2:79–85.
21. Leibniz to Johann Friedrich, Spring 1679, A I, 2:80.
22. Boineburg to Louis XIV, 20 January 1672, A I, 1:249: 'But as he feared squandering the grace of novelty, and perhaps losing the hope of success forever after an unhappy first encounter, since it is tiresome to hear of a thing again after it has already been rejected or neglected on account of initially unfavorable circumstances—he wanted to use it in this manner, and having explained in the attached note the expected outcome of this proposal, which he claims to demonstrate, he thought it more appropriate to reveal the particular details once he had formed a serious reflection on these preliminary positions.'
23. As we will see, Leibniz would deploy the same argument with Duke Johann Friedrich. Leibniz to Johann Friedrich, Spring 1679, A I, 2:80: 'Now it is a truly damaging thing to the public, to make good proposals unnecessarily and inappropriately, not only because we prostitute ourselves, but also because those who are rejected lose the desire for good, and mainly because a proposal, once it has been rejected, passes for bad and useless; and on account of a very strong prejudice one does not want to hear any more about it. Hence many good things are decried on account of the recklessness or misfortune of those who propose them, and the public is deprived of them for ever. Consequently, those who love the public good will take care not to reveal the truth, so as to not receive an affront and be disappointed unworthily.... Besides that, truth is often received only because of its novelty, and once it has lost or needlessly exhausted that claim, it has lost all its grace and won't be attended to anymore.'
24. Johann Christian von Boineburg to the King of France, 20 January 1672, A I, 1:249.

and fruitful opportunities were rare and unlikely to return any time soon. Leibniz thus cautioned Louis XIV through Boineburg that 'the first year [was] the best and that [they had] all the reasons in the world to fear that in delaying things a little too much, [they would] lose the best opportunity without being able to retain anything from it, except the regretful memory of what [they] could have been achieved'.[25] While circumstances could not be fully brought under control, they could be harnessed to an extent in an attempt to gain some mastery or triumph over 'manifest cases of injustice'.[26] 'Bring[ing] into play the springs of auspicious dispositions' could help ensure 'a happy outcome'.[27]

Keeping the Egyptian plan secret in a highly politicized and cutthroat world of rumour and intrigue was absolutely vital for its successful implementation, at least until the outcome was no longer 'uncertain, or something of outstanding importance [had been] achieved', by which point it would be 'welcome to everyone, and as though sent from heaven'.[28] In this context, secrecy served as an instrument that could afford some degree of protection and, as Leibniz went to great lengths to demonstrate, could effect change and help impose order on contingency. When drafting his proposal for Louis XIV, Leibniz elaborated on the latent power of secrecy, which he described as 'being the soul of [the current] project', capable of being turned into a formidable weapon. Not only would secrecy safeguard designs from being 'sniffed out' and thus imperilled, it would also provide a cloak behind which to retreat and save face 'should everything go up in smoke'. A secret project that would 'unexpectedly break out'—Leibniz uses the word *éclater* twice in the same passage—would take everyone by surprise and inspire admiration by confounding expectations.[29] Most important, perhaps, the potency of secrecy would reveal itself fully in the opportunity it afforded to convey the impression of the predetermination and inevitable successful realization of certain designs, even the impression of omnipotence.[30] Leibniz's decision to pair secrecy with the miraculous is

25. Boineburg to Louis XIV, 20 January 1672, A I, 1:252.
26. The conceptual analogy here with Machiavelli and his understanding of *fortuna* seems patent, yet I have found few explicit references to Machiavelli's works by Leibniz in his early years. Finster et al., *Leibniz Lexicon* (1988), records only one instance of Leibniz's own use of the word *fortuna*, in reference to Cicero's *De divinatione*.
27. Boineburg to Louis XIV, 20 January 1672, A I, 1:252.
28. Leibniz to Boineburg, January 1672, A I, 1:247: 'We can go securely, if we can go secretly.'
29. Boineburg to Louis XIV, 20 January 1672, A I, 1:250, 252.
30. Boineburg to Louis XIV, 20 January 1672, AI I, 1:251. Boineburg too seems to delight in the potentialities that could be leveraged through it: 'Wondrous gossip is

especially revealing here because his use of the word *miracle* is deliberately double-edged, maintaining a degree of confusion between the pragmatic and the supernatural, and in the process establishing an interesting nexus between the two: secrecy is miraculous in its efficiency because it enables one to help gain command over events and historical contingency.[31] This, in turn, helps convey the impression of a—supernatural—control over circumstances. In fact, as Leibniz takes care to specify, the execution of secret plans should 'break out in a bolt of lightning'.[32] Simply put, wielding the tool of secrecy would enable the king to appear god-like.

Fortune, as Leibniz pointed out regularly over the course of his correspondence, in a move conflating flattery with genuine sincerity, favours the powerful and the king of France in particular. Certain great designs had been conceived in the past but had since fallen into oblivion; Louis XIV, as Leibniz emphasized in his proposal to the king, had been earmarked by fate to help realize great projects for mankind and advance its happiness: 'Your Majesty alone has sufficient strength to undertake this project which Providence (in regard to past times, when others have thought of it but were unable to realize it) seems to have reserved in its entirety to Him. I leave the decision up to Him, knowing that Your Majesty, if He wills it, will find it easy to succeed in it, notwithstanding the present circumstances, of which, after God, He is the master.'[33]

Not only would Leibniz's proposal to the King of France help confirm the latter's mastery over the elements, it would help perpetuate the 'Glory of his invincible rule' and inscribe him in the Pantheon of mankind for the rest of eternity.[34] More generally, one's good name or 'renown' acted as a powerful antidote against the 'reversals of fortune' as had, for instance, befallen Boineburg.[35]

An interesting tension between precariousness, unpredictability, and fluidity, on the one hand, and attempts to master forces greater than oneself, on the other, thus animates this early correspondence. Leibniz's plan, as transmitted by Boineburg, offers an interestingly self-reflexive

---

divulged here. Nobody knows even the slightest about you, where you are, or what you are doing. Nor is there need to know. Your efficacy shall be [a] God[send]!'

31. Boineburg to Louis XIV, 20 January 1672, A I, 1:252: 'It will confirm the judgement of those who ... are right to characterize His Majesty's recommendations as the miracle of secret.'

32. Boineburg to Louis XIV, 20 January 1672, A I, 1:252.

33. Boineburg to Louis XIV, 20 January 1672, A I, 1:249, 250.

34. Boineburg to Louis XIV, 20 January 1672, A I, 1:252.

35. Leibniz to unknown recipient, beginning of September 1673, A I, 1:365; Leibniz to Johann Friedrich Schutz von Holtzhausen, 22 December 1672, A I, 1:305.

dimension particularly of Leibniz's early years: as we shall see, the young savant set out to secure his own position and impose his will and vision on the world by promising to hold the key to his prospective patrons' own conquest of contingency and pursuit of prosperity. Boineburg's pitch was successful in the first instance: on 12 February the minister Arnauld de Pomponne replied that the king would gladly hear his 'overtures'. Leibniz would head to Paris a few weeks later.

CHAPTER TWO

Encountering Paris

LEIBNIZ REACHED PARIS in the spring of 1672, at the age of twenty-six, having obtained a leave of absence from the court of Mainz for a short journey, which the prince archbishop approved on the assumption that it served scientific purposes alone. From his lodgings on the rue Garancière in the Saint-Germain neighbourhood on the Left Bank, just behind the church of Saint-Sulpice, a short walk to the Quartier Latin of university faculties and *collèges* and not far from the French administrative and diplomatic heart on the Île de la Cité and at the Louvre, Leibniz was determined to take advantage of all that Paris had to offer.[1] The wide-ranging nature of Leibniz's contacts attests to the breadth of his activities in Paris. We know little about his daily activities during his sojourn, but some information can be gleaned from letters, notes, and later recollections.

Prior to coming to Paris, Leibniz had many correspondents both at court and in the scientific realm. Once on location, he was determined to meet with a number of them, including some of the period's greatest luminaries— people who could be useful to him and his designs, people from whom he was likely to learn. Of particular importance were the savants of the Académie des Sciences at the royal library, Colbert's state Republic of Letters, whose meetings Leibniz attended and was clearly impressed by.[2] These

---

1. Leibniz to Huet, 10 May 1673, A II, 1:369; Leibniz to Johann Friedrich, 21 January 1675, A I, 1:493; Davillé, 'Le séjour de Leibniz à Paris' (1920), 148–49.

2. Leibniz to Johann Philipp von Schönborn, 20 December 1672, A I, 1:297–98: 'The fellows of the Academy are people extremely learned in various fields who could compose an Encyclopaedia of arts and sciences.' Aside from two mentions in the session reports relating to his calculating machine and a clockwork on 9 January and 24 April 1675, respectively, there are few traces of Leibniz at the Académie. This dearth of information can be explained by the society's secrecy in its first thirty years of operation, as well as the loss of the minutes. See Salomon-Bayet, 'Les académies scientifiques' (1978).

included the abbé Gallois, Colbert's 'close servant' who had been made secretary of the Académie, which convened twice weekly in the royal library.[3] There he met the great mathematician, physicist, and astronomer Christiaan Huygens, with whom he would develop a lifelong relationship; the prominent physicist Denis Papin, whose publications on Guericke's vacuum pump Leibniz had read and whom he had met at Huygens's house; the Danish astronomer—and tutor to the dauphin—Ole Roehmer, with whom Leibniz would discuss mathematics and mechanics; the Italian astronomer Giovanni Domenico Cassini; and the doctor and architect Claude Perrault. Finally, and perhaps most crucially, Leibniz corresponded with and met Pierre de Carcavy, to whom he had been recommended by Boineburg.[4] Another major figure was Henri Justel, a canonist and adviser to the king, from whose extensive library Leibniz would regularly borrow books and to whom he had been warmly recommended by Boineburg as 'a most skilled and erudite man in all the sciences'—Leibniz would in fact send Justel an anonymous letter in which he expressed wonderment at the extent of his own knowledge at such an early age.[5]

In Paris Leibniz would attend Justel's weekly meetings at his salon, where erudite matters and new books were discussed. Justel at the time was a well-connected savant who cultivated close relations with the Royal Society, notably through its secretary (and Leibniz promoter) Henry Oldenburg, whose communications Justel helped disseminate from his Parisian home.[6] Justel and Leibniz would remain in epistolary contact long after the former's flight to London and the latter's departure for Hanover. Among the less prominent savants Leibniz befriended in Paris were Pascal's nephew Louis Périer and sister Gilberte Périer, whom Leibniz considered most 'erudite and spiritual'; the scholar Gilles Filleau des Billettes and his friend the duc de Roannez; and later the skilled mathematician and fellow countryman Walther Ehrenfried von Tschirnhaus and the Danish mathematician Georg Mohr.[7]

In Paris Leibniz was also in contact with other scholars, such as André Morell, a royal adviser and secretary to whom he had been recommended by Boineburg regarding the resolution of his financial affairs, and otherwise

3. Klopp, *Die Werke von Leibniz* (1864–84), 4:380.

4. Leibniz to Spitzel, 7 April 1671, A I, 1:133; Klopp, *Die Werke von Leibniz*, 2:141, quoted in Davillé, 'Le séjour de Leibniz à Paris' (1921), 70.

5. Davillé, *Leibniz Historien* (1909), 19; Anonymous to Justel, April 1672, in Klopp, *Die Werke von Leibniz*, 2:133.

6. See Brown, 'Un cosmopolite du grand siècle' (1933).

7. Leibniz to Seckendorf, 11 June 1683, A II, 1:840; Leibniz to Oldenburg, 12 May 1676, A III, 2:22.

a numismatist of great distinction;[8] the abbé de la Rocque, editor of the *Journal des Sçavans* since 1675; the numismatist and antiquarian Nicolas Thoynard; the travel writer Melchisedech Thévenot; and Pierre-Daniel Huet, a student of the eminent philologist Bochart and subpreceptor of the dauphin, with whom Leibniz discussed various questions, especially literary ones. Other scholars and men of letters who attracted his attention were Colbert's son-in-law, the duc de Chevreuse; the bookseller Claude Clerselier, whose collection of Descartes manuscripts Leibniz studied; the abbé Eusèbe Renaudot, future editor of the *Gazette de Paris*; and the Hellenist Jean-Baptiste Cotelier, who with Charles du Fresne du Cange compiled the catalogue of Greek manuscripts in the royal library and produced erudite works that met with Leibniz's appreciation.[9] Other *érudits* and philosophers with whom Leibniz came into contact and sparred intellectually included the Port-Royal Jansenists Antoine Arnauld and Pierre Nicole; the Oratorian Nicole Malebranche; Simon Foucher, one-time canon of Dijon, a philosopher and physicist with whom Leibniz discussed ancient philosophy and physics;[10] Jesuits such as the Père Gamans, the Père Gervais (recommended by Thoynard), and the Père Bertet with whom Leibniz conversed about mathematics on several occasions;[11] and other clergymen, including an Armenian Dominican priest with whom he conversed about an artificial language derived from Latin and used in Mediterranean commerce.[12] Leibniz also made the acquaintance of an eclectic array of provincials and foreigners who happened to be passing through Paris, including René Oudard, canon of Tours, with whom Leibniz discussed music, and the Swede Anke Rolamb, a senator in charge of the Swedish police and finances, whose 'almost universal ... curiosity' especially inclined to mathematics and history.[13] On one occasion, Leibniz recounts, 'when I was in Paris, all the people of this great city ... were convinced that Madame the Duchess of Tuscany had children abducted from the streets, to have them killed to

---

8. See Boineburg's 'Memoriale' for Leibniz, 18 March 1672, A I, 1:262.

9. Leibniz to Seckendorf, 11 June 1683, A II, 1:840; Leibniz to Landgrave Ernst von Hessen-Rheinfels, 13/23 July 1691, A I, 6:234; Leibniz to Boeckler, May 1672, A I, 1:271—all cited in Davillé, 'Le séjour de Leibniz à Paris', 75.

10. 'I used to debate this point, in person and in writing, with the late Abbé Foucher, Canon of Dijon, a learned and subtle man, but a little too obstinate among the Academicians' (*Nouveaux essais sur l'entendement humain*, A VI, 6:374).

11. Leibniz to Landgrave Ernst von Hessen-Rheinfels, 1680, quoted in Davillé, 'Le séjour de Leibniz à Paris', 76.

12. *Nouveaux essais sur l'entendement humain*, A VI, 6:279.

13. Leibniz to Nicaise, 1/11 October 1694, A II, 2:858; Klopp, *Die Werke von Leibniz*, 3:81.

bathe in their warm blood, in order to be cured of a certain disease she was supposed to have contracted. And when I laughed at that tale, I thought those good people were angry at me.'[14] He frequented salons, mixing in the circles of the Prince de Condé and Madeleine de Scudéry, attended plays by Corneille and Molière, and took French lessons—by the end of his stay, Leibniz boasted that he could 'speak in Parisian'.[15]

Leibniz confessed that one of his greatest faults was his dislike and avoidance of 'capricious' court politics and the 'constant worries' deriving from them.[16] He also admitted to another fault that 'passe[d] for particularly grave in the world', namely, 'failing to abide by social norms'—and he recognized his failure to make a good first impression.[17] The fact that he did not 'offer too positive opinion of himself at first sight', which he ascribed to such physical infirmities as short-sightedness, an arched back, and difficulty pronouncing words, is a theme that crops up occasionally throughout his correspondence.[18] The fact, too, that he had a limited knowledge of the French language and French customs may, in his opinion, have contributed to making him sound slow and awkward, thus increasing his embarrassment.[19] Furthermore, as he confided to Habbeus, he did not 'drink to show off' and was 'inconvenienced by a great defect, namely often failing to respect rituals'; for he 'did not like to indulge in the ordinary entertainments' of young people. Instead he preferred 'to content his spirit by producing something real and useful for the public'.[20] Boineburg recommended Leibniz to the French secretary of state as an 'an inexhaustible treasure trove of all the beautiful sciences that a solid mind ha[d] ever been capable of'.[21] His mind was not only prodigious but free.[22] Perhaps surprisingly, however, Leibniz was also prone to joking: 'When I was in Paris, I used to say to gentlemen of the Royal Academy of Sciences that if

14. Leibniz to Ernst August, quoted in Davillé, 'Le séjour de Leibniz à Paris' (1923), 69.
15. Klopp, *Die Werke von Leibniz*, 3:126; Leibniz to Melle de Scudéry, in Foucher de Careil, *Oeuvres de Leibniz* (1859–75), 2:251–52; Davillé, 'Le séjour de Leibniz à Paris' (1923), 51; Leibniz to Melchior von Schönborn, early January 1676, A I, 1:397.
16. Leibniz to Habbeus, 5 May 1673, A I, 1:416.
17. Leibniz to Habbeus, A I, 1:416.
18. Davillé, 'Le séjour de Leibniz à Paris' (1920), 142.
19. Leibniz to Malebranche, first half of 1676, A II, 1:403; Leibniz to Boineburg, February 1672, A I, 1:257.
20. Leibniz to Habbeus, 5 May 1673, A I, 1:416–17.
21. Boineburg to Pomponne, 5 September and 9 November 1672, in Klopp, *Die Werke von Leibniz*, 2:138–40.
22. Letter from Schuller to Spinoza, 14 November 1675, quoted in Davillé, 'Le séjour de Leibniz à Paris' (1921), 144.

the souls who have returned to heaven pass through the stars, it is necessary to give Jupiter's moon to Galileo and to Mons. Cassini, but that it was necessary to leave Saturn to M. Huygens.'²³

A German Publicist and Agent in Paris

Leibniz's initial engagement with France had been primarily of a political nature, yet his plan was made redundant as soon as he set foot in Paris on account of France's declaration of war against the Dutch Republic.²⁴ Boineburg had furnished Leibniz with a letter of introduction to the minister for foreign affairs, Simon Arnauld de Pomponne, describing the bearer as 'a man, who despite appearances, w[ould] be able to do what he promises'.²⁵ Antoine Arnauld also recommended Leibniz to Pomponne, his nephew, in a letter of 12 September: 'The individual who presents you this proposal is a German gentleman with an exceedingly impressive mind, and who is very skilled in all kinds of science.'²⁶ Despite his initial interest in the Egyptian plan, Pomponne soon dismissed the whole idea as ridiculous and anachronistic, especially at a time when French policy inclined towards a strategic alliance with Turkey against the Holy Roman Emperor; as early as June 1672, Pomponne declared that 'holy wars ... ha[d] been out of fashion since Saint Louis'.²⁷ By September 1672, six months after arriving in Paris, Leibniz had still not been able to see Pomponne to discuss his Egyptian plan, towards which the minister might be swayed, if the Great Arnauld could be convinced to lend his support to Leibniz's scheme for the union of churches. In fact it seems unlikely that Leibniz ever met Pomponne, and ultimately nothing came of the project.

Yet Boineburg also had more prosaic reasons for sending Leibniz to Paris. These included the recovery of an outstanding pension owed to him by the French king and the transfer of income from a property in the Ardennes.²⁸ Boineburg had cultivated privileged connections with Parisian learned and political circles, to which he eagerly promoted his

---

23. Journal, April 1696, quoted in Davillé, 'Le séjour de Leibniz à Paris' (1921), 143.
24. Davillé, 'Le séjour de Leibniz à Paris' (1922), 16.
25. Quoted in Antognazza, *Leibniz* (2009), 124. See A I, 1:175; A I, 1:255; and Klopp, *Die Werke von Leibniz*, 2:125.
26. Arnauld to Arnauld de Pomponne, 12 September 1672, in Klopp, *Die Werke von Leibniz* 2:139.
27. Feuquières to Pomponne, 4 June 1672, and Pomponne's response, 21 June 1672, in Foucher de Careil, *Oeuvres de Leibniz*, 5:354–57, 359.
28. Antognazza, *Leibniz*, 124.

protégé.[29] He advised that Leibniz mention him in his correspondence, or 'write letters in [Boineburg's] name' by way of introduction in order to 'snatch favours'.[30] Boineburg's discussion of the Rhetellesian law is fascinating and particularly instructive for the light it sheds on the extensive network he thought he could mobilize and the special relationships he hoped to instrumentalize—including with the French finance officer Morell; with the envoy to Mainz, Du Fresnes; and even with Colbert—in order to obtain a prompt resolution of the matter.[31] But none of these connections was to prove efficacious. As with the promotion of the Egyptian mission, the recovery of the French rent and pension was a slow and arduous process, for access to the French court was limited and Leibniz ultimately remained at the mercy of circumstances and the bureaucratic machine.[32]

In November 1672 Boineburg additionally entrusted Leibniz with supervising his son's education. He requested that Leibniz draw up a programme to which the child should apply himself rigorously.[33] Wilhelm Boineburg came accompanied by Melchior Friedrich von Schönborn, who had been sent to present a new plan to Louis XIV, suggesting that a peace conference to end the hostilities in the Low Countries and northern Germany be convened in Cologne—a plan for which Leibniz's writing skills were once again enlisted.[34] This presented a new opportunity for Leibniz to play a conciliatory role by helping to establish peace with the Netherlands, even if he was not directly involved in the negotiations and the statement he had prepared was never read. Still, even though this attempt

29. Paasch, *Die Bibliothek des Johann Christian von Boineburg* (2003), 16.

30. Boineburg to Leibniz, 7 November 1672, A I, 1:285: 'You can write letters to [Arnauld] in my name as you like. I admire the strength of his argumentation, and his precision in examining everything down to the tiniest speck of dust for the sake of the truth.' Also Boineburg to Leibniz, November 1671, A I, 1:245: 'If you want to write something to [*sc*. Antoine] Arnauld, do it so that I can send it tomorrow by post. I will enclose your letter with mine to Du Fresnay. It is necessary that some references to me be made in it, snatching favours.'

31. Boineburg to Leibniz, 7 November 1672, A I, 1:285; Davillé, 'Le séjour de Leibniz à Paris' (1921), 68; and Boineburg to Leibniz, 7 November 1672, A I, 1:283: 'If Colbert deems it worthy to consider me, I will write to Him then, if it will be necessary. He may do the favour described above.' See also Boineburg to Leibniz, 30 April 1672, A I, 1:268.

32. Antognazza, *Leibniz*, 145.

33. Boineburg to Leibniz, 7 November 1672, A I, 1:285: 'Please stipulate for my son what, before all, he has to observe in his studies with his own diligence, so that something worthy of my expectation comes of it. Before all, take care everywhere and make an effort that he grows accustomed to discipline, and that he also attends to his health at the college.'

34. Antognazza, *Leibniz*, 146. See also Beiderbeck, 'Leibniz's Political Vision for Europe' (2014); Badalo-Dulong, *Trente ans de diplomatie française* (1956).

was unsuccessful, Leibniz, who had long wanted to visit the Royal Society in London, was later permitted to accompany the delegation there to obtain a hearing at the court of Charles II.[35]

During his stay in Paris, Leibniz continued the activities as a publicist that he had begun in Mainz, seeking to contribute to the political and diplomatic affairs of his country from afar.[36] He composed political tracts addressing practical issues, such as a pamphlet on behalf of the Duke of Mecklenburg, arguing for the nullity of the duke's second marriage as a Catholic and the validity of his first as a Protestant. Judging that this issue might be of broader interest among German princes and wishing to act as a good 'patriot', Leibniz set out to prove that the marriages of German Protestants contracted *in gradu prohibito* without dispensation from the pope were valid according to the Catholic Church.[37] Drawing on French law and history, the essay was immediately taken up by Jean de Lauravy in a book on royal power in marriage.[38]

Leibniz followed contemporary political and military events closely, composing in 1674 a treatise 'Des affaires de Suède', in 1673 a Latin poem on the naval victory of the Dutch over the united fleets of France and England, and in August 1674 the 'Relation du combat de Seneff donn. contre les confederes par Monseigneur le Prince de Conde', as well as transcribing a report on political affairs in 1674.[39] In Paris, too, he copied a text originally composed at the Hague by the Baron d'Isola in 1672 during the march of the German troops toward the Rhine that had been intercepted. Particularly concerned with political events affecting his homeland, Leibniz regularly requested news from Paris about the situation in Germany, including from Marshal Schönborn, Morell, and Huet, even questioning the soldiers of Turenne who had just made an expedition on the Lower Rhine in 1672.[40] He also sought and provided information from the courts of Mainz and Vienna on the general political state and the peace of the North.[41]

---

35. Antognazza, *Leibniz*, 148.

36. Davillé, 'Le séjour de Leibniz à Paris' (1921), 19.

37. Klopp, *Die Werke von Leibniz*, 3:127–29, quoted in Davillé, 'Le séjour de Leibniz à Paris' (1921), 20.

38. Leibniz, *De matrimoniorum Principum Germaniae Protestantium, in gradibus solo canonico jure prohibito contractorum validate Leibnitii dissertatio*, in Klopp, *Die Werke von Leibniz*, 3. See also Davillé, *Leibniz historien*, 21–23.

39. Davillé, 'Le séjour de Leibniz à Paris' (1921), 21; Klopp, *Die Werke von Leibniz*, 3:80–84; Bodemann, *Die Leibniz-Handschriften* (1895), 137, 179, 180.

40. Davillé 'Le séjour de Leibniz à Paris' (1922), 21.

41. Leibniz to Lincker, 27 July 1673, A I, 1:357–59; Leibniz to Melchior von Schönborn, early January 1676, A I, 1:397.

Leibniz seized on one particular incident to seek to position himself, by means of his legal acumen, as a major player on the German and European political scene. On 16 February 1674, Wilhelm Egon von Fürstenberg, the bishop of Strasbourg who was also an agent in the service of France in the ongoing struggle with the empire, was abducted by an imperial regiment and taken prisoner to Vienna. Considering this a violation of the rights of nations, Louis XIV ended the peace conference that had begun in Cologne the year before. Leibniz, however, saw this as an opportunity to play a broader role in helping broker the peace between the Dutch, the empire, and France and promptly drew up a report on peace (1674; recipient unspecified) as well as a pamphlet in which he both rebuked Fürstenberg's betrayal of the empire and condemned the emperor's actions and stated that Louis would enter negotiations only upon the release of Fürstenberg.[42] Possibly eyeing a role as a plenipotentiary in Vienna—and evidently viewing different political commitments not as incompatible but as part of a strategy to diversify his activities as much as possible—Leibniz was eager to be included in the peace congress that followed.[43] He was even invited in the spring of 1676 by the abbé de Gravel, France's extraordinary envoy to the court of Mainz, to accompany him. Leibniz's hopes for a genuine diplomatic role were dashed, however, when Duke Johann Friedrich, by then his employer, denied him permission to attend.[44]

Otherwise his legal work during his Parisian sojourn was confined to drafting a project for a commission of justice, probably on behalf of a German prince.[45] Thus although he was determined to play a political role, Leibniz's political activity and influence remained very limited, confined to the role of mediator or messenger, not least on account of his status as a commoner.[46] Most of his political writings, produced for

42. Davillé, 'Le séjour de Leibniz à Paris' (1922), 22; Klopp, *Die Werke von Leibniz* 3:84, 101–11.

43. Leibniz would in fact seek to have Crafft recommend him to Emperor Leopold I on several occasions over the course of the decade. See I, 2, 273. Klopp, *Die Werke von Leibniz*, 3:xxxii.

44. Leibniz to Johann Friedrich, mid-February 1676, Klopp, 3:293–94.

45. Leibniz to Johann Friedrich, 'Vorschlag wegen Aufrichtung eines Commissionsgerichts', 3/13 December 1676, A I, 2:5–7; Davillé, 'Le séjour de Leibniz à Paris' (1923), 52.

46. Scheel, 'Hermann Conring' (1983), 282, notes that 'German scholars of international renown on the political stage in the age of absolutism hardly had the opportunity to distinguish themselves as acting statesmen, even when they made an effort. They resigned themselves to the fact that political decisions were made by the prince himself and a few confidants, and that they were consulted as historico-political advisers only as a second resort on account of their knowledge and expertise. What applies to Conring in

specific occasions, had little effect on the broader diplomatic field. His Egyptian plan must have appeared amateurish and transparent, betraying an inability to escape from a scientifico-philosophical way of thinking, and only served to convey his failure to grasp the French foreign policy of the time—one that consisted in strengthening the alliance with the Ottoman Empire rather than overthrowing it to embark on an adventure with unforeseeable consequences.[47] To be sure, Leibniz's political misadventures can be traced back, at least in some measure, to Boineburg, who, his own position having been compromised, likely overestimated his influence and unwittingly hampered Leibniz—who after all was not acting in an official capacity for his prince—from gaining access to relevant Parisian dignitaries.

In another respect, however, Leibniz's Parisian sojourn was successful, for he found many 'wonderful opportunities' to acquire books.[48] These included books on 'curiosities', the sciences, and history.[49] A bibliophile from the age of six, Leibniz visited Paris's many booksellers, including Jean Cusson, publisher of the *Journal des Sçavans* and scientific pamphlets, and one on the rue Saint-Jacques.[50] He purchased many small *curieux* books and brought back from his brief stint in London the '*fleur des livres d'Angleterre*'.[51] Together with the volumes he had acquired during his formative years in Germany, these French and English books would form the nucleus of his private library. But the encounter with 'an infinity of curious French books in the arts and sciences as well as in matters of state' also prompted him to conceive an ideal library—one that did not stock 'books of little use' but rather either those containing 'inventions, demonstrations, [and] experiments' or 'state records, histories, especially of our time, and descriptions of countries'. Such a library, unlike those

---

Wolfenbüttel also applies to Leibniz's official work in Hanover.' Haase, 'Leibniz als Politiker und Diplomat' (1966), 196, characterises Leibniz's political activity as a 'succession of failures in comparison to his scientific work'. As Haase observes (197), only the princes and their mostly aristocratic ministers had the opportunity to be politically active.

47. 'In Leibniz's agile mind, every political problem immediately became a mental game that could be solved like a mathematical equation.' Haase, 'Leibniz als Politiker und Diplomat', 199.

48. Davillé, *Leibniz historien*, 23.

49. Davillé, 'Le séjour de Leibniz à Paris' (1923), 50.

50. Antognazza, *Leibniz*, 45; Leibniz to Kortholt, 9 January 1711; Dutens, *Gotholfredi Guillelmi Leibnitii* (1768), 5:315; Davillé, 'Le séjour de Leibniz à Paris' (1923), 50.

51. Leibniz to Christian Habbeus, 5 May 1673, A I, 1:417. On the books he collected, see 'Observata philosophica in itinere Anglicano sub initium anni 1673 March 1678', A VIII, 1:3–21.

that pass as 'beautiful and curious', would cost little and provide infinite illumination.⁵²

Leibniz's scientific activity was manifold, extending to philosophy and historical research, and he found in Parisian libraries a treasure trove of accessible manuscripts and other documents. Reminiscing at the end of his life about his time in Paris, he wrote, 'Having come to France in the year 1672 ... I brought from our universities all other knowledge than that of deep geometry. Law and history were my subjects. ... In Paris I burrowed myself in the great libraries; and I was looking for rare items, especially in history.'⁵³ He scoured the principal libraries assiduously for manuscripts and unpublished historical or clerical documents that would assist his composition of various treatises and essays.⁵⁴ At the royal library in particular he studied documents relating to the reigns of the French kings, memoirs recounting the relations of France with the Ottoman Empire, ambassadorial accounts of the rivalry between Austria and France, and documents relating to neighbouring European nations.⁵⁵ He also immersed himself in the political and ecclesiastical history of modern France as well as in documents relating to the church from Saint Louis to Louis XIII and the debates at the Council of Trent; consulted Greek, Hebrew, and even Arabic manuscripts, including those acquired by the Orientalist Vansleben and translated by the philosopher Géraud de Cordemoy; took an interest in cartography; and examined various ancient artefacts.⁵⁶

Political economy had preoccupied Leibniz since at least 1669.⁵⁷ Paris at the time was not only the most intellectually remarkable but the wealthiest European capital, and Leibniz set out to understand the reasons behind its prosperity, recording his observations and collecting as much data as he could, eventually identifying Colbert's measures to develop France's navigation, manufactures, and commerce as a crucial factor.⁵⁸

52. Leibniz to Christian Habbeus, 5 May 1673, A I, 1:417. Palumbo, 'Leibniz as Librarian' (2014), 610.

53. Leibniz to the Countess of Kilmansegg, 13 April 1716, quoted in Davillé, 'Le séjour de Leibniz à Paris' (1922), 139.

54. Davillé (1922), 139.

55. Davillé, *Leibniz historien*, 22–23.

56. Klopp, *Die Werke von Leibniz*, 2:148; Leibniz to Landgrave Ernst von Hessen-Rheinfels, end December 1691, A I, 7:229; Leibniz to Fontenelle, 12 July 1702, A II, 4:64; Leibniz to Bignon, 4 May 1710, A I, 26:138: 'The late Carcavy once showed me a medal on which there was inscribed ΚΡΉΤΗΣ ΑΟΡΛΟΎ and no one knew what this meant. This was in the King's Cabinet which, like His Majesty's library, had been entrusted to [Cacavy].'

57. Davillé, *Leibniz historien*, 592.

58. Davillé, 'Le séjour de Leibniz à Paris' (1922), 25.

Leibniz singled out France as a nation in which manufacture, for the most part, '[was] in the most flourishing state that could be wished for'.[59] Under the impulse of Jean-Baptiste Colbert's 1664 memorandum and resulting state-driven policy, Paris had successfully attracted talented artisans and craftsmen 'from all around' and, by 'spar[ing] nothing' to extract their secrets and inventions and granting them privileges, had blossomed into a hub of creativity—where Leibniz himself would encounter many artisans in connection with the construction of his arithmetical machine.[60]

Deeply impressed with Colbert's 'grand design of commerce and policy', Leibniz sought to collect as much information as possible on it, from researching in books to collecting relevant ordinances and policies to finding artisans and 'pinching their secrets': 'It will be important', he stated candidly, 'to poach from workers here the tricks and refinements of their secrets, which one can do sometimes with a bit of shrewdness combined with a little liberality.' A little cash and skill could go a long way in earning the trust of artisans and getting them to reveal their trade secrets, any one of which might be exploited successfully and reap great benefits under the correct guidance. Gaining insights into the latest technological developments and artisanal skills behind the most marvellous objects and 'curiosities' would be well worth the cost and the effort. Leibniz was a devoted practitioner of what he preached, both in his own interest and in that of his employers, as he succinctly explained to the Swedish resident in Mainz Christian Habbeus:

> For my part, I have had the opportunity not only to come into contact with a large number of good workmen, but also to draw something from them, and had I been willing to spend a little, I would have learned much more. For wanting to have my arithmetic machine built, with which the four species of calculation are carried out without the slightest work of mind—a model made a long time ago and shown in France and England to general praise and which is on the verge of being completed—I had occasion to get to know these people. And I will be able to multiply these contacts if my plan is approved, and the knowledge gained thereby can only benefit the country.

This *métropolitaine de la galanterie* now vaunted the most remarkable and versatile array of craftsmen and artisans. Of particular interest

59. Leibniz to Habbeus, 5 May 1673, A I, 1:417.
60. Bertucci, *Artisanal Enlightenment* (2017), 30–33; Leibniz to Habbeus, 5 May 1673, A I, 1:417: 'Wanting to have my arithmetical machine made', Leibniz explained in 1673, he came to know many artisans in Paris, the 'metropolis of *galanterie*'.

to Leibniz was practical or luxury manufacture like ironwork, metalwork, the production of silks and dyeing of fabrics, the production of porcelain and its varnishes, and 'an infinity of curiosities such as silverware, enamelling, watchmaking, glassmaking, leather factories, pewter pottery, to which most foreigners do not pay attention and yet only one of which would repay its expense if reported'.[61] Ultimately, his encounter with France's manufacturing miracle would make him all the more determined to extract artisans' secrets in order to send them to Germany, and more generally to use the knowledge he acquired of French economic practices for the sake of the German states' improvement.[62]

In his 'Mala Franciae' (c. 1672), however, Leibniz painted the rather different picture of a France whose provinces were miserable and exhausted, impoverished by taxes and constant warfare, by usury and the high cost of merchandise.[63] While 'Paris flourishe[d]', the provinces were 'depleted' and everything that was 'appealing in appearance' was in reality 'ugly and contorted'. Aside from the king, his nobles, and bankers, the rest of the country had been left to be 'consumed by a slow fire' and die of hunger.[64] Even as he celebrated Parisian prosperity and patronage of the sciences, Leibniz's ambivalence about France would only increase over time with Louis's growing bellicosity—an ambivalence culminating most famously, perhaps, with his *Mars Christianissimus* (1683).

61. Leibniz to Habbeus, 5 May 1673, A I, 1:417.
62. Leibniz to Johann Friedrich, 26 March 1673, A I, 1:488; 21 January 1675, A I, 1:493; Davillé, 'Le séjour de Leibniz à Paris' (1922), 27. Leibniz would in fact seek to promote collaborations between scholars and savants on the one hand and tradesmen and practitioners on the other, suggesting in spring 1682, for instance, that physicians in Nuremberg communicate their knowledge freely to tradesmen in the interest of science. It seems doubtful that Leibniz would himself have given up information without any form of compensation.
63. 'Mala Franciae', A IV, 1:516–17 (no. 38). The intended recipient is unknown. Also Davillé, 'Le séjour de Leibniz à Paris' (1922), 25.
64. A IV, 1:517.

CHAPTER THREE

Learning in Paris

EVEN THOUGH ENGLAND'S DECLARATION of war on Holland made the Egyptian plan quickly redundant, the plan helped secure Leibniz's passage to the intellectual capital of Europe, where he would remain until late 1676. Leibniz was keenly aware of and determined to seize the opportunity for intellectual development that Paris represented.[1] As he would confide to Duke Johann Friedrich in January 1675, Paris was a 'place where one can achieve distinction only with difficulty. One finds there, in all branches of knowledge, the most knowledgeable men of the age, and one needs much work and a little determination to establish a reputation there'.[2] Prior to his arrival in Paris, Leibniz had been keeping abreast of French intellectual life, notably through his friend Louis Ferrand, who informed him of the latest theological and scientific controversies and publications. The pietist J. H. Horb had provided Leibniz with names of the directors of the Parisian libraries and information about books, including the grammarian and historian Gilles Ménage's works on the origins of French and Italian, and the German theologian Johann Leyser had reported to him about some of the royal academy's activities.[3]

Leibniz's Parisian stay, especially his first and last years, proved conducive to his metaphysical reflections.[4] He was committed to continuing his project of a philosophical refutation of atheism and of the reconciliation of mechanical philosophy with revealed religion, thereby furthering the work he had initiated with his *Demonstrationes Catholicae* a few years

---

1. Barber, *Leibniz in France* (1995), 4.
2. Leibniz to Johann Friedrich, 21 January 1675, A I, 1:491–92.
3. Johann Heinrich Horb to Leibniz, 1 July 1670, A I, 1:95; Johann Leyser to Leibniz, 13 July 1671, A I, 1:158–59.
4. Antognazza, *Leibniz* (2009), 144.

earlier. In the *Demonstrationes*, he undertook to defend Christian mysteries against the charges of incoherence and inconsistency. These writings reflect Leibniz's enduring preoccupation with the practical applications to be derived from his theoretical insights and the placement of knowledge in the service of the public good. Like the rest of Leibniz's philosophical works of that period, they would remain unpublished during his lifetime.[5] In them, he broached a number of philosophical and theological topics, including divine justice, divine and human freedom, and the compatibility of the latter with the principle of sufficient reason.[6] Evil was ultimately explained as inevitably required by the maximum of positive 'compossibilities' that made up the best possible world.[7] During the autumn and winter of 1672–73, Leibniz composed the *Confessio philosophi*, and from December 1675 to April 1676 a series of metaphysical papers generally referred to as *De summa rerum*, which are often construed as his first major attempt at presenting his metaphysical ideas in a systematic form.[8] In them, he sketched out many elements of his mature philosophy, including the principle of harmony that he would come to define in a 1671 letter to Arnauld as 'similitude in variety, or diversity compensated by identity'.[9] According to this principle, everything is connected with everything else in the universe, and the universe displays the greatest perfection and metaphysical goodness. Another key metaphysical principle found in the *De summa rerum* is the idea that every mind perceives and expresses the universe, albeit in a confused manner.[10]

It was in natural philosophy, however, that Leibniz made his greatest strides. Nowhere, he believed, in a city 'paved with so many learned people' could one better perfect the sciences than in France.[11] When he was not visiting friends, conducting his correspondence, or attending to his official duties, Leibniz was left with a little time to 'use for the research into Nature and Mathematical reflections'.[12] He took full advantage of what Paris had

5. Antognazza, 144.

6. Mercer, 'The Young Leibniz' (1999).

7. See Antognazza, *Leibniz*, 145.

8. *Confessio Philosophi*, Autumn 1672–Winter 1672–73, A VI, 3:115–49; Parkinson, 'Sufficient Reason and Human Freedom' (1999), 199–222.

9. Moll, 'Deus sive harmonia universalis' (1999), 65–78.

10. Antognazza, *Leibniz*, 163.

11. Leibniz to Johann Friedrich, September 1671, Klopp, *Die Werke von Leibniz* (1864–84), 2:9; Leibniz to Ferrand, May 1672, A I, 1:452.

12. Leibniz to Oldenburg, 15 July 1675, A III, 1:119. On one occasion at least, Leibniz complains, 'I owe you two letters, and I ask you not make anything of my silence. For I am usually interrupted a number of times and attend to these endeavours in intervals' (Leibniz to Oldenburg, 28 December 1675, A II, 1:393). See also Leibniz to Périer, 30 August 1676,

to offer in matters of science, attending, for example, an experiment on the movement of water conducted by Foucher in the presence of Mariotte and Dalancé.[13] Leibniz extended his geometrical method to problems of optics and broached astronomical topics with certain members of the Academy, including Huygens, Cassini, and probably Roemer, as well as some English savants.[14] He also pursued his work on mechanics, in particular relating to laws of movement and the collision of elastic bodies.[15]

Whereas he had concentrated on philosophy and jurisprudence before travelling to France, once in Paris Leibniz dedicated himself intensively to the study of mathematics, of which he had hitherto had a superficial grasp.[16] Leibniz was particularly keen to improve his mathematical knowledge and for this purpose sought out mathematicians, many of whom were Jesuits.[17] These included two teachers at the Collège de Clermont, Chalais and the Père Pardies, the author of a theory on logarithms; Deschales, whose algebra, Leibniz thought, left a bit to be desired; and Jean Prestet (like Arnauld an Oratorian priest), whose algebra and *Eléments de mathématiques* for its part impressed him.[18] Leibniz discussed algebraic curves with Jacques Ozanam, a self-taught geometer who was

---

A III, 1:588: 'The great number of distractions . . . do not let me dispose completely of my time.'

13. Letter to Foucher, 23 May 1687, A II, 2:207.

14. Davillé, 'Le séjour de Leibniz à Paris' (1921), 165 no. 2, 70; (1920), 143 no. 8; 'Excerpta ex Hookio contra Hevelium' (1674), Bodemann, *Die Leibniz-Handschriften* (1895), 316.

15. Leibniz to Vincent Placcius, 10 May 1676, A II, 1:407: 'I have studied the mathematical sciences with particular care . . . so that I am deemed—in Paris, among such a great number of scholars—as someone who should not altogether be despised. . . . Thus if I have achieved certainly this, if nothing else, then I have learnt true analysis and genuine arts of proof from eminent examples, and will make if not many, then certainly better contributions to the science of law.' See also Brown, *The Young Leibniz* (1999); Garber, 'Motion and Metaphysics' (1982); and McDonough, 'Leibniz's Philosophy of Physics' (2021).

16. Goethe, Beeley, and Rabouin, *G. W. Leibniz* (2015), 57–58. See also Klopp, *Die Werke von Leibniz*, 3:272–73: 'I am attending here neither to jurisprudence, nor to *belles lettres*, nor to controversies (things that principally occupied me in Germany), and instead I have begun basic studies to understand mathematics, in which I have had the good fortune to discover thoroughly unfamiliar truths, as letters from the most skilled mathematicians of our time will attest.'

17. Davillé, 'Le séjour de Leibniz à Paris' (1921), 71.

18. Leibniz to Gerhard Meier, 15 December 1691, quoted in Davillé (1921), 71: 'I once knew Chalesius . . . the author of a course in mathematics, intimately in Paris, where he taught in Collège Clairemont. He was a good and learned man, and even though he was not going to make much of a mark on the basis of his own merits, he was yet able to expound the ideas of others clearly and methodically.' On the Jesuit mathematicians, see Dainville, *L'Éducation des Jesuits* (1978); Feingold, *Jesuit Science* (2002). On Prestet, see Bodemann, *Die Leibniz-Handschriften* (1895), 314.

preparing an edition of Diophantus, and regularly consulted Jacques Sauveur on matters of analysis and geometry.[19] He also had discussions with Claude Hardy, an 'excellent geometer and orientalist' who had edited a volume on Euclid; with the mathematician, astronomer, and Hellenist Ismaël Bouillau; with the royal counsellor and distinguished geometer Frénicle de Bessy; and of course with the Abbé Mariotte, whose extraordinary talent for unearthing nature's ingenuity Leibniz marvelled at.[20]

During his short trip to England in early 1673, Leibniz attended the meetings of the Royal Society, where he gave a demonstration of his still imperfect arithmetical machine.[21] This visit was successful overall (aside from a mishap with the mathematician John Pell, who was irked by Leibniz's inaccurate assertion to have been the first to interpolate numerical series, something that had already been achieved by François Regnaud), providing Leibniz with precious contacts in the English scientific world, including Robert Boyle, John Wallis, John Collins, and Robert Hooke.[22] Hooke, the curator of the Royal Society, chose to belittle the machine, which was designed to multiply and divide, before later producing a similar one himself.[23] But Wallis, the Savilian professor of geometry who had been impressed with Leibniz's work on physics (notably the *Hypothesis physica nova*)—and whose own *Mechanica sive de motu* (1670-71) would influence the development of Leibniz's theory of dynamics—wrote a glowing report.[24]

The secretary of the Royal Society, Henry Oldenburg, who was keen to encourage new talent in conformity with his pan-European vision of knowledge and to promote scientific activity in his native Germany, had been a tireless promoter of Leibniz's work even before the latter's arrival in London, helping publish and disseminate his 1671 *Hypothesis physica nova* and *Theoria motus abstracti*, and played a crucial role in assisting

---

19. Davillé, 'Le séjour de Leibniz à Paris' (1921), 72 ; Leibniz to Jean Bernoulli, 29 January 1697, A III, 7:293.

20. *Nouveaux essais sur l'entendement humain*, A VI, 6:408 (book 4, chap. 7, no. 4), and 6:489 (book 4, chap. 17, no. 13). On Bouillau's correspondence, see Hatch, 'Between Erudition and Science' (1998). On the Abbé, see Roux, *L'Essai de logique de Mariotte* (2011), 130.

21. See Hall, 'Leibniz and the Royal Society' (1978); Beeley, 'A Philosophical Apprenticeship' (2004). For more on Leibniz's trip to London, see Hirsch, *Der berühmte Herr Leibniz* (2000), 39-49.

22. Hofmann, *Leibniz in Paris* (1974), 299. Oldenburg reported to Huygens on 19 February 1673 that 'Monsieur Leibniz has won a lot of esteem here, as he most assuredly deserves'. Huygens, *Oeuvres complètes* (1888-1955), 7:256.

23. Arthur, *Leibniz* (2014), 21.

24. Beeley, 'Learned Discourse' (2018), 27.

Leibniz in the English and Parisian intellectual scenes to establish a reputation as a promising young scholar.[25] By withholding, without the author's knowledge, a letter that Leibniz had intended for Hobbes, in which he had laid out some of his early projects regarding a universal language and his plan to reduce jurisprudence to a demonstrative system, Oldenburg probably shielded Leibniz from the antipathy with which many at the Royal Society, including Wallis, viewed Hobbes.[26] Following Leibniz's election to the society, Oldenburg did not fail to remind its newest member of his promise to send a working exemplar of his arithmetical machine to London once the design was complete, and of his obligation to communicate news of the latest publications and scientific developments in his native country.[27] Leibniz had committed himself to 'bring before the public' those matters which he or 'others in Germany' had 'pursued by reflection and experience in physics or mechanics'.[28]

Ironically, perhaps, Leibniz's visit to London and reception within English scholarly circles proved decisive for the remainder of his Parisian sojourn and served as a major impulse behind some of his discoveries.[29] Even though he was elected to the Royal Society in April 1673, Leibniz was made painfully aware of the limitations of his mathematical knowledge there and sought to remedy this situation immediately upon his return to France, where the Dutch physicist Christiaan Huygens (1629-95) took him under his wing after receiving a glowing recommendation of the young scholar from Oldenburg.[30]

Under Huygens's benevolent mentorship, Leibniz would develop into a highly able mathematician who, by the end of his time in Paris, had invented the calculus, even though this discovery would not be disseminated for several years.[31] By 1660 Huygens had achieved a position of considerable renown and influence in the Republic of Letters, especially within Parisian scientific circles.[32] So it is not surprising that France's all-powerful controller-general of finances, who had already consulted the omnicompetent and practically minded Huygens on a number of technical projects,

25. Beeley, 'A Philosophical Apprenticeship', 68, 48.
26. Beeley, 46-47.
27. Beeley, 69.
28. Oldenburg to Leibniz, 20 April 1673, A III, 1:79.
29. Hofmann, *Leibniz in Paris*, 23-35; Bos, 'The Influence of Huygens' (1978). For an account of Leibniz's second short visit to London, see Beeley, 'Learned Discourse'.
30. Salomon-Bayet, 'Les académies scientifiques' (1978), 164.
31. See Hofmann, *Leibniz in Paris*, on this remarkable transformation.
32. See Taton, *Huygens et la France* (1981); and more particularly Mesnard, "Les premières relations parisiennes' (1981), 39.

entrusted him with the organization and direction of the new Académie des Sciences in 1666.[33] Inspired by the Royal Society's strong experimental and observational bent, Huygens set out to shape the new programmes of research at the Académie along experimental lines, placing the pursuit of a non-Cartesian 'useful' science at the heart of its activities—in contrast to the excessively rhetorical cast that had plagued and ultimately doomed the private Parisian societies, such as the Académie de Montmor, in the 1660s.[34] His influence within the Académie des Sciences, however, would remain limited in a number of ways by the intellectual dominance of some of its members.[35]

The first meeting between Leibniz and Huygens probably took place in autumn 1672 and concerned the topic of the summation of series. Huygens posed the problem of the sum of the reciprocal triangular numbers, which Leibniz had succeeded in solving, thereby gaining the former's praise.[36] Huygens recognized Leibniz's talent and potential but also discerned the numerous lacunas in his mathematical knowledge. While possessing extraordinary natural talent, Leibniz had come late to mathematics, and Huygens would act as a mentor to him, especially upon his return from London, introducing him to key mathematical literature (including the work of Pascal, Grégoire de St Vincent, Descartes, Sluse, Cavalieri, Guldin, Torricelli, Wallis, and others) and suggesting problems.[37] Huygens seems not only to have had a decisive influence on Leibniz, but to have taken a genuine interest in his student's work and research interests. The two met regularly, Huygens stimulating Leibniz's mathematical interest through discussions, teaching him analytical geometry, supplying him with the tools with which he would later attack Cartesian motion, and setting him a standard of careful mathematical work.[38] In general, Huygens seems to have sought to steer Leibniz in the right direction, helping him overcome his imperfections, without dampening his eagerness and enthusiasm.[39]

33. Hahn, 'Huygens and France' (1980), 62; Roger, 'La politique intellectuelle' (1981), 44: 'Monsieur Colbert refers most inventors to me to know if they can contribute anything substantial, which is very rare.' On the founding of the Académie, see Taton, *Origines de l'Académie* (1966); and Hahn, *The Anatomy of a Scientific Institution* (1971).

34. Taton, 'Huygens et l'Académie' (1981).

35. Roberval in mathematics, Auzout in astronomy, and Mariotte in physics: see Hahn, 'Huygens and France'.

36. Hofmann, *Leibniz in Paris*, 44.

37. Hofmann, 1, 47-50; also Bos, 'The Influence of Huygens' (1978).

38. See Leibniz to Jacob Bernoulli, April 1703, A III, 9:187 (Vorausedition); Bos, 'The Influence of Huygens', 65-66.

39. Hofmann, *Leibniz in Paris*, 47-50; see also Heinekamp, 'Christiaan Huygens' (1981), 202, on Huygens as a 'stimulator and model' for Leibniz.

Leibniz and Huygens were bound by a strong and mutually shared relation of respect and admiration.[40] Leibniz consulted Huygens on various matters and welcomed his criticism, and it was through the latter that the former quickly gained a solid reputation as a mathematician. With time, Leibniz became more autonomous, taking the initiative rather than waiting for Huygens to set a problem, and their divergent styles and research interests became more pronounced: Leibniz developed special interests in algebra and number theory, while Huygens's interests remained strongly geometrical and problem-oriented.[41] Over the next few months, Leibniz's determination to perfect Cartesian analysis and develop a new algebraic method of calculus led him to various discoveries, including a method for summing infinite series, his 'transmutation theorem' whereby a 'characteristic triangle' is deployed to deduce the transformations of quadratures, his 'quadrature of the circle' (which Huygens passed on to the Académie), and finally in October 1675 his infinitesimal calculus (which allowed the mathematical treatment of all kinds of known curves).[42] To be sure, Huygens accepted the advantages of the calculus only reluctantly, but Leibniz's great discovery was communicated to various figures, including Oldenburg.[43]

Even before arriving physically in Paris, in fact, Leibniz had sought to leverage his mathematical discoveries to gain entry to the Parisian mathematical community, dedicating his *Theoria motus abstracti* (1671) to the Académie, communicating his various advances to Jean Gallois (himself a mathematician of note), and trying to have his article 'Accessio ad arithmeticam infinitorum' (in which he set out to reinterpret the infinite in terms of an infinite progression within finite limits) published in the *Journal des Scavans*.[44] Having entrusted, when he departed for Hanover, a manuscript on the quadrature to the mathematician Soudry for publication in Paris, Leibniz now hoped that his calculus would help him secure a position at the Académie.[45]

More generally, Leibniz conceived his algebraic discoveries as part and parcel of a more general universal language, a *characteristica universalis*.

40. Huygens to Leibniz, end of 1676, A III, 1:429.
41. Bos, 'The Influence of Huygens', 67.
42. See on this Hofmann, *Leibniz in Paris*; Jones, *The Good Life* (2006); Crippa, 'Impossibility of Squaring the Circle' (2019).
43. See Leibniz to Oldenburg, 28 December 1675, A III, 1:331.
44. Leibniz to Carcavy, 22 June 1671, A II, 1:209; Leibniz to Gallois, end of 1672, A III, 1:2-20; Antognazza, *Leibniz*, 170.
45. Introduction, A III, 1:lxiv-v; Hofmann, *Aus der Frühzeit der Infinitesimalmethoden* (1965), 290-305.

In the same letter of 19 December 1678 to Gallois in which he mentioned his *analysis situs*, Leibniz reiterated that his conception of the *characteristica* marked a considerable advance on previous attempts to create an artificial universal language.[46] This was owing to the algebraic and arithmetic models on which the *characteristica* drew and which, with its use of characters, made it possible 'to fix our vague and fleeting thoughts' and hence to 'assist discovery and judgement'.[47] This method would be instrumental in forestalling disputes and controversies, including in metaphysics and morals, because it would organize thoughts in such a way as to 'illuminate' the thought process instead of confusing it.[48] The identification of primitive concepts and their combination according to predetermined rules would thus help pave the way for a type of logical calculus in which each concept would be assigned an appropriate 'sign' or 'symbol' to represent it. This 'combinatory characteristic' would help perfect the mind, providing it with a *filum meditandi* that would render thinking mechanical by making all knowledge accessible and logically discoverable and thus serve as the basis for an all-encompassing encyclopedia of sciences aimed at promoting human improvement.[49]

46. Antognazza, *Leibniz*, 246.
47. Leibniz to Gallois, September 1677, A II, 1:569.
48. Leibniz to Gallois, 19 December 1678, A I, 1:242.
49. Leibniz to Oldenburg, 28 December 1675, A II, 1:393; 1673–76, A II, 1:378–79; Antognazza, *Leibniz*, 172.

CHAPTER FOUR

Engaging with Cartesianism

LEIBNIZ FULLY IMMERSED HIMSELF in the controversies of his time, cultivating relationships with philosophers of various stripes. The years 1672 to 1679 correspond to the height of some of the debates surrounding Cartesianism and scepticism. Leibniz's Parisian life provided the ideal environment in which to apprise himself of contemporary intellectual currents while maintaining relatively detached, sounding out various hypotheses and ideas for the benefit of his own intellectual and philosophical development.[1] Some of Leibniz's close friendships acted as a propaedeutic to his own thinking, providing the backdrop against which he could clarify his own views, and they turned out to be instrumental in shaping many of his most important writings, not least through the discussion of objections.[2] Leibniz thrived on the intellectual stimulation to seek answers, and he displayed a predilection for 'ingenious badinage' with the aim of 'perfecting' and refining his own philosophical positions.[3]

In Paris, Leibniz engaged more thoroughly than before with Cartesian philosophy, with the aim of identifying its weaknesses and errors.[4] His

---

1. See Popkin, 'Leibniz and the French Sceptics' (1966).

2. Popkin's investigation into Leibniz's relationships with apparently contrary types of philosophers led him to acknowledge that Leibniz, after all, was not as far from the sceptics as is commonly supposed: 'And while Leibniz was certainly no sceptic, nor a man who was particularly concerned with *la crise pyrrhonienne* of the seventeenth century, he was regarded as a closer friend intellectually by the sceptics of his age than any of the other metaphysicians of the period' ('Leibniz and the French Sceptics', 228). Leibniz's affinity with scepticism can be understood in the context of his rejection of dogmatism.

3. Desmaizeaux, *Recueil de diverses pièces* (1740), 1:xxvii.

4. Leibniz to Foucher, 1675, A II, 1:388–89: 'I admit that I have not yet been able to read his writings with all the care that I have undertaken to take. . . . However, what I do know of the metaphysical and physical meditations of Mons. des Cartes I have gleaned

criticism of Cartesian doctrines had begun early, and doubtless emboldened by his successes in mathematics—which he believed exposed the superiority of infinitesimal calculus over Cartesian analysis—Leibniz would develop this criticism in his exchanges with the Parisian learned community.[5] To this end he borrowed manuscripts across Paris, including those of Pascal and Descartes, consulting those available at the home of Clerselier, compiling extracts on mathematical, philosophical, anatomical, and medical topics.[6] While conceding Cartesianism to be 'the antechamber' of the truth, he took issue with most Cartesian doctrines and set out to refute some of Descartes's key tenets.[7] Leibniz judged Descartes's mechanics 'full of errors', his geometry 'limited', his physics 'too hasty', and his metaphysics 'all of these traits at once'.[8] In particular he objected to Descartes's conception of motion, which banished final causes, and he sought instead to relocate the final reasons underlying mechanism and the laws of movement in metaphysical substances, in which connection he later immersed himself in Plato's works.[9]

'Le grand Arnauld'

A particularly decisive and important relationship that Leibniz forged in Paris was with Antoine Arnauld (1612–1694), the Jansenist theologian and philosopher who would later prove to be one of the most penetrating contemporary critics of Leibniz's developing system in the 1680s.[10] Arnauld

---

from reading more familiar texts which report on his opinions.' Letter to Malebranche, 27 December 1694, A II, 3:4. See also the introduction to A I, 2.

5. See the introduction to A II, 1:xxxv–vi, lii; and Leibniz to Foucher, 1675, A II, 1:388–89.

6. Leibniz to Jean Bernoulli, 29 January 1697, A III, 7:292; Bodemann, *Die Leibniz-Handschriften* (1895), 44.

7. Leibniz to Christian Philipp, early December 1679, A II, 1:767; Leibniz to Malebranche, 13/23 January 1679, A II, 1:678: '[Descartes] has only laid the groundwork without getting to the bottom of things: it seems to me that he is still far removed from true analysis and from the art of invention in general.'

8. Leibniz to Malebranche, 13/23 January 1679, A II, 1:455.

9. Davillé, 'Le séjour de Leibniz à Paris' (1923), 57; see also Leibniz's extracts from the *Phaedo* and *Theaetetus*, March–April 1676 (?), in A VI, 3:283–310.

10. Leibniz, 'New System of Nature' (1871 [1695]), 209: 'It is now some years since I conceived this system, and entered into communication about it with several learned men, and in particular with one of the greatest theologians and philosophers of our time.' On completion of the *Discourse on Metaphysics*, Leibniz wrote from the Harz mountains to Landgrave Ernst von Hessen-Rheinfels on 11 February 1686, A I, 4:399: 'Finding myself in a place where for some days I had nothing to do, I have composed a small discourse on metaphysics, about which I would be very pleased to have the opinion of M. Arnauld.

had acquired a reputation for logical acuity and intellectual rigour, choosing to tackle complex philosophical problems, and corresponded at great length with the leading scientists and thinkers of his time, including the philosopher Nicholas Malebranche, whose three-volume treatise *De la recherche de la vérité* Arnauld later attacked. In 1669, taking advantage of the lull in hostilities between the crown and the Jansenists, Arnauld had published with Pierre Nicole the first volume of his *La Perpétuité de la Foi catholique touchant L'Eucharistie*, in which he defended the Catholic faith against the Calvinists. This volume had found its way to the court of Mainz, where Leibniz's employer, Johann Christian von Boineburg, a convert to Catholicism who shared Leibniz's irenicism, suggested that Leibniz write to Arnauld on the basis of his previous work on religious reconciliation. Leibniz, of course, knew about Arnauld and his writings on transubstantiation and the Roman Church.[11] He mentioned to his nephew, Simon Löffler, Arnauld's indictment of Jesuit morality and the controversy that had ensued in the wake of Antoine and Isaac Lemaître de Sacy's 1667 translation of the New Testament into French.[12]

Leibniz had also been deeply impressed by Arnauld's *Logique* and *Nouveaux éléments de géométrie* (1667) and was keen to enter in contact with the most outstanding philosophical and theological controversialist of the time, one whose approval 'would carry a great weight', as Boineburg remarked to Leibniz.[13] Arnauld had been hailed as *le plus grand homme du Royaume*, whose 'golden quill', in the words of Boileau, excelled in various domains.[14] He was seen as a moral leader of the Church of France, like the king himself. Leibniz's friend Louis Ferrand had also provided Leibniz with journals and works pertaining to Arnauld, and Leibniz had developed a special affinity for Port-Royal writings, discussing them with his pietist friends Horb and Spener.[15] These seemed to resonate particularly with his own intellectual preoccupations and ethos,

---

For the questions of grace, the concourse of God and creatures, the nature of miracles, the cause of sin and the origin of evil, the immortality of the soul, ideas, etc., all are addressed in a manner that seems to open up new possibilities for the elucidation of very great difficulties. I have enclosed the summary of the articles it contains, for I have not yet been able to have a fair copy made. I therefore beg Your Supreme Highness to have this summary sent to him and that he be requested to give it a little consideration and to state his opinion.'

11. Horb to Leibniz, 1 July 1670, A I, 1:96, and 2/12 Mars 1671, A I, 1:128; Spener to Leibniz, 10/20 January, A I, 1:112, 16/26 February, A I, 1:122, and 8 March 1671, A I, 1:131; Ferrand to Leibniz, 13 February 1671, A I, 1:119–21.

12. Leibniz to Simon Löffler, 20/30 April 1669, A I, 1:75.

13. Leibniz reported this to Duke Johann Friedrich in autumn 1679: A II, 1:751.

14. Quotations from Orcibal, 'Leibniz et l'irénisme d'Antoine Arnauld' (1978), 15.

15. Orcibal, 15; Rodis-Lewis, *Lettres de Leibniz à Arnauld* (1952), 3.

namely, an attachment to the purity of internal inspiration and the linking of apologetics to an ideal of mathematical rigour and more formal art of reasoning.[16]

By the time Leibniz made contact with Arnauld, the latter had thus already lived a full life with its lot of struggles, controversies, and polemics, not least the affair of *fréquente communion* in 1643. His subsequent refusal to submit to papal authority and accept Jansen's text as heretical had resulted in his removal from the faculty of the Sorbonne in 1656 and forced him to go into hiding after the 'crisis of the Formulaire' in 1661, in which Jesuits and Jansenists had battled over the orthodoxy of Jansen's propositions.[17] Now rehabilitated after the so-called Peace of the Church, which had helped consecrate him as an illustrious representative of French thought—his family's fame would be hailed in the *Journal des Sçavans* of September 1675—Arnauld was at the height of his fame as a philosopher and a theologian when Leibniz, encouraged by Boineburg to present his ideas for theological reconciliation, first wrote to him in early November 1671, intent on securing his approval on metaphysical questions.[18]

Beyond the project of reunification of the churches, Leibniz discerned a more urgent threat to Christianity in Cartesianism: he took particular issue with the Cartesian dogma that extension constituted the essence of matter, which, according to him, imperilled the doctrine of the Eucharist by declaring it impossible for one substance to assume the physical characteristics and extension of another and yet retain its identity, as the Catholic doctrine of transubstantiation held.[19] Already in his *Catholic Demonstrations*, Leibniz had set out to establish the possibility of mysteries, and in particular that of transubstantiation, which had recently become the subject of a controversy when Arnauld defended the doctrine of the Real Presence against the Calvinist minister Jean Claude.[20]

In his letter Leibniz claimed to have found a solution to what had turned into a highly explosive issue for the Cartesian Jansenists in France, after the contradiction between the Cartesian philosophy and the Trentine doctrine of transubstantiation had led Descartes's works being put

---

16. Rodis-Lewis, 4.
17. Sleigh, *Leibniz and Arnauld* (1990), 27–28.
18. Boineburg to Leibniz, November 1671, A I, 1:245.
19. See Goldenbaum, 'Transubstantiation, Physics and Philosophy' (1999); Leibniz to Arnauld, early November 1671, A II, 1:277.
20. Leibniz to Johann Friedrich, October 1671, A I, 2:163.

on the Index in 1663.²¹ This solution would be compatible with mechanical principles while maintaining the mysterious nature of the sacrament and the indestructibility of the mind.²² Leibniz drew here on his work on motion, which he conceived as inextricably connected to his broader philosophical and theological meditations.²³ In his *Theoria motus abstracti* and his *Confessio Naturae*, Leibniz had redefined matter as motion, the mind as *conatus*, its moving principle, and the body as a *mens momentanea*, that is, a momentary mind: thus 'the substance of one and the same body [could] be in many different distant places, or, what is the same, amongst numerous external appearances'.²⁴ From the 'philosophy of movement' to 'the science of the mind' there was only 'one step'. Leibniz therefore bypassed the difficulties posed by transubstantiation and established the possibility of the physical Presence of Christ.²⁵ This approach also allowed him to challenge the Calvinist refusal to accept the Real Presence and justify his own support for the Catholic doctrine of transubstantiation, with its proximity to the Lutheran doctrine of consubstantiation.²⁶

Arnauld was one among several high-profile thinkers, alongside Hobbes and Spinoza, to whom Leibniz wrote in his bid to ingratiate himself with renowned figures in the learned world, and to gain access to elite intellectual circles.²⁷ By writing to Arnauld, Leibniz was seeking the approval of an established thinker—perhaps the most established philosopher and theologian in France—in the hope of gaining support for his programme of religious reconciliation. Arnauld, however, seems to have been reluctant to enter into conversation with Leibniz and left the letter unanswered. Leibniz presumably had great expectations of Arnauld

---

21. See Goldenbaum, 'Transubstantiation, Physics and Philosophy'.
22. Leibniz to Arnauld, early November 1671, A II, 1:277.
23. Nadler, *The Best of All Possible Worlds* (2010), 91–92.
24. Leibniz to Arnauld, early November 1671, A II, 1:277. See also Leibniz's letter of 5 May 1671 to Lambert van Velthuysen, A II, 1:164: 'Just as the physical can be explained by spaces and movements, so the mental can be explained by stimulations and strivings [*punctis et conatibus*].' Although Leibniz appropriated Hobbes's term *conatus*, he firmly rejected the latter's denial of minds. This concept of a 'moving mind' provided the foundation on which Leibniz would later develop his theory of substance as self-standing and self-acting. Leibniz to Arnauld, early November 1671, A II, 1:277 (and repeated in his letter to Johann Friedrich, 21 May 1671, A II, 1:108).
25. Merely demonstrating the possibility of the mystery would be sufficient, Leibniz explained to Arnauld: 'Just as one must consider that a geometer has really solved a problem, when ... he has demonstrated its possibility and verified it with other already solved problems.' Leibniz to Arnauld, early November 1671, A II, 1:277.
26. Goldenbaum, 'Transubstantiation, Physics and Philosophy'.
27. Nadler, *The Best of All Possible Worlds*, 70.

because the latter had engineered a first attempt at reconciliation in 1668 with his *Treatise on the Perpetuity of the Faith* and thus would likely be instrumental in a broader reconciliation of the churches. Arnauld's non-response notwithstanding, Leibniz would eventually meet him three or four times during his first six months in Paris.[28] As he confided on 26 March 1673 to Johann Friedrich—his last remaining German supporter after both Boineburg and Schönborn had died—Leibniz was deeply impressed with Arnauld, whom he held to be one of the few theologians capable of helping engineer a religious rapprochement:

> The famous Arnauld is a man of the most profound and wide-ranging thought that a true philosopher can have; his aim is not only to illuminate hearts with the clarity of religion, but, further, to revive the flame of reason, eclipsed by human passions; not only to convert heretics, but, further, those who are today in the greatest heresy—the atheists and libertines; not only to vanquish his opponents, but, further, to improve those of his persuasion. His thoughts, then, come to seeking how, so far as it is possible, a reform of abuses, frankly widespread among dissidents, would overcome the cause of the division. In this design, on several points of importance, he has made the first step and, as a prudent man, he proceeds by degrees. I am distressed that we have lost von Boineburg just when I have struck up an acquaintance with Arnauld; for I had hoped to bring these two minds, so similar in their honest soundness, on the road to a closer agreement. The Church, as well as the fatherland, has sustained a loss with this man.[29]

We do not know with certainty what was discussed during those meetings, but it seems highly probable, on the basis of his first letter to the great theologian, that Leibniz would have attempted to steer the conversation towards religious matters, and in particular to broach the subject of a potential religious rapprochement. In Paris, Leibniz was still very much preoccupied with theological problems which he conceived as inextricable from his scientific pursuits.[30]

Arnauld, it seems, entertained other ideas—of a more scientific order. He himself later reported to Landgrave Ernst von Hessen-Rheinfels that

---

28. Antoine Arnauld to Arnauld de Pomponne, 12 September 1672, Klopp, *Die Werke von Leibniz* (1864–84), 2:139.

29. Leibniz to Johann Friedrich, 26 March 1673, A I, 1:487–88. Quoted in Sleigh, *Leibniz and Arnauld* (1990), 15–16.

30. Leibniz to Johann Friedrich, Autumn 1679, A II, 1:761: 'My meditations were primarily of a theological nature.'

Leibniz would come to 'see him regularly' when in Paris.³¹ Arnauld lauded the young man's abilities, notably in mathematics, and even set him a problem on triangulation in December 1675.³² Arnauld was a mathematician of repute in his own right, publishing the *Nouveaux éléments de géométrie* (1667), inspired by Euclid's *Elements* and intended as a new method of teaching geometry, as well as a *Traité des proportions* (1683).

During those meetings held at Arnauld's house, Leibniz was introduced to Arnauld's friend and collaborator Pierre Nicole (1625–1695), to the mathematicians Jacques Buot and Ignace Gaston Pardies, and to Carcavy.³³ The exchanges between Leibniz and Arnauld seem to have been concerned primarily with scientific matters, especially at a time when Leibniz was taking a keen interest in the work of Blaise Pascal and his arithmetic triangle, and working towards his own mathematical and physical discoveries, including the calculus and a new mechanical theory.³⁴ Leibniz acquired the reputation of a 'mathematician by trade', and in this domain Arnauld was more forthcoming.³⁵ Looking back on his time in Paris, Leibniz confided his respect and admiration for Arnauld, adding that he often 'conversed with M. Arnauld on the matter of sciences, as he was no less a geometer than a theologian. He meditated something rather beautiful on reasons and proportions'—a recollection that Arnauld would later qualify in 1686 when he claimed, 'I have applied myself to these matters only occasionally and during wasted hours, and it has been over twenty years since I have seen any of these books.'³⁶

As to whether Leibniz embellished the extent of his conversations with Arnauld, we do not know; but it is likely that Arnauld sought in later years to minimize his earlier association with Leibniz.³⁷ Leibniz, nonetheless,

---

31. Arnauld to Ernst von Hessen-Rheinfels, 10 May 1683, in Arnauld, *Oeuvres* (1775–83), 2:255.

32. Arnaud, 2:255: 'He has a very fine mind and one that is very knowledgeable in mathematics. I wonder whether he has managed to build those two wonderful machines that he claimed would be of unrivalled exactness.' Also Leibniz to Arnauld, 12 December 1675, A III, 1:311.

33. Antognazza, *Leibniz* (2009), 141. It is interesting in this regard that Leibniz mixed equally with Jesuits and Jansenists.

34. Sleigh, *Leibniz and Arnauld*, 15–16.

35. Leibniz to Johann Friedrich, Autumn 1679, A II, 1:761.

36. Leibniz to Landgrave Ernst von Hessen-Rheinfels, 27 April/7 May 1683, A I, 3:286; Arnauld to Leibniz, 28 September 1686, A II, 2:98.

37. In a letter to Malebranche of 27 December 1694, A II, 3:4, Leibniz would also claim to have conversed about free will with Arnauld, which the latter allegedly found 'quite impressive'.

cautiously persisted in his attempts to engage Arnauld theologically.[38] In or around 1673 he sent Arnauld a copy of the *Confessio Philosophi*, but to no avail: despite Leibniz's best efforts not to shock him, Arnauld seems to have avoided being drawn on theological topics, and Boineburg's death may have further impeded Leibniz's overtures. Arnauld's caution is understandable, for Leibniz had come to Paris on the eve of a fresh wave of persecution against Jansenists—in 1679 the pope and Louis XIV would mend their relationship formally, thus ending the 'Peace of the Church'. This persecution would eventually result in the destruction of the Port-Royal monastery, the dispersal of its members, and the exile of its leaders, including Arnauld.

In his letter dated 26 March 1673, Leibniz conceded to Johann Friedrich that he had not engaged with Arnauld on religious matters or controversies: 'As I was proceeding with all possible circumspection, so as not to put myself at a disadvantage, Boineburg's death struck ... [removing the possibility] of explaining myself to Mr. Arnauld.'[39] He confirmed this in another letter of 4/14 March 1685, when he wrote about Pierre Nicole's anonymous anti-Calvinist tract *Les Prétendus réformés convaincus de schisme* (1684): 'I had the honour of talking to Mr. Nicole sometimes at Mr. Arnauld's house, but we did not get to religious matters, nor did Mr. Arnauld.'[40] Arnauld trod cautiously with Leibniz, no doubt discerning the latter's aim of seeking to enlist him to his cause: the French theologian would not have wanted to enter into religious controversy again and risk jeopardizing the Peace of the Church by allying himself to a Lutheran.[41] Nonetheless, despite what Huguette Courtès calls their 'impossible dialogue', Leibniz and Arnauld seem to have enjoyed a mutual respect, and Arnauld remained deeply impressed with the young German, writing in 1676 to a Capuchin in Hanover that he 'lacked only the true religion to be one of the greatest men of his age'.[42] On his side, Leibniz's esteem for Arnauld continued unabated after his departure from Paris: he mentioned the French thinker regularly in his correspondence with third parties, was keen to probe Arnauld for his reflections, and even sought his cooperation with a philosophico-religious reflection that Leibniz would then ideally have published.[43]

---

38. In a letter of 22 June 1679, A II, 1:477, Leibniz referred Malebranche to a dialogue on free will that he had written and shown to Arnauld.

39. Leibniz to Johann Friedrich, 26 March 1673, A II, 1:359.

40. Leibniz to Ernst von Hessen-Rheinfels, 4/14 March 1685, A I, 4:352.

41. Courtès, 'Arnauld et Leibniz' (1998).

42. Quoted in Müller, Schepers, and Totok, *Studia Leibnitiana* (1978), 20.

43. Leibniz considered having his correspondence with Arnauld published twice: in

Leibniz's departure from Paris marked a break in their correspondence, which would resume only in 1686 through the intermediary of Landgrave Ernst von Hessen-Rheinfels on the occasion of Leibniz's composition of an outline of his philosophical views leading up to his *Discourse on Metaphysics*, which he had asked the landgrave to forward to Arnauld for comment. Arnauld was not, at first, favourably impressed with Leibniz's essay, discerning some theological dangers in his ideas and finding 'so many things that frighten me and that almost all men, if I am not mistaken, will find so shocking, that I do not see what use such a work can be—it will clearly be rejected by everyone'.[44] Particularly objectionable to him was proposition 13, which seemed to him to imply a complete fatalism: 'Once God decided to create Adam, everything that has happened since and will ever happen to the human race was and is obliged to happen with a more than fatal necessity.'[45] Despite Leibniz's best attempts to keep it going—and he was the more active correspondent, writing eight of the twelve letters exchanged between them—the correspondence eventually fizzled out, Arnauld deeming that Leibniz's 'metaphysical speculations could not be of any use to him or others'.[46]

Leibniz pursued his refutation of matter as mere extension in his brief exchange with Nicholas Malebranche, whose *Recherche de la verité* he had read (albeit hastily, as he admitted to Malebranche) and whose doctrine of occasionalism had affinities with Leibniz's own doctrine of pre-established harmony, whereby each monad simultaneously reflected all changes taking place in the universe, both thinkers once again engaged in a common reflection.[47] Although they did meet, and both sought to harmonize philosophy—in Malebranche's case, Cartesianism—with theology, their philosophical dialogue remained limited.[48] Still, butterflying from one intellectual niche to another as was his wont, Leibniz frequented anti-Cartesians and Cartesians alike, discussing the master's ideas with them.

1695, the year after Arnauld's death, and then a bit later in 1708, first within the framework of the reconciliation of the churches and then in his writings against Bayle.

44. Arnauld to Ernst von Hessen-Rheinfels, 13 March 1686, A II, 2:9.

45. Leibniz, 'Discours de métaphysique' (early 1686), A VI, 4:1546: 'The individual notion of each person includes once and for all everything that will ever happen to him', so that 'one sees in [that notion] a priori proofs of the truth of each event, that is, why one [event] has occurred rather than another.' Arnauld to Ernst von Hessen-Rheinfels, 13 March 1686, A II, 2:9.

46. Arnauld to Ernst von Hessen-Rheinfels, A II, 2:9.

47. A II, 1:254–57; Leibniz to Malebranche, 13 January 1679, A II, 1:455: 'I have had a look at your *Conversations Chrestiennes* ... I have better grasped your feeling that I had not spent enough time poring over your *Recherche de la Verité*.'

48. Hirsch, *Der berühmte Herr Leibniz* (2000), 70–72.

He was, for instance, particularly receptive to the architect and anatomist Claude Perrault's rejection of Cartesian mechanism with regard to the impenetrability of bodies and causes of physical phenomena.[49] Like Leibniz, Perrault disputed a strictly mechanistic character of life: while mechanism provided a tentative explanatory model, it could not account for the nature of life itself, which Perrault understood in terms of an omnipresent inorganic soul that regulated all movement.[50]

Leibniz would eventually prefer his theory of pre-established harmony and 'little perceptions' to Perrault's animism and concept of 'neglected and confused thoughts'.[51]

Foucher

In a predominantly Cartesian decade, when sceptical thinkers were being denounced as a grave threat to the very foundations of rational thought, overthrowing all certainty in natural and revealed knowledge, Leibniz found much inspiration in his interactions with such thinkers.[52] Among the many philosophers he came to know particularly well during his stay in Paris was Simon Foucher (1644–96), whose critique of Malebranche would give him food for thought.[53] The canon of the cathedral of Dijon,

---

49. See Azouvi, 'Entre Descartes et Leibniz' (1982).

50. On mechanism, Perrault wrote in 'Des sens exterieurs' (1727), 514: 'One cannot say that we are assured of knowing how the organs of an animal function, in the way that it is certain that one knows how a counterweight or a spring make a watch run; we do not know what causes a dog who has lost its master to stop eating, even though we are quite sure that it is not chagrin that stops a watch from running.' On the soul, Perrault, 'La mécanique des animaux' (1727), 329, noted: 'By *animal* I mean a being that has feeling and is capable of exercising the functions of life through the soul; that the soul makes use of the organs of the body, which are true machines . . . and that although the disposition that the pieces in the machine have with respect to each other does nothing differently by means of the soul than it does in pure machines, still the entire machine is moved and driven by the soul.' In a note dating from the end of 1675 or the beginning of 1676, Leibniz reported that it was 'the opinion of Mons. Perrault . . . that the soul is equally present throughout the body, and that feeling occurs in ipso sensorio in the eyes as well as the feet.' Bodemann, *Die Leibniz-Handschriften* (1895), 118–19. For more on Perrault, see Borowski, 'Perrault, Claude' (2021).

51. Wright, 'The Embodied Soul' (1991); Azouvi, 'Entre Descartes et Leibniz'; Herrmann, *The Theory of Claude Perrault* (1973). Bertrand, *Mes vieux médecins* (1904), 188, even believes that Perrault's ideas on the topic 'prefigure Leibniz'.

52. Popkin, 'Leibniz and the French Skeptics', 238.

53. A few years later Leibniz would provide a short account of a chance encounter between himself and Foucher when, looking for Foucher's *Critique de la recherche de la vérité* (1675) in a Parisian bookshop, he was ridiculed for referring to it as a work of metaphysics. Foucher had appeared just at that moment, praising Leibniz and prompting the bookseller to change his attitude. See Antognazza, *Leibniz*, 170; and Bodemann, *Die*

Foucher had come in contact with Cartesianism when he attended a series of weekly lectures on physics given by Jacques Rohault, and these had prompted his interest in experimental physics and led him to publish a little pamphlet called *Nouvelle façon d'hygromètres* in 1672.[54] Still, his deeper interests lay in philosophy, and he devoted much of his life to the revival of academic scepticism, the philosophy of Plato's Academy in its later years, which he perceived as a more rigorous methodology for searching after the truth, one devoid of dogmatic prejudices. He sought to steer a middle course between dogmatism and extreme Pyrrhonism and to provide the new scientific spirit of the time with a suitable philosophical and metaphysical underpinning.[55] By seeking to revive Academicism, Foucher sought to offer a method that would avoid the pitfalls of Cartesian dogmatism while also acting as a 'useful and important propaedutic for Christian belief'.[56] He rehabilitated the method of Academicians, with their moderate exercise of doubt, to arrive at a certain number of useful and indubitable truths, namely, the 'evident truths' (*veritez évidentes*) and 'certain knowledge' (*connoissances certaines*) that led to new knowledge. From incontestable first principles such as mathematical and conceptual truths, we could then seek out further knowledge concerning the soul, God, and the existence of the world.[57] For Foucher, Academicism was the only philosophy capable of overcoming the dead ends of Cartesian dualism in terms of its epistemological implications, and he emerged as one of the most vocal critics of the new Cartesian dogmatism.[58]

In his *Critique de la Recherche de la verité. Où l'on examine en mème-tems une une [sic] partie des principes de Mr Descartes* (1675), his response to Malebranche's *De la recherche de la verité*, Foucher began what was to be a polemic of many years between himself and the Cartesians, including Malebranche and Dom Robert Desgabets (who rallied to Malebranche's defence). The *Critique* conducted a detailed critical investigation into the principles and claims of Cartesianism (even if Foucher incorrectly assumed that the published volume of Malebranche's *Recherche* was the whole work) and set forth a series of interesting arguments in response to Malebranche which helped spur on Leibniz's own

---

Leibniz-Handschriften, 339: 'I have never before noticed so clearly', Leibniz later recollected, 'the power that prejudice and appearance have over human beings.' For more on Foucher, see Watson and Grene, *Malebranche's First and Last Critics* (1995).

54. Garber, *Leibniz* (2009), 272.
55. See Popkin, 'Leibniz and the French Skeptics'.
56. Schmaltz, *Radical Cartesianism* (2002).
57. Schmaltz.
58. Charles, 'Entre réhabilitation du scepticisme' (2013).

philosophical development.[59] Foucher criticized Cartesian principles of knowledge, principles on which Malebranche had 'wholly grounded himself', taking particular issue with the presumptions of the Cartesian position regarding what was known. He challenged our ability to have any knowledge of a world of external bodies especially considering that the first principles on which all knowledge rested remained unknown and 'encumbered with nearly insurmountable difficulties'. All ideas, ultimately, were states of mind, 'only modes of our soul', and granted us no access to the external world 'because these objects have nothing in themselves *like* what they produce in us, for matter cannot have modes that are *like* those of which the soul is capable'.[60]

Leibniz met Foucher in Paris probably through Jean-Baptiste Lantin, *érudit* and councillor to the Dijon parliament.[61] Lantin was familiar with Leibniz's early writings on physics and encouraged him early on to apply his 'dynamical meditations' to metaphysics.[62] By 1676 Leibniz considered that he had already known Foucher 'for a long time' and was consistent in giving his ideas a careful and detailed examination.[63] Though not a philosophical sceptic himself, Leibniz shared Foucher's concern about the insufficient demonstration of the first principles in philosophical reasoning.[64] Both agreed on a whole raft of philosophical issues and objections to Descartes—so much so that Stuart Brown speaks of a 'philosophical alliance' between the two of them.[65] This included their common rejection of the distinction between primary and secondary qualities, of the Cartesian doctrine of the essence of matter as extension, of the contradictions incurred by the mind-body dualism, and of Descartes's arguments for the existence of God. They shared the more general conclusion that the doctrine of clear and distinct ideas was inadequate to discover the truth. Both,

---

59. Foucher published his refutation of Malebranche before the latter had published the concluding portions of his book (only the first volume had been published), leading Malebranche, in the preface to the second volume, to enjoin readers to read books in their entirety before refuting them.

60. Quoted in Garber, *Leibniz*, 270.

61. For more on Leibniz's relationship with Foucher, see Popkin, 'Leibniz and the French Sceptics'; Brown, 'Foucher's Critique and Leibniz's Defense (1983); Olazo, 'Leibniz and Scepticism' (1987); Mendonça, 'Leibniz vs. Foucher' (2010); and Pelletier, 'Leibniz's Anti-Scepticism' (2012).

62. The term 'dynamics' itself ('dynamica') doesn't get introduced into his vocabulary until the late 1680s.

63. Quoted in Müller, *Leben und Werk* (1969), 41; Popkin, 'Leibniz and the French Sceptics', 75.

64. Garber, *Leibniz*, 163.

65. Brown, 'The Leibniz-Foucher Alliance and Its Philosophical Bases' (2004).

too, were committed to the revival of ancient philosophy—Leibniz had produced abridgements of Latin translations of the *Phaedo* and the *Theaetetus* while in Paris and hoped Foucher would make available selections in French translation relating to Academic scepticism as a philosophical methodology.[66]

It was probably through reading Foucher's critique of Malebranche in 1675, and discussing those issues with him in Paris, that Leibniz began to consider more seriously the question of how we can know the existence of the external world beyond phenomenal appearances. In his first letter to the author of the *Critique de la recherche de la verité*, Leibniz discussed Foucher's pamphlet in some detail, delving further into the question of the possibility of knowledge of the external world, and in particular whether it can be affirmed that something exists independently of its being thought. Leibniz agreed with Foucher—and Descartes—that was it necessary to question all one's epistemological presuppositions in order to establish a sound foundation for thought. From these truths he then deduced, first, the existence of the thinking self, and second, the existence of something outside the thinking self, something that is the cause of the variety in our thoughts.

Leibniz considered Descartes's effort to demonstrate the existence of a world external to the mind to be impressive but insufficient. Noting the 'variety of our thoughts', Leibniz for his part argued that this variety must originate outside the mind, 'since a single thing by itself cannot be the cause of the changes in itself'. Internally caused changes in our thoughts have no reason, which would be absurd: therefore such changes must be externally caused.

Harking back to the criticisms of the Ancient Academy, Leibniz presents the quandary facing us: senses provide us only with appearances, not direct access to the external world. Appearances as such did not prove the existence of objects, but only that something must be causing the appearances. In this manner an 'invisible power' could, for instance, 'take pleasure in presenting us with continuous and well-accorded dreams' in such a way that we would be unable to distinguish them from reality. Conjuring up Descartes's *genius malignus* from the *Meditations*, Leibniz goes beyond Foucher's line of inquiry asking 'what prevents the course of our life from being a long well-ordered dream, a dream from which we could be wakened in a moment?'[67]

66. Brown, 78.
67. Leibniz to Simon Foucher, 1675, A II, 1:387, 388, 390.

Our belief that appearances amount to actual existence, Leibniz tells us, is grounded in a moral supposition rather than a real ability to back up any chain of causality to the source and the origin of the world: only in this manner would the reality of existence be demonstrated once and for all. Reaching such a 'beatific vision' would, however, be very difficult at this stage, especially considering that our knowledge of matter and bodies is normally very confused, and we cannot establish their existence with any certainty.

On a more positive note, and praising Descartes once again, Leibniz reasserts man's certainty 'of the existence of that which thinks' and ultimately 'our power to undeceive ourselves about many things, at least about the most important ones', a point on which he and Foucher would later disagree.[68]

That Leibniz was mainly hypothesizing the existence of an external world of bodies and playing with ideas seems to be borne out in an essay of 15 April 1676, written shortly after his exchange with Foucher. In it Leibniz explores a more radical position, seemingly dispensing with the necessity for an external reality altogether:

> Since what we can judge about the existence of material things is no more than the consistency of our senses, one has a sufficient basis for judging that we can ascribe nothing to matter apart from being sensed in accordance with some certain laws, the reason for which (I admit) remains to be sought. . . . On due consideration, only this is certain: that we sense, and that we sense in a consistent way, and that some rule is observed by us in our sensing. For something to be sensed in a consistent way is for it to be sensed in such a way that a reason can be given for everything and everything can be predicted. This is what existence consists in—namely, in sensation that involves some certain laws; for otherwise, everything would be like dreams. Further, it consists in the fact that several people sense the same, and sense what is coherent; and different minds sense themselves and their own effects. . . . Therefore there is no reason why we should ask whether there exist certain bodies outside us, or whether space exists, and other things of this sort; for we do not explain adequately the terms that are involved here. Unless, that is, we say that we call a 'body' whatever is perceived in a consistent way, and say that 'space' is that which brings it about that several perceptions cohere with each other at the same time. . . . As this is so, it

---

68. Leibniz to Simon Foucher, 1:391.

does not follow that there exists anything but sensation, and the cause of this sensation and of its consistency.[69]

In this manner, by reframing our understanding of an extended reality that he now ties to the consistency of our perceptual world, Leibniz formulates a way out of the epistemological quandary of inferring physical reality independently of the mind. In fact, Leibniz explores the possibility that there may be nothing more to bodies than coherent appearances.[70] Thus he suggests that all there is to the world of body is the sensations we have:

> For he who can finally make predictions with success must be said to have become sufficiently proficient in nature. And so the objections the Sceptics level against observations are inane. Of course, they may doubt the truth of things, and if it pleases them to call the things that occur to us dreams, it suffices for these dreams be consistent with one another, and to observe certain laws, and accordingly to leave room for human prudence and predictions. And granting this, it is only a question of names for apparitions of this kind we call *true*, and I do not see how they can either be rendered or desired to be truer.[71]

The correspondence with Foucher seems to have resumed only in the 1690s, when Foucher helped Leibniz refine his theory of pre-established harmony and his new metaphysics for the *Nouveau système* (1695). Leibniz thus perceived Foucher and scepticism not as destructive threats but as an intellectual platform with which he held much in agreement, especially as to how little demonstratively established knowledge we can have and how restricted our knowledge of the world of experience is. While refusing to entertain a scepticism with regard to reason itself and remaining largely immune to the *crise pyrrhonienne*, Leibniz thrived on the 'dialectical interplay' of ideas to forge his own original position.[72]

---

69. Quoted in Garber, *Leibniz*, 278.
70. Garber, 280.
71. Leibniz, 'Definitiones Cogitationesque Metaphysicae', Summer 1678–Winter 1680–81, A VI, 4:1398.
72. Popkin, 'Leibniz and the French Sceptics', 246.

CHAPTER FIVE

Defending a Particular Conception of the Republic of Letters

LEIBNIZ ON THE *QUERELLE DES ANCIENS ET DES MODERNES* AND THE REPUBLIC OF LETTERS

WHILE THIS STUDY AIMS to shed light on Leibniz's role and activities in the Republic of Letters, it also takes an interest in how he viewed and understood that intellectual citizenship. Indeed, Leibniz was one of the most active and prolific members of the Republic of Letters of his time, mobilizing his network of correspondents to collect and dispense information or advice, obtain books or journals, engage in philosophical and theological and scientific debate, help publicize discoveries—in short, disseminating knowledge among a community of savants.[1] Similarly, he often inquired after new inventions and curiosities in his attempt to sift between the true and the false and contribute to the advancement of knowledge through the collective endeavour of his correspondents. As he established one of the most extensive philosophical and scientific epistolary networks in Europe, Leibniz at times paused to reflect on this particular intellectual institution and what it represented.[2] For Leibniz, as

---

1. The phrase *République des Lettres* appears several times in Leibniz's correspondence, as for instance in a letter from Louis Ferrand to Leibniz, 23 March 1677, A I, 2:260: 'The news within the Republic of Letters is that Mr. Baluze has given us his Capitularies in two volumes. . . . It is undoubtedly one of the most beautiful collections that one can make.'
2. See Hotson, 'Leibniz's Network' (2014). Between 1672 and 1679 Leibniz had more than a hundred correspondents.

for the overwhelmingly majority of his peers, the Republic of Letters was characterized by the free exchange of ideas between equals and by the goal of perfecting the human spirit and improving the human condition. In several of his writings and letters, in fact, he shared his understanding of the Republic, describing it in a 1670 letter to Hermann Conring as the supralegislative 'respublica optima', characterized by the union of intellect, power, and wisdom.[3] Later, in the *Relatio Codicis Juris Gentium Diplomatici* of August 1693, Leibniz commented that the aim of this 'university of minds' lay in the 'highest and eternal felicity'.[4]

The relation between Pierre-Daniel Huet (1630–1721) and Leibniz has hitherto remained largely unexplored. An Orientalist, poet, physician, and prolific writer, and arguably the novel's first theorist, Huet has often been portrayed as a reactionary figure by a historiography of philosophy eager to reconstruct the seventeenth century as overwhelmingly Cartesian.[5] Huet was no stranger to the Parisian intellectual scene and its salons. There he had befriended the likes of Pierre Gassendi, Gabriel Naudé, Gilles Ménage, and Jean Chapelain, the last acting as one of his patrons in Paris and assisting with Huet's receipt of a royal pension in 1663.[6] His critical edition of Origen's biblical commentaries had met with considerable success and helped secure him a respected position in the Republic of Letters after his many years as a promising young scholar.[7] Huet's contacts with Parisian learned circles from the 1650s onwards had also brought him into closer contact with the court, where he caught the attention of the duc de Montausier, governor of Caen, and it was in Caen that Huet cofounded an Académie de Physique in 1662.[8] The duke, employed at the time as the dauphin's governor, continued acting as an attentive protector of Huet, eventually getting him promoted to the position of tutor to the dauphin in 1670. Leibniz had heard of this great figure even prior to his arrival in Paris, when he had had the opportunity to read Huet's article expounding on a new method

---

3. Leibniz to Conring, A II, 1:46. For more on this subject, see Marras, 'Leibniz Citizen' (2011); 'Grundriß eines Bedenkens von Aufrichtung einer Societät', 1671, A IV, 1:531. See also 'De S. Pufendorfii libro cui titulus est: Jus Feciale Divinum, 1695', A IV, 6:321.

4. A IV, 5:83.

5. Leibniz and Conring discussed Huet's theory of the novel and work on Origen. See Leibniz to Conring, 19 March 1678, A II, 1:598. For more on Huet, who nonetheless remains a largely overlooked figure of the later part of the seventeenth century, see Lux, *Patronage and Royal Science* (1989); Huet, *Mémoires* (1993); Guellouz, *Pierre-Daniel Huet* (1994); Volphilhac-Auger, *Ad Usum Delphini* (2000); Shelford, *Transforming the Republic of Letters* (2007); Lennon, *The Plain Truth* (2008); and Laerke, *Les Lumières de Leibniz* (2015).

6. See Maber, 'Colbert and the Scholars' (1985).

7. Shelford, *Transforming the Republic of Letters*.

8. See Lux, *Patronage and Royal Science*.

of desalinating water.⁹ He gradually took the measure of Huet's immense erudition and the wide-ranging nature of his work, aspects of which, from philology to the natural sciences, he discussed with a number of his correspondents, including Hermann Conring, Johann Georg Graevius, Ludwig Veit von Seckendorff, Gabriel Wagner, Thomas Burnett, and Henri Basnage de Beauval. Leibniz's first reference to Huet dates from a June 1671 letter to Johann Georg Graevius in which he deplored the loss, as a result of the war, of an Origen manuscript that later turned up among the possessions of Queen Christina of Sweden and was used by Huet for his edition.¹⁰ In August of that same year, Johann van Deimerbroeck reported to Leibniz the publication of Madame de la Fayette's novel, 'the famous Zaïde' set in medieval Spain, which was preceded by Huet's *Traité de l'origine des romans*.¹¹ As his letters indicate, Leibniz met Huet relatively early during his stay in Paris, and in all likelihood at Henri Justel's salon.¹² By November 1672, certainly, Leibniz had made personal contact with him, relaying to his patron, Melchior Friedrich von Schönborn, 'credible information' shared with him by Huet 'concerning a Dutchman recently conducted to the Bastille'.¹³

Leibniz seems to have grasped quickly Huet's importance in the Republic of Letters and courtly circles alike, and to have been determined to place himself at Huet's service by demonstrating his commitment to him from the inception of their correspondence. Shortly before his departure for England in early 1673, for instance, Leibniz transmitted to Oldenburg Huet's request to inquire after manuscripts of the second-century astrologer Vettius Valens in the Bodleian Library, and Oldenburg in turn forwarded the request to John Wallis, the Savilian Professor of Geometry, to compare the manuscript with the transcript Huet had made of the *Codex Hamburgensis*.¹⁴ In return, Leibniz hoped that Huet would facilitate his access to Louis XIV in his bid to present the political project that he, as the Court of Mainz's diplomatic envoy, had been sent to Paris to promote—the famous *Consilium Aegyptiacum*.¹⁵

9. Cf. [Huet], 'An Extract of a Letter' (1670).
10. Leibniz to Johann Georg Graevius, 7 June 1671, A I, 1:157. See Lux, *Patronage and Royal Science*, 14–15, on Huet's use of the manuscript.
11. Johann van Diemerbroeck to Leibniz, 18 August 1671, A I, 1:167.
12. Huet, *Mémoires*, 131: 'Every day his house was filled with savants conversing about erudite matters.' See also Vernière, *Spinoza* (1954), 106.
13. Leibniz to Melchior Friedrich von Schonborn, 15 April 1673, A I, 1:340.
14. Beeley, 'A Philosophical Apprenticeship' (2004). See Huet to Leibniz, 20 March 1673, A II, 1:357; and Beeley, 63 no. 33. There is a Vettius manuscript among the Selden papers in the Boldeian Library, Oxford (Bodleian MS Arch. Selden B.19).
15. Leibniz to Boineburg, 25 November 1672, A I, 1:289.

Huet's appointment as tutor to the dauphin had earmarked him as one of the most eminent intellectual figures of the kingdom and as the natural academic director, so to speak, of Montausier's design to produce a large, annotated digest of Latin literature for the use of the dauphin in the first instance.[16] In a letter from the spring of 1673, Leibniz sent a more detailed description of Huet and this project to the secretary of the Royal Society, Henry Oldenburg:

> Perhaps you are not unaware that Huet has been moved to the Dauphin's studies. . . . By the order of Montausier, and under the guidance of Huet, this programme has commenced, [which will be] very useful for improving education, and for recalling, so to speak, the vanishing erudition of antiquity. For men of proven learning have been given the task to organize the old Latin authors in a manner different from that hitherto employed—with a brief and lucid paraphrase added when necessary, so that the texts of the ancients may be easily imparted to a young man.[17]

From 1674 to 1691 a series of volumes would thus be published, with the intent of imparting major works of the Latin antiquity to the prince, and by extension to the cultured public. In addition to the original Latin text itself, each of these works included an introduction of a generally political or historical nature, a paraphrase (*interpretatio*), and notes (*notae*)—all in Latin—designed to facilitate comprehension of the work. The collection's emphasis was decisively pedagogical rather than scholarly and intended for the *honnête* homme rather than the erudite savant. It set out to initiate a renaissance in the study of classical works and provide the dauphin, but also the broader public at large, with models of instruction, eventually serving as the impulsion behind such seminal works as Jacques Bénigne Bossuet's *Discours sur l'histoire universelle* (1681) and Huet's own *Demonstratio Evangelica* (1679). To this end Huet recruited a wide range of collaborators, including first-rate humanists such as André and Anne Dacier, as well as the classicist Jean Hardouin and the philologist Jean-Baptiste Suchay, to whom he gave a precise set of instructions. By spreading the taste of classical letters among the educated public, this edition would act as bulwark against the forces of barbarity: 'Never have the Roman language and antiquity received such solid aid, and such a

---

16. For the conception and realization of *Ad Usum Delphini*, see Volphilhac-Auger, *Ad Usum Delphini*.

17. Leibniz to Oldenburg, 26 April 1673, A I, 2:84.

strong preservative against ignorance and barbarism.'[18] Emmanuel Bury has analysed Huet as a 'man belonging to another generation closer in his formation to a Guez de Balzac', having 'lagged behind in the history of ideas, by attacking Cartesianism at a moment when it seemed to triumph'.[19] Huet belonged to a generation that favoured learning as it had been preserved in books rather than as it was challenged by 'new critics' such as Joseph Scaliger, whom he accused of 'disfiguring ancient authors'. At a time when the ancients had come under attack, it appeared more necessary than ever to uphold the centrality of memory and erudition, and the immutable truths they had preserved over time. It was with this aim that Huet undertook this humanistic project of memorialization against those new philosophies which threatened man with the prospect of a tabula rasa and amnesia. By positioning himself as a staunch defender of those classics which had recently fallen in discredit and had been 'banished from the polite world' and 'relegated to the dust and darkness of a few cabinets', Huet countered Cartesianism's claim to certainty and its faith in the strength of human reason by upholding the equal validity of memory and history.[20] Only memory could provide an antidote to the natural weakness of the human mind.[21]

Like Huet, Leibniz feared that the Republic of Letters would dissolve into a state of barbarity, and he thus greeted the project, and Huet's position more broadly, with considerable and genuine enthusiasm. It was incumbent, he wrote, in 'these times of contempt' to 'recall learning [*Lettres*] in disarray, to revive the nearly extinguished light of an almost moribund Antiquity, and to give a third life to the best authors', those precisely whose 'eyes had begun to close as tired of living'.[22] In his letter of April 1673, Leibniz praised Huet's effort to safeguard the texts of ancient thought while inveighing against Cartesianism. Some people, he continued, had misused the greats to destroy ancient wisdom:

> Indeed, I see some [of our contemporaries] taking advantage of the opinions and complaints of great men, Bacon, Galileo, and Descartes, to put down ancient wisdom and conceal their own ignorance, to the point of seeming rightly to despise a knowledge unworthy of being known. They deprive themselves, and to a certain extent the world, by

---

18. Quoted in Lopez, 'Huet pédagogue' (1994), 223. See also Bury, 'L'humanisme de Huet' (1994).
19. Bury.
20. Leibniz to Huet, 15 April 1673, A II, 1:363.
21. Bury, 'L'humanisme de Huet', 207.
22. Leibniz to Huet, 15 April 1673, A II, 1:363; Laerke, 'Ignorantia Inflat' (2013), 26.

dispensing with all the benefits and trials of so many centuries . . . as if a kind of great restoration, if we are to believe Bacon, or to wipe the slate clean, if Descartes is to be believed, were needed in order to think correctly; but one should perhaps forgive the indolence of those who made discoveries more freely in their own times, and the idle idolatry of those who accept dogmas, for they produce from their own resources important findings comparable to the discoveries of the ancients. One should not recuse those who produce their own experiments, even if in general they judge other disciplines unfairly.[23]

Leibniz proceeded to describe the new sectarianism which had taken hold and even come to threaten holy doctrine: 'Certain disciples, who are just as sectarian as before since they have only changed masters, violently protest against all the dogmas of antiquity. . . . That they consider a man sufficiently philosophical, quite erudite, preferable to Aristotle . . . who explains the phenomena of nature only by means of subtle matter, and . . . vortices—all this clearly ends in the ruin of holy doctrine.' 'Youth', in particular, had fallen under the spell of this new 'ignorance'. Confining ourselves only to the 'experiences and productions of the language of our time' imperilled religion whose truths had been confirmed in antiquity. To 'this disease of the century which creeps in so perniciously,' Huet had 'opposed a remedy which [would] be effective for the second or third generation, for if you imbue a prince with such great hope, destined for such great things, with mysteries that touch the heart of doctrine,' as Huet had done, 'you will have worked towards posterity for many centuries.' 'The world had grasped from the example of [the sixteenth-century ruler] Francois I, whose effects endure today, how much power the encouragement of a single prince wielded', and 'the fate of letters [was] linked to his will, which direct[ed] the flow of valuable spirits'. The nobility, too, could contribute to this undertaking, especially once Huet had smoothed the way and made it possible for everyone, 'by the power of reading alone, without the tedious step of going to beg for help elsewhere, to penetrate into the sanctuaries to which only the greatest minds could hope to gain access thanks to painful work in the past'.[24]

While Leibniz's political proposal was ultimately unsuccessful, Huet recognized in Leibniz an erudite scholar and enlisted him in this

23. Leibniz to Huet, 15 April 1673, A II, 1:363.
24. Leibniz to Huet, 15 April 1673, A II, 1:363. Leibniz's antidogmatism assumed that different intellectual traditions could be reconciled. See Knecht, *La logique chez Leibniz* (1981), 27–30.

monumental undertaking.²⁵ Significantly, with Graevius, Leibniz was the only foreign scholar enlisted in the project.²⁶ Both Graevius and Leibniz were recruited in 1673 with the explicit ambition, as stated in the privilege, to search for letters 'inside the kingdom and outside it' (*intra and extra regni fines*).²⁷ Crucially, these savants would help disseminate the collection abroad and help lend it a scientific imprimatur: Leibniz was a universal scholar whose skill in *belles lettres* Huet recognized, and Graevius one of the best Latinists of the time.²⁸ In his 15 April 1673 letter, Leibniz commented on the great honour that had been bestowed upon him in being invited to partake in the project:

> That eminent men have been chosen to contribute to your monument, no one who knows you doubts. I am all the more amazed at the benevolence of your judgement, you who thought recently that a man like me could be of any use. I first attributed this idea to your natural benevolence . . . but when I saw you stand firm—you whose judgement is so penetrating that I could not doubt it—upon reflection I found a way to reconcile my scruples with your wishes. I admit, in fact, although I do not claim either the mind or the knowledge, that I have nonetheless sought, through my zeal, to win the praise of impartial censors: and what else could be expected from a native of Germany, a nation whose sole gift of the spirit is perseverance?²⁹

Initially commissioned with drafting a new edition of Vitruvius, Leibniz finally offered to translate Martianus Capella's philosophical allegory *De nuptiis Philologiae et Mercurii* instead, judging the first work too technical, especially upon hearing of Perrault's forthcoming translation.³⁰ Nothing came of this, however, owing to the slow progress of the enterprise and the pressure of other demands on Leibniz's time.³¹

---

25. Robinet, *Leibniz* (1994), has noted that Huet had co-opted Leibniz's translations of Plato in his plans for educational reform.

26. In his *Mémoires* (114), Huet mentions Graevius, 'whose name was famous in the *lettres* . . . because of his editions of classical authors', and considered it 'a great honour' to enter in an epistolary relationship with him. Curiously, Huet does not mention Leibniz in his *Mémoires*.

27. Huet, 118.

28. Leibniz to Huet, 19 April 1673, A II, 1:366.

29. Leibniz to Huet, 15 April 1673, A II, 1:365.

30. Leibniz to Oldenburg, 26 April 1673, A III, 1:84–85.

31. See Volpilhac-Auger, *Ad Usum Delphini*, 115–18. Leibniz did send Huet a sample commentary on Martianus on 10 May 1673 but hinted that the work was preventing him from attending to his scientific studies: see A II, 1:369, and for the sample, A VI, 3:189–202.

Placing Erudition and History in Defence of Religion

The correspondence between Leibniz and Huet resumed only in 1679 after the latter's publication of his *Demonstratio Evangelica*. In his introduction, Huet clearly laid out the work's aim: 'This book is intended to attack impiety, which is making new progress day by day, and to prove that common sense should make it a duty to believe in the Christian religion, that so many people have the unreason to give up.'[32] Although he began to compose the *Demonstratio* in the 1670s, Huet had already conceived of it in the early 1650s in Amsterdam when, debating with Menasseh Ben Israel, he claimed that he would prove the fulfilment of the Old Testament prophecies in the New Testament with as much certainty as in a Euclidean geometric proof.[33] While the demonstrability of Christian truth was not by any means a new idea, Huet's choice to couch this ancient apologetic argument in terms of a 'geometrical' method was, as he himself recognized: 'The argument offered in this book is not new; but the way of presenting it is novel.' Huet defined his geometrical method in the following manner in the opening passages of his volume: 'Geometers usually begin by explaining and defining an object with a definition. Then, they ask that we concede certain points to them which are generally agreed upon by all, and which cannot be denied without causing harm; they then establish their statements as something manifest and accepted by everyone. This is also what I am going to do so that our proof can proceed in orderly fashion, and that we can conclude scientifically.'[34]

By organizing his argumentation according to the *mos geometricus* and thus structuring his *Demonstratio* as a geometric proof, which at the time represented the archetype of demonstrative knowledge, especially in Cartesian circles, Huet hoped to contest the Cartesian monopoly over what constituted the criteria for truth and to achieve a similar level of evidence to that 'achieved by any geometrical demonstration' for the sake of moral and religious certainty.[35] By adopting this format, Huet 'forced humanist erudition in an Euclidian mould and reworked Christian apologetics to make them conform with geometric form'.[36] He hoped to erase what he perceived as an artificial divide between humanist erudition and Cartesian geometry by deploying the geometrical method to establish the

---

32. Huet, *Demonstratio evangelica* (1679), 1–2.
33. Shelford, *Transforming the Republic of Letters* (2007), 153.
34. Huet, *Demonstratio evangelica*, 6.
35. Huet, 2–3. Laerke, *Les Lumières de Leibniz* (2015), 150.
36. Shelford, 'Thinking Geometrically' (2002), 601.

authenticity and historical accuracy of scripture.³⁷ In his *Demonstratio Evangelica*, Huet thus set out to show how, according to historical data, most Old Testament prophecies had indeed been fulfilled. By having recourse to the geometrical method, Huet hoped to establish the Christian faith on knowledge obtained by reason, and in particular on knowledge of a historical order, at a time when faith was coming under increasing pressure to provide rational foundations for itself.³⁸ He envisaged his *Demonstratio* mainly as a response to the growing tide of anti-erudition and biblical criticism, explicitly presenting it as a contribution to the pan-European controversy over the biblical lexicon, drawing on his knowledge of ancient languages and civilizations as well as his skill in biblical exegesis to combat impiety, 'which every day makes new inroads, and to prove that common sense dictates that we believe in the Christian religion, which so many people have the unreason to abandon'.³⁹

It seems that, during his stay in Paris, Leibniz may even have sought to guide Huet's pen in the *Demonstratio*, which promised to offer a refutation of Spinoza's *Tractatus theologico-politicus* (1670).⁴⁰ Indeed, having failed to enlist the Orientalist Gottlieb Spitzel, Leibniz had turned to Huet to refute the Dutch philosopher's 'abominable' work, which had denied the Mosaic authorship of the text and undermined the authority of the Old Testament.⁴¹ By advancing a novel type of biblical exegesis that seemed to deny the divinity of scripture, Spinoza's treatise had produced a scandal of unprecedented magnitude in the Republic of Letters.⁴² Unlike Richard

---

37. See Shelford, 601; Laerke, *Les Lumières de Leibniz*, 108–10.

38. See Quintan, 'Le statut de l'apologétique chrétienne' (1994b), 95–96.

39. Huet, *Demonstratio evangelica*, 2. On his skill in exegesis, see Shelford, *Transforming the Republic of Letters*.

40. In his *Mémoires* (176), Huet recounted the various discussions he had with friends during the composition of his book: 'While I was working on my *Demonstratio Evangelica*, I would sometimes discuss it with my friends; I shared its plan, method and proof with them enthusiastically'. Leibniz confirmed to Huet, March 1769, A II, 1:695: 'I often reminded you, Sir, when I was in France, and enjoyed your learned conversations, with what eagerness I awaited the publication of the beautiful work whose prospectus you had kindly shown me.' Some scholars, such as Vernière, *Spinoza et la pensée française* (1954), even assert that it was Leibniz who brought the arguments of Spinoza's *Tractatus* to Huet's attention. Laerke, *Les Lumières de Leibniz*, 119: 'The evangelical demonstration [would contribute] to the controversy over the principles of biblical exegesis conducted by the works of Lodewijk Meyer and Spinoza in the Netherlands, which shortly after the publication of the *Demonstratio*, will continue in France with the quarrel around Richard Simon's *Critical History*.' See also Shelford, 'Of Sceptres and Censors' (2006), 161–81.

41. Huet, *Demonstratio evangelica*, 140–44, attacks Spinoza's *Tractatus* as 'impious' and aligns it with Hobbes's *Leviathan*.

42. Laerke, 'À la recherche d'un homme égal à Spinoza' (2006), 387.

Simon, who also challenged the Mosaic authorship of the Pentateuch in *L'histoire critique du Vieux Testament* (1678), Spinoza, however, was no biblical expert, and his audacity, in Leibniz's view, had not been matched by the requisite erudition. Despite reading the *Tractatus* twice, in 1670 and in 1675, the second time closely at the instigation of his friend Tschirnhaus, Leibniz never felt he quite possessed the necessary philosophical and linguistic skills to take this work on and write a critique of the Spinozian exegesis, undertaking instead a search for a 'man equal to Spinoza' who would be capable of carrying out an effective refutation of the perilous work.[43] In Huet, Leibniz believed he had found a thinker who was equal to the task.[44]

Refuting Spinoza was but one aspect of the *Demonstratio Evangelica*, however, and Huet did not make it the focus of his work, despite subsequent scholars' interpretation of it. He intended it primarily as a rebuttal of Jansenist Cartesianism, in particular its claim that only geometrical proofs could yield certain knowledge, which he felt had led to a devaluing of faith and historical knowledge as sources of reliable knowledge.[45] Huet set out to condemn the 'absolute intellectualism' of Descartes by seeking to emphasize the limits of human reason—and deny it the ability to obtain, in any field whatsoever, any evidence proper.[46] Nonetheless, this did not stop Huet from using the geometrical approach himself.

In response to the *Demonstratio Evangelica,* Leibniz sent to Huet in 1678 and 1679 two remarkable and still largely unexplored letters in which he extolled the virtues of erudition and history in relation to the defence of religion. In his first letter, despite voicing certain misgivings about Huet's methodology and argumentation, he expressed his satisfaction at the publication of Huet's 'great and immortal work'.[47]

Erudition, memory, and history were not only worthy fields of enquiry in themselves, but they could be put to theological use by helping to authenticate the claims of scripture. For Leibniz, as for Huet, who in his *Demonstratio Evangelica* had himself explicitly placed erudition in the

---

43. Leibniz to Gottlieb Spitzel, 10/20 February 1670, A I, 1:85: 'In order that the victory might be perfect and that the mouths of the impious be forever shut, I never tire of wishing that one day there arise a learned man ... in all kinds of erudition, who will clearly establish the harmony and beauty of the Christian religion, and unequivocally dispel the innumerable objections levelled against its dogmas, text and history.'

44. See Laerke, 'À la recherche d'un homme égal à Spinoza', 388–90.

45. See Shelford, 'Thinking Geometrically' (2002), 599–601.

46. Niderst, 'Comparatisme et syncrétisme religieux' (1994); Laerke, 'À la recherche d'un homme égal à Spinoza'.

47. Leibniz to Huet, 18/28 October 1678, A II, 1:641.

service of God and history, historical scholarship allowed us to vindicate the authenticity of the Bible as it had been foretold 'centuries before his birth' and help ensure that it had been passed down uncorrupted and unadulterated, thereby refuting recent attacks on scripture as fraudulent. Therein lay the 'greatest advantage that [could] be expected from history and erudition'. 'The study of antiquity', Leibniz stated, 'seems to me almost to have no other use than to put ourselves in a position to observe and preserve faithfully the ancient titles of our happiness, and if I may so express myself, our nobility, which we must, after our regeneration by baptism, trace back to Jesus Christ.'[48]

Huet and Leibniz thus agreed on the usefulness of history and the recourse to historical criticism in biblical exegesis.[49] This, however, required meticulous historical work, which took the shape of compiling 'inventories': 'But it cannot be proven that all these things happened as they are described, without having solidly established all universal history, sacred and profane; and to be able to establish it firmly, we need collections of manuscripts, medals, inscriptions, and all the other elements which compose the trove of the savants. For it is from this that history derives its credentials.' This task required a particular type of scholar who would be 'willing to work hard to compile, as it were, the most accurate inventory possible of all those precious remains of antiquity which we still possess'. While the German classical scholar Marquard Gudius on the one hand promised an account of inscriptions, and Ezechiel Spannheim on the other hand, along with Carcavy, one of medals, Leibniz still hoped for a 'history of manuscripts, containing a list of the best codices still extant in Europe, especially those from which copies have been made and which are unique or at least rare'.[50]

The chief purpose of the 'whole study of antiquity' consisted in 'the elucidation and the confirmation of sacred history', one which could only be undertaken by a scholar who applied 'the principles of the art of criticism ... and is well versed in the properties of languages, the genius of centuries, and their chronology'.[51] For this task, Huet stood out as the natural candidate, for 'such a work was easy only for [him] alone, perhaps because [he] alone in our century, possess[es] the vast foundation of erudition and philosophy which it demands.'

---

48. Leibniz to Huet, 1:641. For more on Leibniz's recourse to ancient history in the defence of scripture, see Cook, 'The Young Leibniz' (1999), 103–22.

49. See Laerke, 'À la recherche d'un homme égal à Spinoza'.

50. Leibniz to Huet, 18/28 October 1678, A II, 1:642.

51. Leibniz to Huet, 1:641–2. Leibniz was intent on reconciling ancient and modern thought.

In fact, 'who ever deserved better than this glorious title [of scholar] in the sense that we are going to define it' than Huet? Still, as Leibniz further specified, 'when it comes to demonstrating the truth of the Christian religion, much more material and research is needed. For the issue here is the fall and restoration of the human race, the differences among nations, the most ancients texts, and this analysis requires not only a philosopher, but a scholar.'[52] A true scholar was

> a man ... who has learned the greatest things in the world from the beginnings of human memory ... a man who is acquainted with the principal phenomena of heaven and earth, the history of nature and the arts, the emigrations of peoples, the revolutions of languages and empires, the present state of the universe, in a word, who possesses all knowledge which is not purely from genius, and which is acquired only through the very inspection of things and the reports of men; and that is what accounts for the difference between philosophy and erudition; the first is to the second what a question of reason or law is to a question of fact.[53]

Erudition, for Leibniz, consisted in the exploration of the truths of fact, of that which had actually happened, the *historiae rerum gestarum*, rather than the exploration of the underlying principles behind the phenomena in the world. It should serve towards the contemplation of man's perfection and the *scientiam felicitatis* through the deep study of universal history and the immersion in texts of great men.[54]

While history alone 'was very useful for providing posterity with beautiful models, to arouse men so that they too could be inspired to undertake actions that would immortalize their memory, to describe the ends of empires and bring to a close illustrious controversies in the Republic', it should, according to Leibniz, necessarily be paired with criticism: 'One of the essential uses of history and criticism is to establish the truth of religion. For I have no doubt that if the art of criticism perished once and for all, the human instruments of the divine faith would perish at the same time, and that we would have nothing solid to show to a Chinese, Jew, or Mahometan in order to prove the truth of our religion.'[55] It was necessary to exercise an 'art of criticism' in order 'to distinguish the imagined from

52. Leibniz to Huet, March 1679, A II, 1:696.
53. Leibniz to Huet, 1:696.
54. Laerke, 'À la recherche d'un homme égal à Spinoza'; Laerke, *Les Lumières de Leibniz* (2015), 107.
55. Leibniz to Huet, March 1679, A II, 1:697.

the true, and the fable from historical fact': 'If we were to recall these famous times only through the accounts of the mythologists ... there would be no more certainty regarding the facts; and far from being able to prove that scriptures are of divine origin, it would not be possible to establish that they are of such a time and such authors.'[56]

This art of criticism, and in general the study of antiquity, which helped establish the truth in matters of fact, had been 'flourishing' at the beginning of the century, but more recently had 'degenerated into open warfare' and entered a period of decline. This inspired 'sorrow' in Leibniz who deplored that 'this class of scholars, whom we call critics, and which, in the Republic of Letters is prepossessed to the safeguard of monuments, diminish every day, to the point of making one fear that it will disappear altogether in the not too distant future.'[57] This shift had been compounded by the 'revolution' that had occurred in scholarship, which had further challenged the 'old' Republic of Letters and its intellectual methods, projects, and spirit.[58] Leibniz explicitly aligned himself with Huet in his attempt to rescue a Republic of Letters that had become beholden to 'contempt for erudition', laziness, and the glorification of scholarly ignorance, thus endangering its enlightened ethos.[59]

In this, Leibniz singled out the new philosophers and the sectarian admiration for Descartes—in particular those who had been egregious in their ignorance of history, antiquity, and metaphysical matters, confining themselves instead to natural philosophy and mathematics.[60] He rejected the Cartesians' abusive deployment of hyperbolic doubt to denigrate even solidly established knowledge in favour of often frivolous opinions.[61] More generally, Leibniz inveighed against those who claimed to overturn the past, and he emphasized the necessity of incorporating

---

56. Leibniz to Huet, 1:697; Leibniz, 'Discours touchant la méthode de la certitude et l'art d'inventer', August 1688–October 1690, A VI, 4:953. Leibniz disputed Huet's claim that all certainties carried the same weight: while moral certainty derived from experience, mathematical certainty derived from logic. See Leibniz to Thomas Burnett, 1 February 1687, in Foucher de Careil (1875–90), 3:191: 'This is the certainty to which we can aspire when we address the truth and antiquity of facts, the genuineness and divinity of our sacred books.'
57. Leibniz to Huet, March 1679, A II, 1:698.
58. Leibniz to Huet, 1:698; See Shelford, *Transforming the Republic of Letters*.
59. Laerke, 'Ignorantia inflat' (2013), 27.
60. Leibniz to Philipp, end of January 1680, A II, 1:790; A II, 1:775; Leibniz to Malebranche, 22 June 1679, A II, 1:718–19; Leibniz, 'Contemplatio de historia literaria', early 1682 (?), A VI, 4:462.
61. Leibniz to Malebranche, 22 June 1679, A II, 1:718–19; Laerke, *Les Lumières de Leibniz*, 96.

new discoveries while preserving the achievements of the past, of 'combining the two philosophies, and starting the new one where the old one end[ed]'.⁶² This is crucial to his philosophical undertaking: Leibniz believed that truth could be attained not by opposing but drawing on both ancient and modern knowledge. Accordingly, he explained, 'whenever I discuss matters with the Cartesians, certainly, I extol Aristotle where he deserves it and undertake a defence of the ancient philosophy, because I see that many Cartesians read their one master only, ignoring what is held in high esteem by others, and thus unwisely impose limits on their own ability'.⁶³ Ever conciliatory, and against Conring's claim that nothing substantially new in doctrine seemed to have been established by experimental philosophy, Leibniz emphasized that the new empiricism, for all its faults, should be embraced and celebrated in the spirit of 'favour[ing] the ingenious minds of all ages' and 'especially ... [those of] our own century'.⁶⁴ Rather than presenting the Republic of Letters as a state of knowledge, as he would later do in his *Guiliemi Pacidii plus ultra, praefatio* (1686), Leibniz in his early days focused on its 'intellectual ethos', one that placed concern for intellectual partnership and the public good—rather than hostility or coercion—above everything else.⁶⁵

62. Leibniz to Conring, 19 March 1678, A II, 1:606.
63. Leibniz to Conring, 1:606.
64. Conring to Leibniz, 16/26 February 1671, A II, 1:141; Leibniz to Conring, early May 1671, A II, 1:154.
65. Laerke, *Les Lumières de Leibniz*, 68–69, 75.

CHAPTER SIX

The Struggle for the Heart of the Republic of Letters

SCIENCE AND THE STATE

LEIBNIZ LIVED AT a transitional moment when the 'old' Republic of Letters was being challenged.[1] In his view, it had recently become plagued by 'audacity' and the 'disdain for genuine erudition', which had been brought about by an 'excessive liberty to philosophize' and 'the arrogance of men'.[2] The quest for personal fame and glory, rather than the public good, had proved particularly nefarious to the Republic of Letters, prompting Leibniz to recommend that intellectuals be placed 'under administration' to Emperor Leopold, since 'most scholars do not care about augmenting human knowledge' but were like mercenaries who 'work for money or out of vanity . . . [but not] for the glory of God or common good'.[3] Obviously Leibniz implicitly excluded himself from this description.

In 1675, towards the end of his Parisian stay, Leibniz composed a text loosely modelled on the famous moment in the *Odyssey* when Hermes shows Odysseus moly. This unfinished text, though striking, has hitherto evaded scholarly scrutiny.[4] In it he set forth an altogether different

---

1. Shelford, *Transforming the Republic of Letters* (2007).
2. Quoted in Laerke, *Les Lumières de Leibniz* (2015), 95. Leibniz, 'Dialogue entre un habile Politique et un Ecclesiatisque d'un pieté reconnue', 1679–1681, A VI, 4C:2242: 'Unfortunately, these ways of reasoning take hold of the clearest minds, because they imperceptibly flatter the pride of men and that natural inclination we have to debauchery.'
3. Leibniz, 'Aufzeichnung für die Audienz bei Kaiser Leopold I', 1688, A IV, 4:22.
4. See A IV, 1:133–216. In the introduction to this volume, on xix and xxiv, Paul Ritter characterizes Leibniz's musings in his 1675 Relations as 'a confused mix of remarks and reflexions'.

understanding of the Republic of Letters, one far removed from any ideal of equality, brotherhood, or the circulation of new knowledge or 'practicalities of life' (*commodités de la vie*), as he was otherwise wont to emphasize. In fact, far from extolling practical matters and realizable projects, Leibniz, this time, engaged in an allegorical meditation on the topic of the Republic, transporting his reader to fabulous lands and mythical ages.[5] While Leibniz had used and would continue to employ the image of sending colonizing explorers to unknown lands to extend the map of human knowledge, here he gave the image a satirical twist to account for what he believed had devalued and cheapened the Republic of Letters, juxtaposing this Republic with its ideal incarnation. Leibniz presented us with a traveller with the traits of 'a certain Greek adventurer of the nation, named Pythagoras', who amidst the 'infinity of worlds to be discovered' has founded 'a colony of the other world'. In draft B, this Pythagoras resolves on populating the colony by sending a small fleet whose members, selected after a harsh apprenticeship of five years, would 'speak only once Pythagoras had opened their mouths'.[6]

In draft C, Leibniz provides a variation on the previous draft. Here we find that Pythagoras, after clearing part of the land, has 'planted a certain drug there that Mercury had shown him and which we call glory'—and which in fact is none other than the 'real nectar of the gods and the liqueur of immortality'. Trade flourished between this colony and our world until eventually 'savages from the land of ignorance and misery' brought the colony's self-subsistence to an end, leading it to grow dependent on 'refreshments called pensions'. Through the indiscriminate distribution of praise, the inhabitants of the colony began sending a surplus of products of more or less dubious value until a great prince, scandalized at the profanation of the celestial gift of immortality through which 'men had been made into gods', ruled that only real heroes should be 'rewarded for their works by receiving true glory and immortality'.[7]

In his first and lengthiest text (draft A), entitled 'Pensées pour faire une Relation de l'Estat present de la République des lettres', which he dedicated to Louis XIV, Leibniz himself assumes the role of a traveller 'to a country in which all great men aspire to live'. He leads us on a journey in

---

5. Robinet's interpretation in *Leibniz* (1994, 285) that Leibniz here expounds a project based on a 'true and eternal friendship', which 'responds to a political philosophy of wisdom and love', is not supported by the text.

6. 'Relation de l'état présent de la République des Lettres', 1675, A IV, 1:568–71. See the appendix to this book. It is not clear that any of the drafts was circulated.

7. A IV, 1:571, 570.

time by framing his depiction of the Republic of Letters as a future—and fundamentally 'foreign' (*aliter*)—projection seemingly at odds with the text's title, which promises an account of the current state of the Republic of Letters. By offering 'conjectures as to its future development', Leibniz set out to provide correctives to what he perceived as the many imperfections of his own time: in this ideal future state, reason would finally be reconciled with revelation at a time when, in defiance of those who 'had too wicked an opinion of the Holy Scriptures', 'rigorous geometrical demonstrations *de Deo* and *Mente*' would be available. The path opened by Descartes would be explored, the art of machines exhausted, the 'secret qualities of natural bodies' exploited, and Hobbes would be as known for his moral teachings as Descartes for his physical teachings. This future republic would inaugurate a new age of physiological, mathematical, and medical discoveries—indeed, the state of medicine would be vastly improved and more new drugs made available than in all previous ages. In fact, this age would be so perfect that 'the soul, if the dreams lasted a long time, [would] finally be able to reason perfectly.' After this detour via the future, Leibniz returns to the present to consider the tasks that still remained to achieve this ideal and eradicate contemporary ignorance: 'The people are ignorant everywhere, and I never would have believed, that in a city as refined as Paris, children would be kidnapped today to bathe the sick in their blood.' Chief among the tasks is the necessity of arguing 'rigorously' against Cartesian intuitionism and of upholding the 'rigour of demonstrations.' By providing an insight into the current state of affairs and appealing to a different future, Leibniz hoped precisely to 'excite [the] devotion' necessary to help realize that future.[8] When he wrote his 'Pensée', Leibniz was hoping to become a member of the French Académie des Sciences, and in it he directly addressed Louis XIV, even dedicating the piece to him. In this connection he referred to 'Glory', as he was to do several times during his first few years in Paris, and especially in his writings to the king. His concept of glory would undergo an interesting transformation in the span of a mere few years, between 1672 and 1675.

While he initially adopted a more 'classical' understanding, whereby glory consisted in the 'most intense passion of the soul', and in his 1672 written address to the king extolled this 'heroic' model of glory, he would gradually come to decry it: 'An integral glory cannot be acquired without converting one's own power into the wealth of others. One who makes himself well-known through crime and slaughter is judged to be powerful

---

8. A IV, 1:568–70.

but not wise.'⁹ He even considered Louis XIV's patronage to exemplify the wrong kind of glory, which he compared to 'earthquakes, floods and other public calamities'.¹⁰ Already in his 'Pensée', Leibniz had begun to distinguish true from false glory, identifting the former with intellectual advancement for the benefit of humanity and the latter as a form of symbolic capital in the pursuit of worldly advancement. The colony had declined, he explained, because the inhabitants, having lost their autonomy, were forced to 'sell' their glory to obtain 'bursaries'.¹¹

The multiperspectivism that contributed to the exchanges and vitality of the Republic of Letters should not, Leibniz maintained, degenerate into acrimony. While it 'affected novelty', audacity, more often than not, ended up spreading a spirit of sectarianism, strife, and complacency, which impedes the collective discovery of truth and progress.¹² In their eagerness to disparage and make a name for themselves, *novatores* and amateur philosophers were keen to undermine any progress, causing the Republic of Letters to lose itself in a confusion of disputes and the proliferation of conflicting 'singularities', and hence to stray from the search for truth: 'Instead of holding hands to guide each other and to assure we are on the right track, we run around haphazardly and in all directions and we run into each other, rather than help and support each other' and 'with their heads full of empty subtleties without demonstration, and fighting out of caprice and passion, people miserably lose the precious time they could have used to advance solid knowledge'.¹³

Leibniz henceforth associated 'glory' with what was created, transmitted through time, and enduring. The proliferation of writings and opinions and the lack of a centralized control over them contributed to creating a 'labyrinth of letters' that would soon be impossible to unify and would make any collective action or thought impossible:¹⁴ 'This horrible mass of books, which always increases . . . the multitude of authors that will shortly become infinite, will expose them to the danger of falling into oblivion, and the desire for glory which animates many scholars will suddenly cease; it will perhaps become shameful to be an author when it

---

9. A IV, 1:247–48.

10. Piro, 'The Sellers of a Sweet Powder' (1999); Leibniz, 'Conversation du Marquis de Pianese et du Père Emery Eremite', 1679–80 (?), A VI, 4:2275–76.

11. Piro, 'The Sellers of a Sweet Powder'.

12. Leibniz, 'Recommendation pour instituer la science générale', 1686, A VI, 4:695.

13. See Garber, 'Novatores' (2018) ; Laerke, *Les Lumières de Leibniz*, 66, 96; and Leibniz, 'De la philosophie cartésienne', 1683–85 (?), A VI, 4:1482, in Laerke (2015), 67, and (2013), 26.

14. Piro, 'The Sellers of a Sweet Powder'.

was once honorable.'[15] The discoveries and heterodox opinions encouraged by this new 'philosophical age' had culminated in a 'kind of creative disorder'.[16]

Leibniz thus seemed to suggest that the ideal, true Republic of Letters was born out of the transmutation of the contingent into the perennial. The Republic thus described by Leibniz is one in which those 'marvellous things' of each century have passed into posterity—and more precisely, reached the supratemporal state in which the past has been transcended and contingency overcome.[17] (He did not, however, elaborate on the requisite criteria of supratemporality.) Here we come to the crux of the issue in these three short texts and Leibniz's conception of the Republic of Letters: Leibniz strikes an unusual chord not only by his choice of genre, the literary device of allegory and fable, but by his emphasis on the passing of time and the judgement made by history, which determines what subsists throughout the ages and what falls into oblivion: 'We erect great buildings on ideas which will be overthrown one day. But something solid and admirable will remain.'[18]

Like true glory, which attached to the eternal, 'time' for Leibniz was 'like the King who abides by the maxim to only reward those who are truly singular in each species'.[19] Time's passage presented itself all the more acutely in the realm of the *belles lettres*, for whereas scientific progress appeared linear and incremental, the value of the humanities (as we would now call them) lay in their exemplarity and conformity to perennial standards. As Leibniz stated, 'a beautiful musical tune that will always be sung is like a freshly discovered geometrical theorem'.[20] This distinction lay at the heart of the ongoing *Querelle des anciens et des modernes*, which culminated in the 1680s and in whose terms Leibniz framed his understanding of the Republic of Letters. Only by a recourse to a past—that is, to the values that had defied time to become supratemporal and would

---

15. A IV, 3:789–95; see also A VI, 4:698. Interestingly enough, in his *Praefatio operis ad instaurantionem scientiarum* of 1682 (A VI, 4:440), written only a few years later after his 'Pensées', Leibniz presented a similar allegory, but this time concerning 'erudite' knowledge. The issue here was not the quality or excessive quantity of the goods, but their lack of organization and a proper inventory. The inability of the *érudits* to produce a cohesive scientific body had resulted in their irrelevance and oblivion in the past.

16. Leibniz, *De Republica Literaria*, 1681, A VI, 4:438; Leibniz to Arnauld, 1671, A, II, 1:171; Piro, 'The Sellers of a Sweet Powder'.

17. A IV, 1:570.
18. A IV, 1:569.
19. A IV, 1:569.
20. A IV, 1:569.

ideally continue into an equally supratemporal future—could one hope to escape the vagaries of contingency and mutability. Since medals 'are for all places and time', and 'inscriptions, similarly, are for all times', it was necessary to 'find the secret to stabilizing the French language so that it does not evolve—something that only belongs to the dead or decaying languages whose peoples no longer exist or whose empires are in decline'.[21] For the same reason, owing to the 'abundance of good books' written in the vernacular, those written in Latin are likelier to survive.

For Leibniz, these visions and projects were eminently political, and contemporary France was best placed to realize them. As Leibniz had already sketched out in the more utopian plans for academies that he had drafted in Mainz and reiterated in various letters to Conring and Chapelain, the crisis brought on by political and religious turmoil, as well as by the wave of new scientific discoveries, had made it necessary to rethink the role of intellectuals in society and the articulation between power and knowledge.[22] As seen earlier in this book, Leibniz's ideal political state both supported learning and was informed by it. In this state, a large number of the citizens of the Republic of Letters would hold official positions in courts, academies, and universities, helping to guide and administer the state. His ideal Republic of Letters, the goal of which he would pursue, was thus simultaneously political and scientific. For only when it was achieved would 'something solid and admirable' remain.[23] That is, only a properly functioning Republic of Letters, paradoxically restored by political fiat to its original autonomy, could ensure the immortality of humanity's great achievements.[24] As he sought to indicate in his allegorical tale, Leibniz was persuaded that true glory derived from a mutually beneficial alliance of power and wisdom, especially in the political cultivation and sponsorship of the arts and sciences.[25] Accordingly, he presented the king as being able to salvage the Republic and give it new life.[26]

Some of his early proposals thus reveal Leibniz to be a radical political thinker and visionary who, by weaving science, economy, philosophy, and theology together, sought to redesign society as a whole. By sketching

---

21. A IV, 1:569.
22. Marras (2011, 68): 'The Republic of Letters for Leibniz is inevitably more than an ideal, imagined republic; it was a concrete one.'
23. A IV, 1:569.
24. Piro, 'The Sellers of a Sweet Powder'.
25. Griard, 'Le meilleur régime selon Leibniz'.
26. Piro, 'The Sellers of a Sweet Powder'.

out far-reaching scientific, socioeconomic, administrative, and educational reforms, he aimed to 'establish a *res publica* of the entire human race', under the guidance of an 'authoritarian welfare state' headed by philosopher-king figures uniting in themselves wisdom and power and driving the rational reform of their society.[27] By setting up such metaphysical technocracies, such princes would reap not only financial benefits but also eternal fame by contributing to God's glory.[28]

'Funny Thought'

Leibniz expanded on the nexus between knowledge, science, and power in his 'Drôle de Pensée' of 1675, a reverie that took the shape of a wishlist of the most extraordinary sights and curiosities. In it he envisaged a completely unprecedented kind of scientific academy that would bring together experts and *honnêtes gens* alike, and turn scientific demonstration and experimentation into a public spectacle within the broader 'theatralization' of scientific conceptuality. Within a society reorganized by science as a permanent and ubiquitous encyclopedia, several 'Academies of Games' spread about throughout the city of Paris—and ideally later throughout cities in Europe—would showcase 'all things imaginable', and even the seemingly impossible, with self-playing instruments and vessels that could navigate against the wind.[29]

Inspired by the spectacle of an automaton in the form of a man that had been made to run across the surface of the Seine from the quai de la Grenouillère, Leibniz composed a small text in which he imagined a vast array of the most extraordinary and beautiful representations.[30]

---

27. See Schneiders, 'Sozietätspläne und Sozialutopie' (1975), 58–80; and Ramati, 'Harmony at a Distance' (1996), 430: 'All of his outlines were intended to further the same utopian ideal: political power should be given to wise and pious men, who are best suited to guide mankind into a more ordered and rational phenomenal reality because they are able to combine scientific theory with practice.'

28. See, e.g., 'Grundriß eines Bedenkens von Aufrichtung einer Societät', 1671, A IV, 1:531. It is unknown whether these drafts were sent to Louis or anyone else, but it is difficult to avoid the impression that these visions expressed the essence of Leibniz's ideal scientific society (with him as a member), whether the Académie or another.

29. Robin, 'L'Académie des plaisirs de Leibniz' (2004); 'Drôle de Pensée', September 1675, A IV, 1:562–68. As Michael Kempe perfectly sums up, 'Funny Thought' 'embodies the promise ... to accomplish things previously deemed impossible using research and technology ... [in] an era of renewal during which inventions, machines and new constructions offer the promise of surpassing the limits of reality and open up a space of unsuspected possibilities'. Kempe, *Sept Jours dans la Vie de Leibniz* (2022), 31.

30. Rateau, 'La "Drôle de pensée"' (2023), 152.

Giving free rein to human ingenuity and imagination, the 'alternative to the [Parisian] scientific academy' presented in 'Drôle de Pensée' encompassed all kinds of displays, shows, inventions, games, artefacts, and natural phenomena, including 'magic Lanterns, kites, artificial meteors, all manner of optical marvels; a representation of the sky and the stars . . . fireworks, jets of water, vessels of strange forms; Mandrakes and other rare plants . . . rare and extraordinary animals', as well as a 'Royal Machine for races with artificial horses', not to mention 'speaking trumpets'. Leibniz imagines that 'the representation could be combined with some sort of story or comedy', and that this performance might include 'extraordinary tightrope dancers [and] perilous jumps'. The public might observe 'a child who raises a great weight with a thread', and there would be an 'anatomical theatre', as well as a 'garden of simple [elements]'.[31]

In their seminal work *Leviathan and the Air-Pump*, Steven Shapin and Simon Schaffer have examined the production of matters of fact through the use of performance and rhetoric.[32] This increasingly involved making the domains of theatre, opera, and spectacle privileged modes of legitimation and dissemination in a bid to make the new scientific discourse more palatable and ensure its widest possible dissemination.[33] The realignment of scientific discourse with the canons of Parisian *honnête* sociability took the shape of the exploitation of the essentially demonstrative and performative nature of the new science, one which appealed more to the senses than to pure intellect.[34] What Jacques Truchet describes as the later seventeenth-century's 'triumph of theatricality', in which spectacle and theatricality assumed privileged roles in the processes of cultural legitimation and diffusion, therefore also came to encompass scientific conceptuality.[35] It was epitomized by the Parisian scientific, cultural, and social scenes that Leibniz came to inhabit.

Leibniz further exploited an interesting confusion of genres which contributed precisely to the poetic of ambiguity characteristic of his age and its production of norms of truth. By deliberately borrowing a term from the lexicography of drama, he emphasized the naturalization of theatricality and performance underpinning the production of scientific—and

31. A IV, 1:563.
32. Shapin and Schaffer, *Leviathan and the Air-Pump* (1986).
33. Robin, 'L'Académie des plaisirs de Leibniz', 179; see Jones, *The Good Life* (2006), 184–87.
34. Robin, 'L'imaginaire scientifique' (2003), 151, speaks of 'the intimately theatrical essence of scientific experimentation'.
35. Truchet, 'Les arts du spectacle' (1992); Robin, 'La théâtralisation' (2007), 335: 'The new science also undertook to raise itself to the rank of a performing art.'

social—norms.³⁶ As he shows, representation lies at the heart of *honnête* sociability, and, conversely, *honnête* sociability lies at the heart of representation. Only under these conditions—whereby cheating in gambling, unless banished by agreement, might be permitted—would the new science be widely disseminated and 'ennobled'.³⁷

Calling this institution an 'academy of representations' captured the performative nature of scientific experimentation. And by showcasing science and learning as a profoundly empirical experience, 'opening the eyes' of its audience, Leibniz offered a further refutation of the Cartesian tendency to neglect sense-experience.³⁸ By turning the experimentation at its heart into play and spectacle, this new science would appeal to the 'taste' of Parisian high society by 'drawing on ... what was fashionable and had the appearance of quality'.³⁹

Leibniz thus rendered the 'dissemination of wisdom' contingent upon the acquisition of the imprimatur of polite sociability.⁴⁰ Enlisting illusion, strategic deception, wonder and excitement for the 'curious and monstrous', this extraordinary spectacle would not fail to further scientific understanding and even foster further discoveries.⁴¹ In fact, Leibniz imagined that 'all honest men would want to have seen these curiosities, so that they would be able to speak about them'.

'Even women of quality', he adds, would wish to see them. In this interdisciplinary wonderland of 'new representations' in which the acquisition of knowledge and curiosity converged with entertainment and pleasure, 'one would always be encouraged to push things further'.⁴² In this manner, one could 'heal' the world by exploiting audiences' weakness for entertainment and educating it surreptitiously.

36. See Robin, 'L'imaginaire scientifique', 156: 'In Richelet, three entries out of twelve on "Représentation", "Représenter", or even "Se représenter", belong to the comedian's vocabulary.' In his entry on *representation*, Furetière (*Dictionnaire universel*, 1690) includes references to acting: 'The Poets in their Tragedies make vivid representations of the incidents of History and of the passions of the Heroes ... TO REPRESENT. To make an image or painting of an object, which lets us know it as it is; is also said of what conveys things by words, & by gestures.'

37. Robin, 'La théâtralisation', 340–41.

38. Wiedeburg, *Der Junge Leibniz* (1970), 646. Leibniz would in fact declare several years later that he 'prefer[red] a Leeuwenhoek who tells me what he sees, than a Cartesian who tells me what he thinks.' Leibniz to Huygens, March 2, 1691, A III, 5: 62–63.

39. A IV, 1:567; Jones, *The Good Life* (2006); Bertucci, *Artisanal Enlightenment* (2017).

40. A IV, 1:567.

41. Wiedeburg, *Der Junge Leibniz*, 1005, 621.

42. A IV, 1:565.

Leibniz's reverie includes a most curious and unexpected passage that further highlights the ambiguity of this 'academy of pleasures': 'There would be several houses or Academies of this nature throughout the city. These houses or rooms will be built in such a way that the master of the house can hear and see everything that is said and done, without being noticed by means of mirrors and pipes. Which would be a very important thing for the state, and a kind of political confessional.' This 'academy of games' would serve to educate its audiences through play, entertainment and illusion; at the same time it would help collect information useful to the state on those very same audiences. Leibniz here hints at the versatility of such a panopticon—including for disciplinary purposes—hoping no doubt to ingratiate himself with the political authorities.[43] The very same apparatus that would distract audiences and lull them into a sense of wonder would provide the state with a ubiquitous gaze with which it would be able to patrol their lives. Leibniz here placed the senses, in particular that of sight, in the service of political power. That which served to experience, observe, verify, probe, and, finally, elucidate could, conversely and disconcertingly, be deployed to achieve the very opposite, namely, used to monitor and possibly deceive from the shadows.[44] In Leibniz's conception, transparency and deceit could thus perfectly coexist. His long-held dream of a ubiquitous gaze in the service of philosophical, logical, and experimental illumination has now acquired a more sinister sense here by doubling as a 'political confessional' for an unwitting public.

We find here, therefore, little distinction between science, pedagogy, profitability, and political control. The philosophical and scientific are coupled with the playful, which itself doubles as the pedagogical and, potentially, the disciplinary. Similarly, Leibniz was one and multiple people at the same time, whose various identities did not exclude each other but in fact often overlapped: passing seamlessly from one function and register to another, Leibniz the projector and experimenter regularly doubled with Leibniz the political operator, agent, and informant. Epistemological fluidity coincided with a degree of social fluidity.

Leibniz's interest in political control, or what may be termed the state's disciplinary apparatus, has remained relatively unexplored by scholarship.[45] And yet, from—and perhaps especially at—the very beginning of his career, Leibniz offered his services to aristocratic employers in a wide

---

43. Robin, 'L'Académie des plaisirs de Leibniz'. It remains unknown, however, whether anyone in power ever saw this vision.
44. Knecht, *La logique chez Leibniz* (1981), 16.
45. Kempe, 'Dr. Leibniz' (2015), is an exception in this regard: Kempe's work has

range of capacities, from the legal to the commercial to the theological. Although he was committed to a number of utilitarian projects for the public good, he obviously remained beholden to his employers. In fact, in what perhaps seems to contradict more widespread perceptions of Leibniz, especially his well-publicized irenicism, he regularly advised or sought to advise on military matters, drafting political and military memoranda, designing guns, or, as we saw, recommending war against the 'barbarian' and calling for the conversion of the 'infidel'.[46] Leibniz would retain a lifelong desire to set up a kind of virtuous absolutist rule which would crystallize the marriage of power and reason and promote the perfection of man and society.[47]

While seeking to erect a new rational social order and progress towards 'greater culture' was theoretically a noble goal, its implementation was not unambiguous. As with many of his schemes, Leibniz did not necessarily work out the concrete details or real-world implications of many of his great projects and abstract ideas, including those aimed at the greater good and welfare of mankind. Leibniz, furthermore, operated within the constraints of the prevailing political absolutism of the time, and while he rejected arbitrary rule, he seems to have been more concerned with ensuring the public welfare, the improvement of living conditions, and the virtue of citizens rather than the dissemination of liberal values of liberty, rights, and representation, such as were being propounded by English political thinkers.[48]

Achieving the common good—including furthering God's plan—according to Leibniz would require a certain degree of coercion (and willful blindness) if necessary, which would appear fully justified in the light of reason.[49] Overlooking the exploitation and brutality of colonialism, he speculated that once indigenous peoples had been awakened to their power of reason, they would welcome their conversion to the true

---

touched on Leibniz's military science in which play and games play an essential role as a way of acquiring knowledge.

46. Wilson, 'Leibniz on War and Peace' (2016), 13–16. Leibniz followed news of developments in the firearms industry with great interest, from gun powder experiments to the discussion of the production of bullet-proof harnesses with Martin Elers (Martin Elers to Leibniz, Berlin, 20/30 December 1681, A III, 3:525–27) and even the production of a form of machine gun. Kempe, 'Dr. Leibniz, oder wie ich lernte, die Bombe zu lieben. Zum Verhältnis von Wissenschaft und Militärtechnik in Europa um 1700' (2015).

47. Riley, *Leibniz's Political Writings* (1988), 24.

48. Riley, 24–25

49. 'The common good was accordingly widely agreed in the 17th century to be compatible with natural or contractual slavery. In addition, it justified the brutal handling of the "barbarian" world outside of Christian Europe.' Wilson, 'Leibniz on War and Peace', 16.

faith.[50] A war directed towards 'taming' the 'beasts (that is, barbarians)' was therefore 'just'.[51] Similarly, in a 1671 addendum to the Egyptian plan, Leibniz even envisaged the enslavement, if necessary, of young boys in the style of Janissary armies, the training of an army of warrior slaves, who would welcome defending European Christendom once they had been trained and presumably civilized.

Ultimately, too, irrespective of his schemes for philosophical languages and religious reconciliation, Leibniz remained throughout his life first and foremost in the employ of princes whom he needed to placate—and this points to an ambiguity at the heart of his vision. To sell his services to the French state would simply have been a logical next step for him, especially given that he had actively courted Colbert and the king even before arriving in Paris.[52] The confusion of genres and coexistence of different registers in his envisioned academy attested to the nexus between science and political power at the heart of Colbert's vision of learning, one in which Leibniz was eager to participate.

Yet, radically, this 'Academy', for which Leibniz would obtain royal approval, would also be the first of its kind to be completely self-funded, being constituted as a business venture in such a manner that public good and profitability would coincide.[53] In evident contradiction to the academy's surveillance function for the purposes of state control, Leibniz thus also imagined it as an autonomous and intellectually independent body, free to pursue its own research objectives. No wonder, then, that he struggled at times to navigate the tension between scholarship and power, and to reconcile the ideal of unfettered research with the reality of subservience to power. As he would find out quickly enough, willing as he was ready to place his knowledge in the service of power, power did not in turn feel obliged to implement his schemes or cater to his vision of learning.

50. Daniel Cook sees Leibniz as a 'colonialist *malgré lui*'. Cook, 'Leibniz on "Advancing toward Greater Culture"' (2018), 178.

51. As Cook writes (165): 'The only conclusion that can be reached (though Leibniz never says so explicitly) is that their conversion to Christianity and the consequent enlightenment of their posterity trumped their present suffering, thus justifying European expansionism: in effect, the sins of the slavery, the colonialism and the rise of imperialism in his day will be washed away by the future enlightenment and salvation of the "barbarians" and (perhaps most importantly) their posterity, thus aiding in "advancing toward greater culture" in keeping with God's plan for His world stage in the ultimate *"perfection of mankind"*.'

52. See Moll, 'Von Erhard Weigel zu Christiaan Huygens' (1982); Soll, *Information Master* (2009).

53. A IV, 1:565.

CHAPTER SEVEN

Infiltrating Colbert's State Republic of Letters

EVEN BEFORE SETTING OUT TO PARIS, Leibniz had set his sights on the Académie Royale des Sciences and more broadly, on the state-backed Republic of Letters. The figure who had now for the past ten years masterminded the economic as well as the cultural, intellectual and scientific revival of Louis XIV's France was Jean-Baptiste Colbert (1619–83), the king's finance minister. As controller of finances, he worked relentlessly to improve the state of French manufacturing and restore the government's finances from the brink of bankruptcy through a fiscal refoundation, notably a series of financial reforms at a time of increasing economic competition from European neighbours. Crucially, he implemented the *dirigiste* policies codified in his *Memorandum on Trade* (1664), which aimed at ensuring France's economic recovery and independence through the encouragement of commerce and manufacturing in a wide array of fields.[1] More generally, the authorities established new industries, protected inventors, invited specialized workers from foreign countries, and prohibited their French counterparts from emigrating in a bid to confine talent to France. Little escaped Colbert's purview in his quest to establish France as a worthy economic rival to its more successful European counterparts—from encouraging major public works projects to allocating monopolies, setting tariffs, promoting the French merchant marine, and

---

1. See Bertucci, *Artisanal Enlightenment* (2017), 30–40. This included, among others, cloth and glass production with the foundation of the Manufacture royale de glaces de miroirs in 1665 to supplant the importation of Venetian glass (forbidden in 1672, as soon as the French glass manufacturing industry was on a sufficiently sound footing) and encourage the emulation of Flemish technical expertise in cloth manufacturing.

instituting the French East India Company. Key to this economic, cultural, and scientific transformation and success was Colbert's newly erected centralized state information system.[2] To set it up, Colbert turned statecraft into a 'practically oriented state research institute',[3] as he built a network of state scholars who placed their erudition, methods, and skills in the service of the state, from its policy-making to its quotidian administrative functions. His was a learned administration in which political administration and scholarship went hand in hand in what Arthur de Boislisle called *erudition d'état*.[4]

Part of Colbert's strategy concerning the state Republic of Letters was the recruitment of foreign scholars and scientific figures such as Huygens, Cassini, the German Hermann Conring, and the Polish-Lithuanian Protestant astronomer Johannes Hevelius. In return for royal largesse and pensions, these scholars were unambiguously expected to place their particular skills and mobilize their intellectual networks in the service of the state and to engage in propaganda on its behalf.[5] His project went beyond bringing the instruments of learning under state control and thus regulating the public marketplace of ideas. By co-opting learning for the service of the French absolutist state, Colbert's administration epitomized the confusion between state and scholarship, which came to coexist in an ambiguous symbiotic relationship that was simultaneously nurturing and repressive.[6]

By setting up this 'state Republic of Letters' and bringing the political and learned realms together, Colbert set the tone for the kind of research that should be carried out by favouring research projects of a practical,

2. See Cole, *Colbert* (1939).

3. Soll, *Information Master* (2009), 3; see also 'The Antiquary and the Information State' (2008).

4. Boislisle, *Correspondance des contrôleurs généraux* (1874–97), 1:xxi–xxii; see also Briggs, 'The Académie Royale' (1991).

5. Maber, 'Colbert and the Scholars' (1985).

6. Dew, *Orientalism* (2009), and Soll, *Information Master*, offer sophisticated accounts of Louis XIV's chief minister co-optation of scholarship—including, as we have previously seen, Oriental studies—on behalf of the state. To quote Soll: 'The case of Colbert's information system shows the extent to which a public sphere and Republic of Letters coexisted in a symbiotic and competitive relationship with the growing sphere of state information and knowledge. It also shows the growing role of experts and how the state played a central and innovative, as well as repressive, role in the growth of modern information culture' (12). Colbert's administration does not accord with Jürgen Habermas's well-known interpretations of the public sphere as a hub of civic opposition to state power. In this case the interconnection between state and scholarship was much more complex and ambiguous. On the limits of openness in the Republic of Letters, see Malcolm, 'Private and Public Knowledge' (2004), 299; cf. Habermas, *The Structural Transformation of the Public Sphere* (1989).

empirical, and industrial bent.⁷ Scholars were no longer independent figures but state operatives, with many now employed in the service of the crown as natural philosophers, historians, librarians, archivists, as well as diplomats and propagandists.⁸

The centrepiece of this edifice was perhaps the Académie des Sciences, the creation of which in 1666 had institutionalized science and formally linked it to power.⁹ Research in natural philosophy—strictly separated from *belles lettres* and political and religious questions—was henceforth to be conducted no longer by amateurs funded by private patrons, but by remunerated scholars 'living for and through science' and working for the state towards the production of technical and practical applications.¹⁰

In Leibniz's view, Paris and in particular the Académie would provide him with financial security. It would also afford him the freedom to devote himself to his research and the 'calm state' to which he had aspired since his days in Mainz and in which he could devote himself undistracted to his own scientific research,—indeed there was 'nowhere better opportunity than in France' since the king supported with pensions scholars 'from whom something [was] to be expected'.¹¹

Already as a young scholar, Leibniz was animated by the dream of meeting an enlightened and generous prince who would leave him to pursue his ideas for the public good,¹² and to this end Leibniz sought to place his learning and science in the service of the French state. One of the

---

7. See Colbert, *Lettres, instructions et mémoires* (1861–67), vol. 5, containing his correspondence with cultural figures. See also Collas, *Un poète protecteur des lettres* (1912); Ranum, *Artisans of Glory* (1980), 188–96; Stroup, 'Nicolas Hartsoeker' (1999); Rothkrug, *Opposition to Louis XIV* (1965). On Colbert's relationship with the poet Jean Chapelain, whom he enlisted to guide the king's distribution of pensions to scholars, see Soll, *Publishing 'The Prince'*, 48–50; and Chapelain, *Lettres* (1880–83), 2:275.

8. See Stroup, 'Science, politique et conscience' (1993).

9. Salomon-Bayet, 'Les académies scientifiques' (1978). Even though we do not possess direct comments, it is fair to assume that Leibniz grasped the highly controlled nature of court and science in Paris, and that, like Colbert, he had a particular appreciation for systems.

10. Salomon-Bayet.

11. Salomon-Bayet, 160; Leibniz to Johann Friedrich, first half of October 1671, A II, 1:165; Leibniz to Pierre de Carcavy, early November 1671, A II, 1:183; Leibniz to Jean Gallois, end of October 1682, A II, 1:530; Leibniz to Johann Friedrich, second half of October 1671, A II, 1:268.

12. Leibniz to Johann Friedrich, January 1675, A I, 1:492: 'I am disinterested and seek only the protection of a great prince ... a man as myself whose only concern is to acquire glory through significant discoveries in the arts and sciences and to oblige the public through useful works should find a great prince able to penetrate to the bottom of things in order to assess their value and who is guided by generous feelings and the pursuit of glory, provided that the state of affairs allows him to favour wonderful endeavours.'

first strategies he devised was writing to the eminent academician Jean Chapelain (1595–1674), who had come to occupy a position of unrivalled influence in the official direction of literary life in France.[13] Transformed from a '*curieux* who informed other *curieux*' into Colbert's political agent, adviser, and informer,[14] Chapelain placed his numerous epistolary contacts from the Republic of Letters in the minister's service and drew up lists of *gens de lettres* for Colbert to decide which scholars deserved patronage and royal gratifications. A keen discerner of talent, Chapelain introduced the figures he identified to relevant and vital intellectual circles.[15] He singled out the scientist Nicolas Steno, offered d'Herbelot a place in the new academy of Oriental languages in order to lure him back to Paris, ingratiated himself with the scholar Isaac Vossius by informing him that Louis XIV himself had taken a personal interest in his work, and even asked Hermann Conring to work for the French crown by assembling historical documents that could then be used as French propaganda.[16]

After previously failing to win the approval of the scientific community with the dedication of his *Theoria motus abstracti* to the Académie in 1671,[17] Leibniz now wrote a letter not short on intellectual ambition to Chapelain to recommend his work on legal reform. Legal knowledge had hitherto proved unfailing in its ability to ensure Leibniz's access to the highest political circles, earning him a promotion from the elector of Mainz to the position of *Revisionsrat* at the supreme court of appeal (*Oberappellationsgericht*),[18] and he was therefore relatively confident that his project of legal reform and of the 'rational ordering' of the entire body of law would help him replicate beyond Germany the success he enjoyed in Mainz. Leibniz's letter to Chapelain thus marked the culmination of several years spent closely working on the reform of the *corpus juris* at the Mainz court, work that he now sought to instrumentalize to penetrate the French intellectual and scientific scene at the highest possible level, even before he had set foot in Paris.[19] Leibniz envisaged a reform of legal thinking and practice along philosophical lines: jurisprudence would be

13. Maber, 'Colbert and the Scholars'.
14. Collas, *Un poète protecteur des lettres*; Martin, *Livre, pouvoirs et société* (1999), 668.
15. As, for example, Huygens to the Académie de Montmor: see Roger, 'La politique intellectuelle de Colbert' (1981).
16. Soll, *Information Master*, 127; Dew, *Orientalism*. For more on the use of scholars for political gain see also Montcher, *Mercenaries of Knowledge* (2023).
17. A VI, 2:27–262.
18. Antognazza, *Leibniz* (2009), 195.
19. For more on Leibniz's legal thinking and his project for a *jurisprudentia rationalis*, see Schneiders, 'Respublic optima' (1977); Berkowitz, *The Gift of Science* (2005); Armgardt,

reconceived as a branch of a broader logico-philosophical project in which all disciplines were inextricably connected and erected as a science of law based on the geometric method, uniquely apt, according to him, to ensure certainty.[20] With this universal method, it would be possible to deduce new legal truths by 'bringing only attention and patience to bear' and simply 'following the rules available'. This 'kind of systematic art', whereby the whole corpus of Roman law could be reduced to a few first principles and a limited number of propositions that could then be combined to infinity, would help solve all 'expressed and unspoken problems', especially when it came to the countless mutual complications and exceptions that had emerged between natural and positive legal norms. France, where Roman legal norms were already prevalent, could lead the way in this legal reorganization and help homogenize the legal system across Europe.[21]

Leibniz's determination to use his project for legal reform to advance his own career, not just in Germany but now across Europe, is attested by his eagerness to publicize it to the major thinkers of his time. In fact, around the same time he wrote to Chapelain, Leibniz sent out similar letters to others, including Hobbes.[22] He also sent news of his project on the codification of Roman law on the one hand to Arnauld, and on the other to Johann Georg Graevius, a German classical scholar based in Utrecht.[23]

Leibniz also reported the news of his uniquely imaginative and intrepid endeavour of a 'rational jurisprudence' to Duke Johann Friedrich of Hanover,[24] and even to Emperor Leopold—since it affected the Roman common law as the basis of imperial law. With this project, it would be possible to grasp the whole 'geographical map of science' at 'a single glance' and to 'walk through' its 'individual provinces' through the

---

'Leibniz as Legal Scholar' (2014); Meder, *Der unbekannte Leibniz* (2018); and Johns, *Leibniz's Geometric Method* (2019).

20. Antognazza, *Leibniz*, 61, 85; Johns, *Leibniz's Geometric Method*; Leibniz to Jean Chapelain, early 1670, A II, 1:87.

21. Leibniz to Jean Chapelain, early 1670, A II, 1:83–89.

22. Leibniz to Hobbes, 13/23 July 1670, A I, 2:57: 'Four years ago, I started to work out a plan for compiling in the fewest words possible the elements of the law contained in the Roman Corpus . . . so that one could, so to speak, finally demonstrate from them its universal laws.'

23. Leibniz to Arnauld, early November 1671, A II, 1:277: 'I am thinking of summarizing the elements of Roman law in a short table which presents, at a single glance, a few clear rules whose combination can solve all cases, and furthermore, new measures for abridging lawsuits. . . . In addition to these, I am planning to collect in a short book the elements of natural law, from which everything will be demonstrated from definitions alone.'

24. Leibniz to Johann Friedrich, 21 May 1671, A II, 1:105.

combination of basic 'elements'.[25] Little did Leibniz know, however, that by the time he drafted his letter, the French legal reform movement had lost much of its momentum, and Chapelain's authority was waning.[26]

Leibniz's poor grasp of the situation did not, however, stop him from making other attempts to infiltrate the French establishment. Later he would ask Huet to recommend him directly to the highly esteemed Bishop Bossuet, whose theological designs accorded with his and with whom he sought to forge a mutually beneficial relationship at a time when Christian doctrines were under attack from various sides, including the new science and Cartesianism.[27] In 1675 he even sought to join the dauphin's tutoring team, and in late 1675 he drafted a fascinating letter to the fourteen-year-old dauphin himself.[28] Presumably intended for Montausier, the dauphin's governor and Huet's superior, the letter expressed Leibniz's commitment to a particular educational ideal—to be elaborated four years later in his better-known 'Portrait of a Prince'—and interspersed extravagant praise of his prospective employer with passages on the proper education of princes and the kinds of virtues they ought to cultivate to govern justly.[29] Whether the letter was actually sent remains unclear, but in it Leibniz delivered an allegorical exposition, which he may have deemed more appropriate to convey his message and more appealing to a teenage boy, in which he sketched out the history of philosophy from antiquity up to the 'new world of the mind . . . discovered in our age by Galileo, Bacon, Descartes, and other Argonauts'. In a vivid account that reads rather like a martial epic, replete with onslaughts, retreats, defeats, invasions, ambushes, victories— no doubt calculated to stir the imagination and appeal alike to the head tutor and France's future king—Leibniz proceeded to retrace philosophical progress and the 'histories of the struggles of the human mind', a gap

---

25. Leibniz to Kaiser Leopold I, August (?) 1671, A I, 1:60.

26. See Hammerstein, *Jus und Historie* (1972). Leibniz's writing to Chapelain, instead of to the jurist Guillaume de Lamoignon (who had been recruited to prepare the codification of French laws), is another example of Leibniz's imperfect grasp of a situation: see Brockliss, *French Higher Education* (1987), 277–334.

27. Leibniz to Huet, 1/11 August 1679, A II, 1:737.

28. Leibniz to the Dauphin, end of 1675, A II, 1:394–98. This letter has hitherto remained untranslated. It is reproduced in translation in the appendix.

29. See 'Portrait of a Prince' in the original French in Klopp, *Die Werke von Leibniz* (1864–84), vol. 4, and in English in Riley, *Leibniz's Political Writings* (1988). Riley, 85: 'Since the order of states is founded on the authority of those who govern them, and on the dependence of peoples, nature. . . . Thus princes must be above their subjects by their virtue, and by their natural qualities, as they are above them by the authority which the laws give them to reign according to natural law and civil law.'

in the dauphin's education that he felt uniquely capable of remedying.³⁰ After devoting much of his narrative to Aristotle's rise in the history of philosophy, Leibniz then recounted how the Aristotelian reign had later come under attack, especially with the rise of mechanical philosophy and new discoveries in physics and chemistry, before finally having to concede defeat and withdraw on several fronts. This state of affairs had been exacerbated by the rise of Cartesianism, for the French philosopher had sent out his 'legions' and engaged in a full-fledged assault that threatened philosophy itself. Philosophy had now reached a precarious state, and although the Cartesian invasion had been temporarily repulsed, new dangers had become manifest. Descartes had struck up new alliances with Paracelsus and Van Helmont, and these threatened new hostilities. In this overview, Leibniz presented history and erudition as extremely valuable, especially for their ability to attest to the authority of scripture and theology. It was necessary, he wrote, to expound the history of that war which man conducted with nature. Yet this remarkable attempt to ingratiate himself with Bossuet or Montausier was evidently fruitless.

Louis Ferrand and the Pursuit of Abulfeda's Geography

Louis Ferrand (1645–99), was a young *provençal* scholar and Arabist who had been sent to Paris to study medicine but whose 'own Genius irresistibly carrie[d] him another Way, *viz.* to Oriental Studies'.³¹ In Paris, Ferrand spent much of his time in the royal library, where he had caught Colbert's eye and come into his employment. The latter had recently spearheaded a royal patronage of *lettres* and Oriental studies as part of a programme of cultural influence and mythologization of Louis XIV, which would earn him a reputation as a patron of the 'République des Lettres' even during his own lifetime.³² Ferrand, with his work on the history of the Crusades and Saint Louis, as well as his later involvement in the controversy against the Protestants in the 1680s, fit the bill perfectly. Ferrand had demonstrated his 'assiduity' and come into contact with 'learned persons', and he had come to be known for his prodigious proficiency in Oriental learning

---

30. Leibniz was a staunch believer in education through play and imaginative stimulation: see Jones, *The Good Life in the Scientific Revolution* (2006).

31. Pocock, *Theological Works* (1740), 1:66–67 (paraphrase of a letter of F. Vernon to Pococke, 12 November 1671), quoted in Dew, *Orientalism*, 26. Ferrand had earlier written to Pococke to ask for references on Arab historians of the Crusades.

32. Dew, 27, 29, 30.

among men of science. In the 1660s he had begun a new translation of the Hebrew Bible and met Leibniz in Mainz.[33]

The relationship of Leibniz and Ferrand was one of mutual assistance from the outset. Ferrand was also an esteemed scholar in his own right, referring to himself as 'the last of the learned men in letters', and he relished the kind of erudite correspondence that Leibniz made possible, offering to share some of his work with the latter, requesting Leibniz's help on some finer points of erudition, either directly or on behalf of his network of contacts.[34] Perhaps Ferrand's most valuable asset was his knowledge: helping Leibniz to familiarize himself with the Parisian learned world even before he set foot in Paris, Ferrand revealed himself to be a gold mine of information. In particular, he apprised Leibniz of the king's general design 'to restore all things to a better state, and to make the whole reign shine with new splendour' with the Académie des Sciences. There, 'inventions [were] proposed' and 'secret things revealed with one word', and the 'most excellent things' in science were discussed and 'puzzled out', yielding a 'great light ... for the sciences'.[35]

In Paris Ferrand acted as Leibniz's first French contact, and of all Leibniz's correspondents, he was perhaps the most resourceful in keeping the German abreast of news in the Republic of Letters, regularly reporting on major debates raging at the time (such as that between Arnauld and Claude) and on theological and scientific developments, whether connected with notable centres of intellectual activity (such as the salon held at Rohault's house) or new publications (such as Pascal's *Lettres Provinciales* of 1656–57 and Michaele Baudrand Parisino's *Ferrarius Locupletatus*).[36]

33. On Ferrand in Mainz, see Dew, 123–24.
34. The correspondence began after Ferrand's return to Paris. Louis Ferrand to Leibniz, 13 February 1671, A I, 1:121: 'I will ask the most illustrious Abbé Gravel to share his copy [of this summary] with you. . . . I come now to uncertain things which I had promised to place before you for clarification. What is, I beg of you, that village which the Rabbi calls Nossa . . . what is the outer wall near Cologne that the Rabbi, in Hebraic, writes of as Volkenbourg and elsewhere Voltzkenbourg or Folkenbourg; what is the place near Mainz that the Rabbi calls Garmaissa'; Ferrand to Leibniz, 1672–76, A I, 1:454: 'A friend of mine asked me to find out if at the Nuremberg Academy [or in its vicinity] . . . there would not be some scholar Astronomer with whom one could cultivate a correspondence regarding the observation of the stars. . . . I await your answer.'
35. Ferrand to Leibniz, 13 February 1671, A I, 1:120.
36. See for example Ferrand to Leibniz, 3 June 1671, A I, 1:152; Ferrand to Leibniz, May 1672, A I, 1:452–53; Ferrand to Leibniz, 11 September 1671, A I, 1:167: 'What is more, I had extensively discussed the business of the Republic of Letters of the French, and will talk more about this.' Ferrand to Leibniz, May 1672, A I, 1:435: 'I received news from

Knowledgeable about the Republic of Letters, Ferrand advised Leibniz to write to Arnauld, even offering to deliver the letter himself (albeit not without a warning), and he recommended Pierre de Carcavy to Leibniz as 'a person whom [he] ought well to cultivate, for he [was] all powerful around Mgr. Colbert in everything concerning letters.'[37]

In Ferrand, Leibniz seems also to have found a well-intentioned agent with whom he had much in common.[38] Once in Paris, he would visit Ferrand frequently.[39] For his part, Ferrand appeared to have genuinely valued Leibniz, peppering his correspondence regularly with marks of esteem and on more than one occasion acting on Leibniz's behalf by praising his talents to notable figures of the Republic of Letters in Paris, such as the mathematician Père Berthet, who, in turn, had shared his opinion on Pater Grandamicus's experiment on magnetism.[40] Crucially, Ferrand sought to promote Leibniz to Carcavy, reassuring the German savant in July 1671 that Carcavy had formed a high opinion of him.[41] And in December of that year, Ferrand even requested that Carcavy promote Leibniz to Colbert.[42]

Germany yesterday, but there is nothing specific to report. The abbé de Gravel sent me a little book entitled *Breviarium Chronologiae cum sacrae tum profanae* printed in Virzburg in 1672. If you want to read it I would be happy to share it with you.' Ferrand to Leibniz, 13 February 1671, A I, 1:117: 'But now, in order to satisfy your wish on all sides, I will touch on new publications. . . . Various books have recently come out here, among which certain stand out.'

37. Ferrand to Leibniz, 25 January 1672, A I, 1:179: 'It would not be out of place, if I am not mistaken, for you to write to this most famous man and to instruct me to deliver him your letter'; Ferrand to Leibniz, 3 December 1671, A I, 1:175: 'I have delivered your letter to Arnauld, he has given me no response to your letter; and he will not, I believe. For when that learned man is immersed in books, he responds to almost no letters, as I hear. I will take care of the remaining things you have instructed me to do in the next few days'; Ferrand to Leibniz, 3 June 1671, A I, 1:153.

38. Ferrand to Leibniz, 3 June 1671, A I, 1:153. It is interesting to note the distaste of both towards ceremonies and other social rituals: 'I implore you that there should be no 'ceremonies' . . . that trite, indeed barbaric word.'

39. Ferrand to Leibniz, May 1672, A I, 1:453: 'You will always do me a great deal of honour when it pleases you to come to my house, and I will always count this as a new grace that you will do me. So, Sir, you will be welcome whenever you wish to give yourself the trouble.'

40. Ferrand to Leibniz, 3 June 1671, A I, 1:153: 'You have delivered outstanding examples to me both of your benevolence and your learning in your last letters, and to say it with one word: you have satisfied my wish in every aspect; I thank you, Most Honourable Lord, in both regards; and I would like you to believe that the memory of your merits towards me will never be forgotten'; Ferrand to Leibniz, 1 October 1671, A I, 1:170.

41. Ferrand to Leibniz, 28 July 1671, A I, 1:160.

42. Ferrand to Leibniz, 3 December 1671, A I, 1:175.

Leibniz was particularly active in helping funnel books to and from various correspondents across Europe in a bid to establish himself as an indispensable member of the Republic of Letters, as well as to ingratiate himself with established and prospective patrons. The young scholar, Ferrand explained, could prove his worth to Carcavy and the newly formed Académie by initiating a book trade between Mainz and Paris and purchasing books in Germany for the royal library:[43]

> [Carcavy] collects books from everywhere in the world at the King's command and has already brought together innumerable volumes of books in all sciences and languages.... He told me that you would oblige him greatly if you could send to Paris outstanding books that will appear ... [in the] Frankfurt [book fairs] and also inform us of any other ... library, museum ... or curiosity for sale. I would be delighted if ... you consented to this business and Carcavy's wishes; in this way ... we could establish the greatest book trade; concerning money, Carcavy would pay you fairly and nothing should be lacking in this area.[44]

Ferrand thus used Leibniz as a key intermediary in the purchase of books that were then dispatched to France primarily for Carcavy's benefit.[45] Ferrand acted as a go-between between Leibniz and Carcavy, exchanging correspondence and delivering instructions, especially regarding the acquisition of books.[46] In particular, the abbé de Gravel, the French envoy in Würzburg, acted as an intermediary between Leibniz and his Parisian contacts, not without leveraging Leibniz's eagerness in order to obtain books in particular for Carcavy and Ferrand.[47]

---

43. Ferrrand to Leibniz, 13 February 1671, A I, 1:120.

44. Ferrand to Leibniz, A I, 1:120.

45. Ferrand to Leibniz, 3 June 1671, A I, 1:152: 'Please indicate the price of the books which you have sent, and the money will be paid immediately.' Ferrand to Leibniz, 13 February 1671, A I, 1:121: 'Meanwhile, Carcavy asks for Patinus's works about which you talk in your letter; all of Petrus Lambeccius's commentaries of the Caesarean Library; the edition of Joseph which has been completed by Andreas Bose; the Bishop of Paderborn's chronicle of his diocese; and Marius Nizolius's books ... whose edition you have undertaken. If you would deliver these books, I would be most grateful, and fair payment made.'

46. Ferrand to Leibniz, 3 June 1671, A I, 1:152; Ferrand to Leibniz, 1 October 1671, A I, 1:171: 'The most illustrious Carcavy has left for the countryside, and before he left, he instructed me to greet you in his name; and to remind you that he much approves of your plan regarding Zunner, the Book Vendor in Frankfurt, who may still state his terms, and that he will attend to his duties to you in this matter with utmost pleasure.'

47. See Badalo-Dulong, *Trente ans de diplomatie française* (1956).

The Abulfeda episode illustrates well how, under Louis XIV's rule, power and erudition were inextricably linked, and how, while the Ottoman threat was still looming on the horizon, interest in and the dissemination of knowledge about the Islamic Orient reached a peak. In this regard, Colbert had begun sponsoring networks intended to bring books from the Levant to Paris and assemble information about the Orient. Compilations of the collected and translated material, which would then serve as the basis for reference manuals with which to supplement other historical texts, were particularly prized.[48] Still, despite Colbert's best efforts to provide an institutional base for Oriental learning—a royal college—this college remained largely inoperative, and Oriental studies in Paris at the time remained marginal.[49] One of the best examples of these compilations is the Catholic humanist Barthélemy d'Herbelot's *Bibliothèque orientale*, which was published in 1697 and served for more than two centuries as an encyclopedia of Islam.[50]

A key figure within this Orientalist 'culture of curiosity' was the traveller Melchisédech Thévenot (1620–1692), who, in his desire to bring potentially useful and hitherto hidden knowledge (especially from overseas) into public circulation through printed translations, had accumulated Hebrew, Arabic, and Persian manuscripts, especially those with historical and geographical content.[51] Ferrand had become aware of Thévenot's project to edit Abulfeda's early fourteenth-century geographical work, based on Ptolemy's *Geography*, which Arabic authors had supplemented over time, and which contained tables of the latitudes and longitudes of Middle Eastern cities.[52]

This work was the object of widespread scholarly fascination, every major European Orientalist scholar from Guillaume Postel to Edward Bernard attempting and failing to produce an edition of it.[53] An edition

---

48. As Abdel-Halim, *Antoine Galland* (1964), 242, sums up: 'These geographic and historical compilations . . . remain a complete repository of the knowledge of Muslims in these fields. . . . They represent the ultimate evolution of these genres before the centuries of scientific decline; and seeking to make an inventory and synthesis of previous work, they constituted successful and easy-to-use general works. In the West, they appeared—and rightly so—as the most appropriate general sources for further research.'

49. On Colbert's support for Oriental studies, see Dew, *Orientalism*, 25–40; and Bevilacqua, *The Republic of Arabic Letters* (2018), 25–29.

50. For more on this subject, see Gaulmier, 'À la découverte du Proche-Orient' (1969). On d'Herbelot, see Laurens, *Aux sources de l'orientalisme* (1978); and Richard, 'Le dictionnaire de d'Herbelot' (1997).

51. For more on Thévenot, see Dew, 'Thévenot's Collection of Voyages' (2006).

52. Dew, 126.

53. See Dew, 108–9

and translation of Abulfeda's *Geography*, fragments of which had already appeared in Thévenot's *Relations de Voyages*, fell squarely within the Thévenot circle's remit as an important means of 'illustrating' and perfecting geography 'for the public benefit'.[54] Editing and translating Abulfeda would prove to be an all-consuming and ultimately unsuccessful venture lasting many decades—the work would not be published before the nineteenth century. Thévenot's role included listing it in 1687 among manuscripts to be purchased in the Ottoman Empire by the French ambassador Girardin, and hiring Antoine Galland to produce a French translation.[55] Thévenot's ultimately unsuccessful attempt to produce an edition of the Arabic text is emblematic of the ambivalent and precarious state of Oriental learning at the time.[56] Although, like many other *curieux*, he was dedicated to helping promote and disseminate Oriental scholarship in Europe, helping in the process to incorporate it to the broader production of knowledge, Thévenot faced an astounding number of obstacles in trying to bring this particular project to completion. After another failed attempt to have the manuscript printed in Holland, he returned empty-handed to Paris in the spring of 1670. He then resolved to acquire the copy made for Wilhelm Schickard, German Orientalist and professor of astronomy, from a manuscript found in the imperial library in Vienna.[57]

As Nicholas Dew has pointed out, Thévenot's inability to get Abulfeda's *Geography* published is representative of the broader fragility of the nascent field of Oriental studies in Paris.[58] Despite the dedication of networks that were to some extent successful in collecting material, Oriental studies were hampered by a lack of necessary skills, printing presses, and funding: only substantial institutional backing would enable larger-scale projects such as the production of a full-scale Arabic edition. State support was limited, however, even hindering erudition after co-opting it. But as restricted as Oriental studies in France remained in the seventeenth century, Leibniz capitalized on the interest in them.

54. That is, according to the anonymous 'Project de la Compagnie des Sciences et des Arts' (1663), as 'work[ing] towards the perfection of the Sciences and the Arts, and search[ing] comprehensively for everything that could be of some utility or convenience to the human race, and particularly to France' (quoted in Dew, 'Thévenot's Collection of Voyages', 46). Quotes from Chapelain to I. Vossius, 12 September 1666, in Chapelain, *Lettres* (1880–83), 2:476n.

55. Dew, *Orientalism*, 125.

56. Dew, 127.

57. For more on the subject of Thévenot's quest to produce an edition of the Abulfdefa, see the excellent account in Dew, 81–130.

58. Thévenot even commissioned at his own expense Arabic printing types.

Ferrand promptly informed Leibniz of the venture and the attempts being made to secure the manuscript before suggesting that an arrangement be made with Magnus Hesenthaler, since 1656 a professor of history, politics, and rhetoric at the University of Tübingen, who possessed the Schickard transcript.[59] This would be an ideal opportunity for Leibniz to demonstrate his utility to the French state: in this respect, providing Oriental material within the Republic of Letters promised to be a fruitful avenue for him. In 1671, for instance, he was tasked by his patron, Boineburg, with acquiring Johann Michael Vansleben's *Conspectus operum Aethiopicum* (1671) through the intercession of the Parisian agent in Mainz, the Abbé de Gravel, and Ferrand back in Paris.[60] (He succeeded.) Oriental scholarship remained dependent on the mobilization of scholarly networks that, in turn, could help Leibniz himself gain patronage and possibly further his career.

Knowledge was indeed shaped by power,[61] but could, conversely, be instrumentalized and serve as a conduit to power. Indeed, Leibniz's quest for the Abulfeda manuscript illustrates how, within the Republic of Letters, knowledge and power could overlap. In the field of Oriental studies, a subset within the broader Republic of Letters, Leibniz identified a niche which, if he played his cards well, could serve as his entrée into Colbert's growing institutional apparatus.

Leibniz sought to instrumentalize to his own advantage Thévenot's attempt to secure a copy of the Abulfeda, discerning in it an avenue by which to make his way to Paris, even before his infamous *Consilium Aegyptiacum*. Securing it, however, turned out to be an arduous process. On one occasion, after having had his hopes raised perhaps too often and upon discovering that there existed a copy of the manuscript as well as a Latin translation in Paris,[62] Leibniz wrote rather despairingly to Ferrand,

---

59. Dew, 123. For a discussion of Leibniz's quest for a translation of Abulfeda's *Geography*, see also Bodéüs, *Leibniz-Thomasius Correspondance* (1993), 343–46. The story can be traced through A I, 1:118, 153, 155, 157, 160, 163–66, 167–68, 170–71, 173, 175–76, 178–79, 189–90, 197, 453; and II, 1:195. There are twenty references to Abulfeda in the first volume alone of Leibniz's *Allgemeiner politischer und historischer Briefwechsel* in the Akademie-Ausgabe.

60. Abbé de Gravel to Leibniz, 20 September 1671, A I, 1:169. The German Orientalist Vansleben (also spelled Vansleb and Wansleben), financed by the duke of Saxe-Gotha, had copied some Ethiopian religious manuscripts in Egypt in 1663–64 but had not been able to reach his intended destination of Ethiopa itself. In 1672–73 he returned to Egypt in Colbert's service. See Hamilton, *Johann Michael Wansleben's Travels* (2018), 1–53.

61. Especially within the French context where Colbert attempted to co-opt all Orientalist networks: see Dew, *Orientalism*.

62. Louis Ferrand, 13 February 1671, A I, 1:118.

'I am quite happy to hear the news concerning Abulfeda. For, to confess the truth, I have been unable to talk to you until now—during the recent uncertainty—without embarrassment [*sans rougir*].'[63] Indeed, the onerous negotiations over the Schickard transcript lasted over a year, with Leibniz acting on behalf of Hesenthaler, relaying a high asking price and his lists of demands back to Gravel. Hesenthaler, who soon realized that Ferrand and Colbert were involved in the venture, was prepared to relinquish the manuscript only in exchange for copies of the printed edition when it appeared, as well as a shipment of some of the most sumptuous Parisian publications, including a complete set of the 'Byzantine du Louvre' (the Jesuit-edited *Corpus Historia Byzantina*, then still in progress), an eight-volume Bible—both works produced by the Imprimerie royale—and Gassendi's works in six folio volumes.[64] After the exchange of the goods at the Frankfurt book fair in March 1672 and verification of the manuscript by one of Gravel's associates, the Schickard Abulfeda eventually made its way to Paris, soon followed by Leibniz, his credit enhanced by his role in the acquisition. Nonetheless, he had other, greater, plans in mind for himself.

Pierre de Carcavy and Leibniz's Calculating Machine

Pierre de Carcavy (1600–84) was the director of the Académie des Sciences, the newly established 'academy of *curieux*', access to which he controlled closely.[65] Carcavy had not only been appointed to the academy, where he was in charge of cartography, at its founding in 1666 but had also served as the custodian of the royal library since 1663. His aim was to make the king's collections the best in the world across the whole spectrum of learning, from historical and antiquarian fields to the natural sciences. He had previously managed Colbert's own library, at the time providing his master with documents pertaining notably to religious rights and exemptions in France.[66] Carcavy was a keen mathematician who had himself worked on the problem of the quadrature of the circle and whose interest had been further encouraged through his friendship with the mathematician Pierre de Fermat. At one point he had tried to

---

63. Leibniz to Ferrand, May 1672, A I, 1:451.
64. Dew, *Orientalism*, 124.
65. Leibniz to Hans Eitel von Diese zum Furstentein, 19/29 June 1671, A III, 2:4; Ferrand to Leibniz, 13 February 1671, A I, 1:120: 'Carcavy, the Royal Librarian, directs this Academy, with Colbert's favour, and he will, if he may be in possession of a rather long life, promote letters and sciences not by a little.'
66. Dew, *Orientalism*. See also Soll, 'The Antiquary and the Information State' (2008).

get Fermat's *Novus secundarum et ulterioris ordinis radicum in analyticis usus* published. A member of the circle of Marin Marsenne, whose role he had effectively assumed at the latter's death in 1648, Carcavy had also become an important member of the Académie de Montmor, corresponding with leading intellectuals and mathematicians such as Roberval, Galileo, and Torricelli. By the time Leibniz contacted Carcavy, the latter had become one of France's most powerful and influential scientific courtiers, amassing an impressive array of contacts throughout Europe, to the extent, in fact, that Henry Oldenburg had once mistaken him for the president of the Académie. Leibniz seems to have been well acquainted with Carcavy's reputation early on. In letters to Peter Lambeck, the librarian at the Imperial court in Vienna—a position he would later covet—and Duke Johann Friedrich of Hanover, Leibniz clearly identified Carcavy, with whom he had been in epistolary contact, as a key player in the Republic of Letters, alongside Ferrand in France, Oldenburg and Wallis in England, and Conring and Guericke in Germany.[67] He was determined to ingratiate himself with Carcavy by acting as an intermediary on his behalf, sending him books, and more generally keeping him informed of literary developments, including the publication of new journals.[68]

In the early 1670s Leibniz had conceived the idea of a calculating machine with the aim of improving Pascal's calculator, the Pascaline, by enabling it to perform multiplications and divisions automatically.[69] Prompted by Ferrand, Leibniz sent a letter to Carcavy with news of his calculating machine and other instruments he claimed to have invented as well as several printed works on natural philosophy. Before his trip to Paris, Leibniz had recently learned the 'details of [Colbert's] great plan

67. Leibniz to Peter Lambeck, August 1671, A II, 1:149; Leibniz to Johann Friedrich, second half of October 1671, A II, 1:160: 'Many *Curiosi*, of whom I have not the slightest knowledge, have replied to my letters with extraordinary courtesy and willingness, among them gentlemen of the French Academy and the English [i.e., Royal] Society, also Mr von Boineburg, H. P. Kircher and Lana in Italy, Messrs Gericken, Linckern, Conringium, Boeclern in Germany, Messrs Graevium, Velthuysen, Diemerbroeck in Holland, Oldenburg and Wallis in England; de Carcavy the royal librarian, Ferrand and others in France, and many other virtuosos, including *chymicos* and *mechanicos*.'

68. Leibniz to Carcavy, 22 June 1671, A II, 1:213: 'I do not doubt that you have received the books which I sent a week ago, and among them the *Paderbornian monuments*, which the most Honourable Baron Boineburg has given to me to pass on to you, with greetings from himself added. I will write to the most famous Ferrandus when I have received the answer about Abulfeda.... You have without doubt seen the new *Giornale de Letterati*, whose publication has begun in Venice. I will write more about letters and sciences also to ... Ferrandus, to whom I send regards most officially in the meantime.'

69. For more on this topic, see Jones, *Reckoning with Matter* (2016).

for commerce and policing (of industry)', a centrepiece which consisted in attracting foreign talent and artisanal skill to France. In Germany, as Leibniz knew, the cameralist Johann Daniel Crafft's talents had been wasted and his endeavours largely unsuccessful on account of the laws by which artisanal guilds guarded their privileges jealously and stifled innovation.[70]

By seeking to cultivate a relationship with Carcavy, Leibniz hoped that Carcavy would grant him not only a privilege for his calculating machine, thus validating his claim to superior expertise and yielding great benefits for the state and market alike, but also, and more importantly, grant him access to Colbert's state Republic of Letters. Leibniz was particularly impressed by the French Académie des Sciences, which, by providing a collective and state-propelled research structure, sought to improve the lot of mankind as a whole:

> What the English Royal Society has done so far, too, is great in itself but scant in comparison to what would be in our power if we wanted it enough. And I believe that you pursue this goal whenever I imagine the excellent mind of Your King, who—as He has begun plans to improve human affairs—will secure a blessed memory of himself for mankind. You will not only win over the English, but all men: if you would make a further noble effort worthy of your grandeur and of a King who seeks his glory in the happiness of mankind.[71]

Carcavy, it seems, was very impressed with Leibniz's project for a machine, comparing it to Pascal's 'machine of a time past', and he hoped to show it to Colbert.[72] In early June 1671, he sent Leibniz a very encouraging letter in which he enthused over Leibniz's 'renown' and 'outstanding merit' and obliquely signalled to Leibniz that the latter might potentially be considered for membership of the Académie.[73] In a letter he sent only a few days later, Carcavy's enthusiasm continued unabated.[74] After much praise, Carcavy was thus already determined to put Leibniz and his projects to good use, enlisting the latter to compile lists of useful German books that Gravel would then have shipped to Paris. Crucially, Carcavy now requested

---

70. Jones, *Reckoning with Matter*, 115. See Leibniz to Carcavy, 22 June 1671, A II, 1:215.
71. Leibniz to Carcavy, early November, A II, 1:288.
72. Carcavy to Leibniz, 20 June 1671, A II, 1:208.
73. Carcavy to Leibniz, 5 June 1671, A II, 1:191–92.
74. Carcavy to Leibniz, 20 June 1671, A II, 1:208. 'Meanwhile, farewell, Sir, and continue to illuminate the letters and sciences.'

that Leibniz send him a copy of his calculating machine, which he believed to be 'meticulous and useful', so that he could present it to Colbert and recommend Leibniz to the minister, emphasizing that 'deeds have more power, especially with him who does more than he says'.[75] In his lengthy response, Leibniz did not offer the required description of his machine, preferring instead to expound his latest physical discoveries, especially as contained in the two treatises that he requested be printed at Carcavy's behest.[76] In November 1671 Leibniz sent Carcavy a most remarkable letter which he requested remain strictly confidential.[77] Possibly emboldened by the prospect of an upcoming trip to Paris in the wake of his diplomatic activities, he undertook '[to] recommend ... [him]self now with increasing boldness': 'I see that you are not only diligent, but also aspire to outstanding and great things and embrace with your mind the progress of the arts and sciences, since the great Colbert, the most apt minister for the true glory of the Great King, encourages such an excellent plan. And truly, what can concern the State more than to raise its own power and human happiness?'[78]

Leibniz proceeded to submit his own proposal for consideration, offering himself to Carcavy as an 'instrument perhaps not useless ... for advertising many outstanding things' from many branches of learning, including chemistry, mechanics, medicine, and mathematics. On account of his extensive correspondence with learned men throughout Germany and of the high opinion in which he was held 'throughout courts everywhere, including among German and foreign ministers', Leibniz had achieved a certain stature not only with Germans but with Italians, Englishmen, and Belgians, all of whom responded to his letters kindly and communicated matters of substance. Moreover, he insisted, he had been offered state positions but had declined them in order to be freer to pursue 'more fruitful scientific research'. Continuing, he emphasized that

75. Carcavy to Leibniz, A II, 1:208.
76. Leibniz to Carcavy, 22 June 1671, A II, 1:212: 'For one thing because I did not think that my *Hypothesis* had been seen then, which I now believe has reached your hands along with other books, as you have ordered me to discuss some problems in physics next to my principles.'
77. Leibniz to Carcavy, early November 1671, A II, 1:290: 'Furthermore, I wish my proposition to be kept secret at least, whatever you may think of it: if you discard it, then because I do not want to be held accountable for things not said; if you lend your ears to it, then because in that case men who could be useful to me might, once the matter is made public, become resistant out of envy.'
78. Leibniz to Carcavy, A II, 1:288.

he had 'never asked for anything more ardently than that he be allowed' to occupy himself quietly with improving the sciences.[79]

A scheme whereby Leibniz could explore 'the secrets, circumstances, and arts by which human power is increased' was in fact in perfect keeping with Colbert's ethos and would not fail to earn Carcavy himself no small amount of recognition and glory.[80] Leibniz, with his customary brashness, thus proceeded to dictate his terms: Carcavy would recommend the German's admission to the Académie des Sciences. In exchange for his membership and adequate compensation, as well as his freedom from other duties, Leibniz would offer a range of services, including obtaining books, compiling indexes, and revealing information about scientific inventions, experiments, publications, and the like.

While there undoubtedly existed 'more learned' scholars than he, Leibniz conceded, it would be difficult to find one 'better prepared for this plan through experience, education', and 'finally through the joy itself which [he had] almost from childhood on received from the variety of manifold acquaintances'. Addressing Carcavy's concerns in the hope of formalizing a working relationship with the Académie as soon as possible, he had proven himself to be not merely a man of words, but also one of action: 'Nobody regretted having recommended me, I have shown that I could achieve something in that realm, not only perhaps by vain words until now, but also sometimes by deeds. And you will not meet anyone who knows me who ... will not agree.' Indeed, he would be able to deliver 'even greater things than expected'.[81]

Carcavy responded to Leibniz a couple of weeks later with marked reserve. After briefly elaborating on Pascal's calculating machine, Carcavy offered to send Leibniz a more detailed description of the machine, or better yet to compare Pascal's directly with Leibniz's own, were the latter to send an exemplar. Carcavy seemed at first to share Leibniz's caution in his previous letter against the dangers posed by an 'excessive imagination', which he credited with the 'invention of many beautiful secrets'. Nonetheless, he strongly distanced himself from Leibniz's position, not, in fact, without a touch of sarcasm, emphasizing that schemes

---

79. Leibniz to Carcavy, A II, 1:289.

80. Leibniz to Carcavy, A II, 1:290: 'If the priests had passed a bill that all inventions by monks should be made public only by the trust of obedience, they would without doubt be the rulers of the world, for Berthold Schwarz's invention of gunpowder for instance could have remained with them alone'; 'There is no mortal who knows this better than the great Colbert.'

81. Leibniz to Carcavy, A II, 1:290.

should be put forward for consideration only once they had been proven to be valid:

> But as it is very difficult to produce riches, and even more so to profit from what [projectors] themselves do not fully grasp, displaying neither order nor method in any of their operations, and quite often persuading themselves that trivial secrets are of great consequence since they invested so much effort conjuring them up in the first place, you will agree with me, Sir, that these schemes should not be trusted until they have been confirmed.[82]

Carcavy's criticism was double-pronged. On the one hand, he rebuked Leibniz again for failing to expound his theories fully and sending 'confused speculations', thus rendering some of his texts difficult to grasp. On the other hand, Carcavy was concerned with the feasibility and possible practical implementation of the many schemes that Leibniz had sent him and his colleagues. While he remained interested in Leibniz's project for a new calculating machine, of which Leibniz had still disclosed very little, Carcavy was nevertheless unwilling, in the absence of tangible results, to accede to the young man's requests for immediate membership of the Académie and a pension in anticipation of his *possible* usefulness to Colbert and the Académie. Since Colbert was only 'satisfied with what was real and solid', Carcavy would, he cautioned, present something to the minister only once Leibniz 'had begun to send something effective' for assessment.

That said, Carcavy was not unwilling to reward Leibniz if the latter sent him something tangible and convincing. In this case the German would, Carcavy assured him, receive credit for his invention, be able to specify the conditions of its use, and be appropriately remunerated. Clearly growing frustrated with Leibniz's tendency to promise much while delivering little, Carcavy concluded his letter by noting the dangers of the *amour-propre* of 'authors' who overestimated the novelty of their inventions. Carcavy was much older than Leibniz and no novice at this game, having had to contend with his fair share of schemes and projectors over the years. The proliferation of schemes had made him wary and more determined than ever to act as a gatekeeper on Colbert's behalf, protecting the latter from potential scams while seeking to attract genuine talent.[83] Even before setting foot in Paris, it seems, Leibniz had begun to squander his credit by leaving Carcavy and members of the Académie des Sciences increasingly

82. Carcavy to Leibniz, 5 December 1671, A II, 1:307.
83. See Jones, *Reckoning with Matter*, 114–16.

unimpressed, wasting their time and goodwill with 'unclear ideas and half-baked projects'.[84]

To be sure, this rebuke did not stop Leibniz from claiming regular contact with Carcavy, and even presenting himself as one of Carcavy's inner circle.[85] In a letter to Albert von Holten, for instance, Leibniz discouraged the Tübingen Orientalist from presenting his 'cylindrical Hebrew grammar' (*Grammaticam . . . Hebraicam in cylindris*)[86]—inspired by Leibniz's own *De arte combinatoria*—to the Royal Society, suggesting instead that more a fruitful avenue might be pursued with the French Académie, where, by intervening on his behalf with Carcavy and Ferrand, Leibniz could (he claimed) provide von Holten 'with the surest route to those Maecenases'.[87]

Over the next half a year, it seems that Leibniz requested Carcavy's assistance on a number of matters and proposed to him, either orally or in writing, a number of projects—of which only a few details are known—but there is no evidence that Carcavy accepted these proposals.[88] In this regard the contrast with the difference in treatment received by Leibniz's friend and fellow projector Johann Daniel Crafft is striking. Having obtained a privilege to produce steel from iron in France, Crafft was left in limbo by various delays and interruptions in the plans to transfer his workshop

---

84. Jones, 114–15.

85. Leibniz to Jakob Thomasius, 21/31 January 1672, A II, 1:320: 'I am used to having exchanges with Carcavy, the king's librarian, who also attends to literary matters [*sc.* the books in the library] on Colbert's orders and I have found by chance, among other things, a mention of the Reinesianarum inscriptionum.'

86. In what was probably a note to himself (A II, 1:no. 96), Leibniz contemplated the usefulness and combinatory potential of Von Holten's cylindrical grammar, comparing it to similar ventures including Kircher's, Wilkins's, Comenius's, and his own *De arte combinatoria* and leading Leibniz to ponder on the possibilities of a logical machine. Similarly, Leibniz muses, it would be possible to construct a cylinder that would provide all possible relations between certain given terms decomposed into simple elements. This could form the basis for an artificial language which could be applied to questions of law. For more on this, see Couturat, *La Logique de Leibniz* (1901), 115.

87. Leibniz to Albert von Holten, 17/27 February 1672, A II, 1:324: 'Carcavius the king's librarian has been given this brief concerning literary matters by Colbert. He has some assistant librarians, learned men, of whom . . . Cotelerius attends to the Greek holdings and Ferrandus the Oriental ones. I have exchanged letters with Ferrandus often and with Carcavius sometimes.'

88. See, for example, Leibniz to Carcavy, early November 1671, A II, 1:290. 'A short while ago I recommended book and machine vendors to you. . . . The book vendor Zunner will diligently compile the prices of books first, and will then proceed to order them if they are requested. . . . This matter, I hope, along with the more important one I executed at Colbert's request as you will recall, will be settled upon my arrival.'

to France, and he finally enlisted Leibniz to act on his behalf.[89] In his response to Leibniz, Carcavy reassured him that, provided Crafft carry out what he had promised to Colbert and agreed with Du Fresne, Mainz's envoy in Paris, both would hold up their own end of the bargain.[90] Crafft was—at this time at least—clearly perceived as a more reliable and effective operator whose ventures offered great potential. Leibniz, from whom Carcavy and Gallois had in July 1671 greeted the prospect of learning,[91] had lost much of his credibility by December and was unable to make headway with the Académie. By mid-1672 his principal value to Carcavy seems to have been reduced to reporting on new discoveries, particularly concerning the progress of Otto von Guericke's experiments on vacuum.[92] In less than a year, therefore, it seems that Leibniz's credit with Carcavy and at the Académie largely evaporated.

Enthralled by the mechanical arts and inventions, from Des Billettes's 'thousand instruments and pretty inventions' to the machine that walked on water,[93] not to mention his own plan for a calculating machine, Leibniz was particularly receptive to France's new policy to attract foreign talent. For Leibniz was a project-maker who could design ingenious solutions to real practical problems, and it was in this capacity that he put himself forward. He was an audacious thinker and a master of discerning the needs and concerns of his patrons or potential patrons and mirroring them back to them. But what Leibniz truly sold when promoting his projects was himself. He presented himself as a superior form of projector, 'upon whom the success of the crown depended' at a time when the royal administration was more than ever intent on having natural philosophical and artisanal knowledge yield practical and technological applications.[94]

When Leibniz arrived in Paris in 1672, he was interested above all 'in understanding and inventing machines',[95] and he began working on producing a functional machine. Alongside his work on mathematics, he

89. Johann Christoph Crafft to Leibniz, 5 October 1675, A I, 1:426–27; Leibniz to Carcavy, 22 June 1671, A II, 1:215.
90. Carcavy to Leibniz, 10 July 1671, A II, 1:229.
91. Johann Leyser to Leibniz, 13 July 1671, A I, 1:159.
92. Compare this to the occasion in July 1671, when the theologian Johann Leyser reported back to Leibniz from Paris that Carcavy, along with Jean Gallois, editor of the *Journal des Sçavans* and secretary of the Académie, wanted nothing more than for the Académie to benefit from Leibniz's learning. Antognazza, *Leibniz*, 123. On the vacuum experiments, see Leibniz to Carcavy, July 1672, A II, 1:340–41.
93. Davillé, 'Le séjour de Leibniz à Paris' (1922), 40.
94. Jones, *Reckoning with Matter*, 87, 120, also 10.
95. Leibniz to Bernoulli, April 1703, quoted in Jones, *Reckoning with Matter*, 66.

intended the machine as his entrance card to Parisian learned circles, and he hoped to sell it to the French crown and have it mass-produced once the king had made it 'fashionable'.[96] Upon his return in 1673 from England, where he had demonstrated an incomplete prototype of his machine, he applied for funding from the Académie des Sciences to bring the twelve-digit machine into practice.[97] After showing a prototype before Colbert and the Académie, Leibniz received a preliminary commission from Colbert for machines for the Royal Observatory and the financial offices in December 1674.[98] However, Leibniz's claim that the production of his machine was well under way, or even complete,[99] could not have been further from the truth.

Transforming Leibniz's blueprint into a functioning model, especially with regard to the mechanisms involved, proved far more difficult than anticipated. The materialization of his ideas could not simply be reduced to the mere imposition of an idea onto 'inanimate matter', but required a process of remediation, from paper to brass, of Leibniz's original idea, especially to accommodate the mechanical problem of propagating the 'carry-over'.[100]

The difficult—and frequently unsuccessful—implementation of many of his schemes during much of his life could in fact be ascribed to Leibniz's 'hylomorphic' or 'mentalistic' conception of inventions.[101] This consisted in the belief in the radical division between an idea and its practical realization. In this scheme of things, the inventor could claim payment and privileges on the basis of the former, whilst the later was left to the artisan as a matter of simple execution. In one eye-opening letter to Louis

---

96. Jones, 115, 119; 'De Machina ad usum transferenda,' quoted in Jones, 114.

97. Mackensen, 'Die Vorgeschichte' (1968), 174.

98. See Denis Papin to Leibniz, 23 July 1705, in Gerland, *Leibnizens und Huygens' Briefwechsel* (1888), 347. Papin regretted that he had not been present when the model was shown but remembered well that Huygens had been 'very content' with it. The presentation of the machine is referred to in an entry of January 9, 1675, in the Procès-verbaux of the Académie. Leibniz to Johann Friedrich, 21 January 1675, A I, 1:492. 'I nevertheless have been recommended to Mr. Colbert and have been urged to manufacture my arithmetical machine at his command.' Also Leibniz to Johann Friedrich, second half of October 1671, A II, 1:267.

99. Jones, *Reckoning with Matter*, 119.

100. As Dumas Primbault shows, the issue was not one of communication between inventor and craftsman but rather 'an epistemological problem relating to what it is possible to conceive in one particular medium . . . [and] then translating this from one medium to the next'. This task furthermore would have been complicated by the particularly hazy and speculative nature of Leibniz's invention. For more on this, see Simon Dumas Primbault, 'Leibniz en Créateur' (2022).

101. Dumas Primbault; as Jones points out, this anticipates the concept of intellectual property. See *Reckoning with Matter* (2016).

Ferrand, Leibniz depicted the enterprise as a veritable battle of the will, comparing his lot to that of Tantalus, the Greek mythological figure who was punished to endure eternal frustration by the gods after being admitted to their table and betraying them by revealing their secrets and stealing their nectar and ambrosia to give to mortals. Like him, Leibniz had to pay the price of what had seemed like a 'century' of silence, isolation, and 'hermitude'—aside from the presence of two labourers—in the midst of all the great men with which Paris was 'paved'. Leibniz was nonetheless prepared to sacrifice himself for this demiurgic feat on behalf of mankind. The business of his machine required 'the whole of his person', especially since he was alone, as he emphasized, and had to attend to an 'infinity of trifles' that detracted him from his labour, even only one of which could send everything 'tumbling down' forcing him, like Sisyphus, to start all over again.[102]

In early 1675 Leibniz proposed to the Académie des Sciences that he be compensated for the delivery of successfully operating calculating machines—remuneration not only for the actual expenses and demands on his time, but also for the financial and reputational risk to which he was exposing himself by focusing exclusively on producing the machines.[103] With no timely delivery of the machines, however, Leibniz's once overconfident language shifted to pushing for the recognition from the crown of his intellectual ownership of his invention.[104]

Leibniz ultimately failed to produce a sufficiently functional machine, much to his would-be patrons' dismay, and in early October he missed yet another appointment to demonstrate the machine in Saint-Germain-en-Laye before the duc de Chevreuse.[105] In fact, he was never able to produce a model for his machine nor ever received a privilege for it.

Leibniz's name was proposed by Jean Gallois, the duke of Chevreuse, and Christiaan Huygens as a replacement at the Académie for the recently deceased mathematician Roberval.[106] But Gallois withdrew his support, having taken offence at Leibniz's reaction to one of his talks.[107]

102. Leibniz to Louis Ferrand, May 1672, A I, 1:452.
103. Mackensen, 'Die Vorgeschichte' (1968), 170; Jones, *Reckoning with Matter*, 74, 97, 118.
104. See Jones, 119.
105. See Salomon-Bayet, 'Les académies scientifiques', 156.
106. See the summary in the introduction to A III, 1:lxix–lxxii; and Leibniz to Gallois, 2 November 1675, A III, 1:306–7. Unlike the Royal Society, the Académie had a *numerus clausus*, so that a new member could be elected only after an existing one had died.
107. Müller, *Leben und Werk* (1969), 167; Antognazza, *Leibniz*, 174; Bos, *Studies on Christiaan Huygens* (1980), 373; and introduction to A III, 1:lxxi.

Huygens, who had been taken ill, sought unsuccessfully to change Gallois's mind.[108] Furthermore, Leibniz's failure to produce his machine in a timely manner, as well as the number of foreigners already present at the Académie, as the duc de Chevreuse intimated to him,[109] likely contributed to undermining his candidacy for a position. To Leibniz's profound disappointment, the seat left vacant by Roberval went to a French candidate instead of to him.

108. Salomon-Bayet, 'Les académies scientifiques' (1978), 167.
109. Barber, *Leibniz in France* (1955), 174.

CHAPTER EIGHT

Finding a New Patron

AFTER LOSING BOTH Boineburg and Schönborn in quick succession in 1673, Leibniz found himself in a precarious financial and professional situation. He had initially obtained from Schönborn a leave of absence to remain in France to 'perfect himself' and 'serve the cause of piety'.[1] But wanting to prolong his stay, he was ready, if necessary, to accept a new scientific or political assignment in order to remain in Paris for as long as possible. He tried to obtain a diplomatic post—to replace Du Fresne, who had just died, as Mainz's resident in Paris—but this effort was scuppered by France's declaration of war on the Holy Roman Empire.[2] More generally, Leibniz hoped to stay in Paris on some sort of scientifico-political mission in the service of an enlightened prince to whom he would report back on the latest political and scientific developments while maintaining a diplomatic presence at the court of Louis XIV. More specifically, he hoped to be hired by the new prince-archbishop of Mainz, who, in addition to agreeing to pay him the arrears on his salary for the past two years, would allow him to remain in Paris, immersed in the French learned world, while serving the interests of the Mainz court.[3]

Such an 'amphibian' existence, enabling him to maintain a degree of independence during which he could attend to his own pursuits before 'finding matter to fix [himself] advantageously', would suit him rather well, as he confided to Christian Habbeus.[4] (He would later make a similar proposal to J. C. Kahm, Duke Johann Friedrich's private secretary, but to no avail.) The new elector, Lothar Friedrich von Metternich, however,

1. Leibniz to Johann Friedrich, Autum 1679, A I, 2:226.
2. Davillé, 'Le séjour de Leibniz à Paris' (1920), 147–48.
3. Antognazza, *Leibniz* (2009), 152.
4. Leibniz to Habbeus, 14 February 1676, A I, 1:445.

had plans of his own, none of which involved giving Leibniz a position as a cultural and political resident in Paris, and consequently he granted Leibniz permission to remain in Paris only 'for some more time . . . without danger to his post'.[5]

Leibniz was confirmed in his duties as a tutor for Boineburg's son, the young Philipp Wilhelm, by Boineburg's widow in 1673, a task that neither tutor nor pupil seems to have found particularly rewarding.[6] Towards the end of December, Leibniz requested from Boineburg's widow the payment of his outstanding salary, for tuition and other services, including the laborious recovery of Boineburg's French rent and pension.[7] When he was dismissed from his duties in September 1674, he was not remunerated for the many additional tasks assigned to him by the late baron.

Soon, the necessity of finding new employment could no longer be avoided, and Leibniz, who was possibly in debt and had already borrowed money from his half-brother, could no longer afford to be as selective in his choice of employment as he had perhaps been in the past.[8] He had previously received two offers of employment, the first soon after his arrival in Paris, as the secretary to the Danish prime minister, Count Ulrik Frederik of Güldenlow, for which he would have received an annual stipend of 400 thalers together with free accommodation and dining rights. Leibniz, however, had then not been ready to give up his freedom to pursue philosophical projects in Paris.[9] Whether he had not found the offer sufficiently attractive or whether his demands, which included being granted the title of royal counsellor and the use of a copyist to help him with his more mechanical tasks, had been considered excessive, nothing had come of it.

In his search for an enlightened prince who would employ him as an adviser while allowing him to pursue his own intellectual pursuits away from court, Leibniz's initial contacts with potential employers did not run as smoothly as he might have wished. Although he was generally careful to model his proposals after the concerns and imperatives of potential new patrons and adhere to the norms of courtly behaviour,[10] he was not reluctant to express his own priorities in making the patronage work

5. Melchior Friedrich von Schönborn to Leibniz, 5 May 1673, A I, 1:349.
6. Davillé, 'Le séjour de Leibniz à Paris' (1921), 169.
7. Antognazza, *Leibniz*, 153.
8. Hirsch, *Der berühmte Herr Leibniz* (2000), 76.
9. Leibniz to Habbeus, 5 May 1673, A, I, 1:415–16.
10. On which see Kettering, *Patrons, Brokers and Clients* (1986), esp. chap. 1; van Houdt et al., *Self-Presentation and Social Identification* (2002); and Fumaroli, *La République des Lettres* (2015).

for him: the patronages Leibniz envisaged were ones of mutual benefit, and he clearly sought patrons who would enable him to implement his own projects. He certainly understood that the cultivation of patrons was governed by certain rhetorical conventions, but his impatience to succeed, and perhaps his very determination to realize his schemes of human advancement, led him occasionally to neglect those conventions or to deploy them in such an exaggerated manner that risked appearing insincere and gauche, thereby undermining his aims. Before finally settling at the court of Duke Johann Friedrich of Hanover, Leibniz sampled the job market, so to speak, in the expectation that his philosophical and scientific standing would automatically secure him a courtly position, an expectation of which he would soon be disabused.

Between March 1675 and March 1676, Leibniz wrote several letters to potential employers, and they are particularly illuminating of his state of mind at the time. One particular noteworthy overture was to Duke Christian von Mecklenburg-Schwerin, whom Leibniz had met in Paris and for whom he had previously drafted legal documents regarding the annulment of the duke's second marriage. In a series of rather remarkable letters, Leibniz attempted to transform this limited commission into the more permanent position of state counsellor, of which he had a rather flexible understanding. In his March 1675 letter to Mecklenburg, for instance, he requested that he not be 'forced to follow his Court, since if [he] had been willing to settle down so soon in a fixed place, [he] would long ago have been in employment as advantageous as [he] might wish'. Despite the dearth of employment offers, Leibniz presented himself to the duke as being in a strong bargaining position and as having freely chosen his predicament—rather than having had to address an increasingly urgent need for employment and salary—and as such, he laid down his conditions. If the reasons just mentioned were not sufficiently compelling, there was another one: his safety, which he linked to the duke's own interests. Since 'the assignments that [he] would be undertaking could come to light in spite of [his] best efforts to conceal them', an official appointment would 'enable [Leibniz] to push the duke's interests with zeal', notwithstanding that the latter's 'enemies were powerful here'.[11] By all accounts and much to his surprise, Leibniz's letter did not produce the expected result. It seems that he initially lacked the self-awareness to appreciate that his brazenness and disregard for social hierarchy, combined with a

---

11. Leibniz to Christian von Mecklenburg, March 1675, A I, 1:477.

lack of tangible results corresponding to his self-panegyrics, were bound to cause friction and irritation.

Upon realizing his misstep, Leibniz promptly apologized for having failed to explain himself properly and causing the duke to misunderstand his intentions. He attempted to rectify his blunder by adopting a more deferential tone and assuring that he expected the duke to appoint him only once he had produced tangible results. This backtracking, however, came too late: the damage had already been done and Leibniz was forced to recognize that he was not considered as special or indispensable as he thought himself to be. This momentary setback did not deter him from attempting to cling to his conditions, citing, for instance, the menacing prospect of unspecified and 'powerful forces' in order to secure an official position and a salary that would 'cover [him] from any pursuits or reproaches'. In exchange, Leibniz would serve the duke as a lawyer and in other capacities, even if he could not commit to being a full-fledged courtier because of his desire to pursue the life of the mind.[12]

Despite his efforts, this second letter failed to restore Leibniz to the duke's good graces, and his third and final letter signals the complete breakdown of the relationship. A clearly irritated Leibniz lashed out at the duke and came close to crossing the line. Leibniz vented his indignation by striking an uncharacteristically unrestrained, reproachful, and direct tone to denounce what he perceived as his ill-treatment: 'There are people of the first order who are kind to me, and who would not approve of such mistreatment.' Ever the lawyer, he proceeded to deploy a wide umbrella of arguments in the space of a paragraph, appealing to the principles of natural justice and the duke's need for reliable and loyal agents, insisting that the force of logical necessity would make his conclusion 'difficult to resist'.[13] His correspondence with Mecklenburg betrays a combination of naiveté and a belief in his exceptionality, which he seemed to think dispensed him from attending to the niceties of courtly etiquette. This is one of many examples of Leibniz's overestimation of his political and social skills. Undeniably gifted as he was, he was 'skilful but clumsy', as Claire Salomon-Bayet puts it.[14]

Leibniz must have recognized his error with Mecklenburg, for he changed tack when dealing with another potential employer, Johann Friedrich. His initial contact with the Hanoverian ruler can be traced back

---

12. Leibniz to von Mecklenburg, 1:478, 479: 'Because of my design, which is to devote myself to science rather than to business, and to cultivate the reputation I have acquired among people of knowledge.'

13. Leibniz to Christian von Mecklenburg, 1:479.

14. Salomon-Bayet, 'Les académies scientifiques' (1978).

to 1671, when, from Frankfurt, Habbeus had drawn the duke's attention to Leibniz by recommending him in the most glowing terms, prompting the duke to ask Habbeus to urge the young jurist to depart immediately for Hanover. Although an offer failed to materialize at that time, Leibniz, following Habbeus's recommendation to keep his options open,[15] began an intermittent correspondence with Johann Friedrich in which the savant reported on his scientific, philosophical, and theological work in Paris and London. From the outset, Leibniz again presented himself as exceptional, but this time in a more muted tone than with Mecklenburg. Against all odds, he explained to the duke, he had succeeded in making his mark through much labour and, after 'an infinite number of imperfect attempts', had finally been acknowledged by his peers to be capable of achieving something extraordinary. In his 'completely new study' in mathematics, he had 'discovered the strangest and most useful truths', as the greatest mathematicians of the time had been able to attest. His arithmetical machine, he claimed inaccurately, had been completed 'and surpasse[d] all expectations' revolutionizing calculation in such a manner that it would henceforth no longer necessitate 'any labour or application of the mind.'[16]

Though motivated fundamentally by financial necessity, Leibniz now presented himself as one in whom duty and natural inclination coincided. He was now careful to insert a leavening of flattery into his terms for the fortunate recipient of his services: he owed 'everything' to the duke, who had showered Leibniz with his 'goodness'.[17] Having 'never manifested any haste for things of this nature'—a claim that many recipients of his letters would have contradicted—he would simply accept the unfolding of the course of events. He could not believe his luck to have met such a 'generous' prince,[18] worthy of a 'particular cult' and to whom veneration and even a kind of tenderness was owed.[19] He was not acting out of necessity or duress but under the most auspicious circumstances. He characterised his success as the product of luck and favourable conjecture—indeed he was 'not quite sure how he had succeeded'. He had been particularly blessed to meet the duke, without whom he would undoubtedly fail to 'produce something great or useful'.[20]

15. Antognazza, *Leibniz*, 87.
16. Leibniz to Johann Friedrich, 26 March 1673, A I, 1:492.
17. Leibniz to Johann Friedrich, 1:492.
18. Leibniz to Johann Friedrich, 10/20 November 1675, A I, 1:501.
19. Leibniz to Johann Friedrich, 11 January 1676, A I, 1:504.
20. Leibniz to Johann Friedrich, 26 March 1673, A I, 1:492, 493.

Only a few days later, Leibniz repeated his determination and readiness to serve Johann Friedrich 'as soon as possible'. He even claimed to have no stronger 'passion than to be' in Hanover.[21] This was certainly disingenuous, given that he would do his utmost to remain in Paris as long as possible, eventually delaying his move to Hanover by nearly a year.

Even when Leibniz did adhere to social norms and respected the distinctions of hierarchy, he was not hesitant to try to leverage better conditions for himself before his employment had been confirmed. He was prepared to divulge an 'auspicious opportunity'—once his terms for appointment had been met—and deserved to be confirmed an official counsellor on account of his invitation to accompany the Abbé de Gravel to a peace conference.[22] In response, Johann Friedrich was generous but inflexible, establishing the pattern that would come to define their relationship—at least from the duke's perspective: when it was appropriate for him to grant Leibniz new marks of distinction, he would happily do so.[23] What emerges from Leibniz's letters, therefore, is a portrait of a remarkably ambitious individual with a strong sense of mission who was not always adept at balancing his intellectual goals and personal expectations with the social realities and power dynamics at play.

Nonetheless, such was Leibniz's determination to remain in Paris that he even wrote to Colbert directly in January 1676, on the very same day that he confirmed his employment in Hanover, in a last-ditch attempt to land an appointment at the Académie. It is fair to say that, throughout his correspondence with French dignitaries, Leibniz was always at least implicitly addressing himself to Colbert.[24] But this time, writing in some desperation, Leibniz was emboldened to recommend himself, which his relatively lowly social status would not otherwise have permitted, purely on the basis of the advancement of the sciences. The strategy of presenting himself as a disinterested scholar, distinguished by his 'loyalty' (as he informed Christan Habbeus), is one that Leibniz was to adopt repeatedly, particularly in exchanges with prospective patrons. Loyalty was especially important in such precarious times, and Leibniz would ensure his to a

---

21. Leibniz to Johann Friedrich, 22 March 1676, A I, 1:513.

22. Leibniz to Johann Friedrich, 10/20 November 1675, A I, 1:501; January 1676, A I, 1:507

23. Johann Friedrich to Leibniz, 11 February 1676, A I, 1:510.

24. Hence Leibniz's frequent references to Colbert, to the latter's power of judgement and personal interests, in letters to others entreating Gallois, for example, to 'keep him in . . . Colbert's good graces'. Leibniz to Gallois, first half of December 1677, A III, 2:297.

patron who was endowed with 'a solid genius, good taste and prompt to lend his support to useful and easily achievable endeavours':

> Of all the qualities which you mention to me, loyalty alone is what I dare to boast, and of which all those with whom I have interacted have been fully persuaded, having been graceful enough to confide in me their greatest secrets without regret ... And as I know that this quality, combined with some effort, and a modest knowledge, can compensate for the failings of others even in greater matters, I begin to regain a little courage, and to console myself for what I lack.[25]

Yet even as he accepted the appointment in Hanover, Leibniz sought to keep his options open.

Learning from France for Germany

Leibniz was to remember his time in Paris fondly, and he sought to keep that memory alive by cultivating an extensive French epistolary network thereafter. The city had provided him with opportunities and contacts that would not have been available to him in Germany.[26] He would continue viewing the French capital as one of Europe's preeminent centres of novelty in the scientific and *galant* worlds alike,[27] which he felt he had experienced at their heights: 'France never boasted more talent than in 1672', Leibniz wrote in 1715, the year before his death, lamenting that in the intervening decades 'erudition had suffered on account of a decline of the public finances, even if they were still superior to those of other nations'.[28] Leibniz's stay in Paris was of decisive importance in determining the rest of his intellectual trajectory. He had been transported from a small German court to the centre of European civilization and come into contact with some of his generation's greatest thinkers. And it was in Paris that he had become a first-rate mathematician, pioneering a new branch of mathematics and mechanics, and had begun to develop some of his key philosophical ideas.

However limited his political influence, Leibniz was able to observe and learn from France's successes, especially in economic matters, with the aim of enabling the German states to emerge from France's shadow

---

25. Leibniz to Habbeus, 5 May 1673, A, I, 1:415.
26. Fischer, *Geschichte der neuern Philosophie* (1855), 150; Davillé, *Leibniz historien* (1909), 147–78, 463.
27. Davillé, 'Le séjour de Leibniz à Paris' (1923), 57–58.
28. Leibniz to Westerloo, 8 July 1715, quoted in Davillé (1923), 58.

and establish their own set of manufactures and politico-scientific structures.[29] Germany would then be able not only to compete with France, but to 'undermine' its commerce 'for ever' and to wage industrial war against it.[30] In Paris, Leibniz revealed a talent for extracting natural philosophical as well as trade secrets, and his notes over the years are replete with numerous kernels of artisanal knowledge concerning materials, techniques, and mechanisms, knowledge that he would seek to direct towards Germany's benefit. Unsurprisingly, then, his admiration for France was coloured by a pronounced ambivalence towards the nation. As an outsider and critical observer, he was not oblivious to its many miseries, which he investigated in some detail,[31] and his admiration for its learned and scientific apparatus coexisted with a deep disdain of its bellicosity and destructiveness. In an 'ode to France' written between 1672 and 1676, he implored it to 'take hold of [its] fortune' and to 'look around [it] at the smoking walls of the world' it had left behind in its ravaging hubris, and instead 'hold fortune in reverence', for 'there are Powers in heaven, and prosperous men are ill-advised to contend with Nemesis'. France did not nor would not always enjoy its present success: in this bold plea, Leibniz urged France to 'offer peace' and 'pardon a supplicant world'. If he had previously presented fortune as subservient to France's destiny and Louis's glory in political and military terms, he now enjoined restraint on France and implored the nation to 'learn how to live among mankind'.[32]

29. Davillé (1923), 60.
30. Couturat, *La Logique de Leibniz* (1901), 520 no. 7, quoted in Davillé (1923), 60.
31. See Davillé, 'Le séjour de Leibniz à Paris' (1922), 25.
32. A I, 1:456–57.

PART II

CHAPTER NINE

Leibniz in His Correspondence with Duke Johann Friedrich (1676–78)

THE SINCERITY OF A PROJECTOR

NO LONGER ABLE TO PROLONG his sojourn in Paris, where he had settled in early spring 1672 and enjoyed several years of relatively unconstrained intellectual and scientific freedom, Leibniz finally took up service at the court of Duke Johann Friedrich in Hanover in late 1676. In Paris, at the time the world's hub of science and scholarship, he had pursued a wide array of scholarly interests and got to know the greatest minds, and so it was reluctantly and with a heavy heart that he had accepted the appointment in Hanover, and only once his hopes of finding a position either in Mainz, Paris, or Vienna had failed to materialize.[1]

Johann Friedrich had recently turned fifty when Leibniz came into his employ. He was the son of Georg (1583–1641), Duke of Brunswick-Lüneburg-Calenberg, and Anna Eleonore of Hessen-Darmstadt (1601–59). A well-educated prince, the duke entertained the notion of setting up his court as a local power and intellectual centre, notably through the recruitment of officials and brilliant minds from all different corners of Europe.[2]

---

1. Rescher, *On Leibniz* (2013), 119; Hofmann, *Leibniz in Paris* (1974), 302. Leibniz's attempts to get appointed as historiographer at the court of Leopold I were unsuccessful.

2. Antognazza, *Leibniz* (2009), 199; Rescher, *On Leibniz*, 121. See also Breger and Niewöhner, *Leibniz und Niedersachsen* (1999); and, more generally on the Hanoverian court, Schnath, *Geschichte Hannovers im Zeitalter* (1976).

The court Leibniz joined upon his departure from Paris was comparatively small. At court Leibniz interacted with the two chamber councilors, Otto Grote, who led the Privy Council (*Geheimer Rat*), the highest administrative body of the duchy; Hieronymus von Witzendorff, who oversaw the cabinet office charged with the general administration of the duchy (*Kammer*);[3] the vice-chancellor Ludolf Hugo, who directed the Chancery or Ministry of Justice (*Kanzlei*); and his direct superior, the valet Johann Carl Kahm. To these would later be added Leibniz's correspondence with the Mining Authority in Clausthal and its president, Friedrich Casimir zu Eltz, as well as with Friedrich Wilhelm Leidenfrost, the government secretary in Osterode during the negotiations with Eltz and the mining office.

Leibniz had been acquainted with Johann Friedrich ever since 1669, when Leibniz had been recommended to him by Christian Habbeus, and got to know him personally in October 1671 during his stay in Frankfurt.[4] Already shortly after they had met, Leibniz had been very candid with the duke about his scientific and theological ambitions, offering an overview of his work and laying out his plans before him by discussing, for instance, his plan for a rational jurisprudence or the strides made by his *Theoria motus abstracti* and *Hypothesis physica nova*.[5] Over the years Leibniz had kept the duke abreast of his work and activities, and upon the unexpected demise of his two patrons, Boineburg and Schönborn,[6] the ruler renewed his offer of employment, which Leibniz finally accepted in 1676. His appointment marked the beginning of a forty-year association between Leibniz and the House of Hanover in which Leibniz would spend much of the rest of his life serving as court librarian and adviser while acquiring a reputation as one of Europe's most eminent philosophers and mathematicians. He had several face-to-face meetings with the duke,[7] which were supplemented by an extensive and wide-ranging epistolary exchange that spanned a little less than three years, from Leibniz's departure from Paris in late 1676 and return to Germany via London and the Low Countries to the duke's premature demise in December 1679. Over this period, he wrote seventy-nine letters of varying lengths to the duke. Whether Leibniz's letters were delivered directly to the duke or passed

---

3. Antognazza, *Leibniz*, 199.

4. Christian Habbeus to Leibniz, 30 November 1669, A I, 1:210; Davillé, 'Le séjour de Leibniz à Paris' (1921), 76.

5. Leibniz to Johann Friedrich, second half of October 1671, A II, 1:260–69; 21 May 1671, A II, 1:170–76.

6. Leibniz to Johann Friedrich, 15/25 April 1673, A I, 1:491.

7. Leibniz to Jean Gallois, first half of December 1677, A III, 2:293.

through his chamberlain, Johann Carl Kahm, and whether all of them reached him at all remains unclear. That some did not would help account for Leibniz's frequent repetitions in the letters.[8]

Leibniz as Court Librarian

Leibniz had been recruited first and foremost as a librarian, the previous incumbent, Tobias Fleischer, having taken up the post of cabinet secretary at the court of Denmark. As such, one of his main tasks was the constitution of a ducal library.[9] This he set out to accomplish diligently by bringing together books, old and new, in order to build an 'exquisite and accomplished' library that would act as an 'encyclopaedia' and 'inventory'.[10] While he was disappointed with his appointment, it was perhaps this very sense of discontent that prompted him to excel at his duties and take possession of this role and make it his.[11] As he confided to Duke Johann Friedrich in mid-December 1676 soon after his arrival in Hanover, Leibniz had no intention of being an ordinary court librarian, but wanted to expand a little 'the usual care of the Library, and therefore the traditional duties of a Librarian'. He was determined not only to secure a more rounded position at court as a court counsellor, but also to leave his mark on a particular field he conceived as being part and parcel of a broader project for the advancement of the human sciences and the reform of learning.[12]

While Leibniz never wrote a systematic treatise on library science along the lines of Gabriel Naudé's widely circulated *Advis pour dresser une bibliothèque* (1627) or Louis Jacob's *Traité des plus belles bibliothèques* (1644), he did commit to paper his thoughts on library organization and

---

8. Many of the letters were reports of Leibniz's activities rather than a two-way exchange, which would account for some repetitions. He also kept copies of most of his recipients' letters, from which it is generally possible to reconstruct much of the exchange.

9. On Leibniz's library work, see Palumbo, 'Leibniz as Librarian' (2014); and Paasch, *Die Bibliothek* (2003).

10. Leibniz to Johann Friedrich, January 1677, A, I, 2:9; May 1679, A I, 2:175: 'My opinion has always been and still is that a library must be an encyclopedia, that is to say that one can be instructed there if necessary in all matters of consequence and practice.' Leibniz himself himself collected books assiduously, bringing back many from France, England, and the Low Countries: see Palumbo, 'Leibniz as Librarian', 610–11, 613–14.

11. Leibniz felt that the library post as such was simply not commensurate with his talents or expectations, writing that the advantages it offered were 'fertile only for the mind, and sterile in everything else' and 'that there [was] no occasion to make friends, nor to acquire the slightest authority or excel at anything'. Leibniz to Johann Friedrich, January 1677, A I, 2:12.

12. Leibniz to Johann Friedrich, A I, 2:12.

improvement in letters and memoranda to various patrons, princes, officers, and ministers as well as fellow librarians and booksellers—to such an extent that he would later come to be seen as an authority on the subject. In fact, shortly after he arrived, Leibniz addressed to the duke a dense *promemoria* for the improvement of the ducal library in which he emphasized the crucial role of the Republic of Letters and its networks of correspondences in helping relay the correct books and learned journals and offered to place his own extensive correspondence network in the service not only of acquiring new books but also of assembling the ephemeral documents of contemporary controversies in learned circles.[13]

Upon his arrival, the library possessed only 3,310 printed books and 158 manuscripts.[14] As Leibniz set out to explain to the duke—and later to his successor Ernst August—the library was the concrete embodiment of the current state of knowledge in the very physicality of its books. In this capacity, it should not be a mere accumulation of volumes, but a carefully selected repository offering the best and most useful books across all disciplines. This general inventory would bring to light the 'seeds of the evident and wonderful truths', 'new realities', 'ingenious observations', and 'remarkable considerations', including in fields conventionally considered marginal.[15] In this respect, in fact, devising methods of organization of libraries as privileged 'storehouses' and 'memories' of humankind would prove a lifelong preoccupation.

During his stay in Paris, Leibniz had imagined a *bibliothèque à sa phantasie*,[16] an ideal library that would contain two categories of books: first, those reporting on scientific demonstrations, and inventions, and second, books on politics, history, and geography.[17] A library was unlike a stone or a metal object but a living being, one which, in the image of the *res publica literaria* and with the help of enlightened patronage, would record past achievements as well as the state of all knowledge at a given moment with its share of beautiful discoveries, experiments, and curiosities. The most important books were not the most 'voluminous' ones but those that, ultimately, would help realize the common good and the true felicity of humankind. For this reason, Leibniz challenged the traditional image of the library—and the role of traditional court librarian hitherto

---

13. Leibniz to Johann Friedrich, A I, 2:15–18.
14. Palumbo, 'Leibniz as Librarian', 610.
15. Leibniz, 'Einrichtung einer Bibliothek' (1680), A IV, 3:349–53, quoted in Palumbo, 613.
16. Leibniz to Christian Habbeus, 5 May 1673, A I, 1:417–18.
17. Leibniz to Johann Friedrich, January 1677, A I, 2:15–16.

entertained. The latter was usually expected to suggest to patrons the acquisition of sumptuously illuminated manuscripts or lavishly engraved folio volumes with the intention of reflecting the prestige of a dynasty or of projecting sovereign political power.[18]

But in Leibniz's view, libraries should place the values of usefulness and learning above those associated with social status and should not be mere courtly ornaments filled with books 'of little use' or 'fashionable ideas'.[19] In keeping with his more general stance towards knowledge, Leibniz was concerned with sifting among the 'awful mass of volumes' that could be detrimental to true *savants* and selecting those precious gems that would help contribute to the advancement of knowledge and science. In this manner, a 'beautiful library' should circumvent social hierarchies in favour of the aims and ideals of the Republic of Letters;[20] and as time passed, Leibniz would elaborate increasingly on the strong link between the library and the Republic of Letters. Writing later in his life to the court librarian Hertel, he defended his understanding of his own role and function as court librarian: 'I prefer thirty small curious books to a voluminous work that contains nothing more than repetitions.'[21] Rejecting for himself the role of traditional court librarian, Leibniz thus proceeded to extend his subversive gaze to the library itself by envisaging it less as a mere display of wealth and status than as the very embodiment of the Republic of Letters. A diligent librarian, Leibniz attended to the library's smallest details and oversaw its expansion, notwithstanding various difficulties created by the need to accommodate the newly acquired Fogel collection.[22] Over the course of his correspondence, he regularly updated the duke about the library's operations and acquisitions, as well as his various negotiations to purchase libraries and books from various booksellers from Hamburg to Amsterdam, including Johann David Zunner in Frankfurt am Main, Thomas Heinrich Hauenstein in Hanover, Daniel Elsevier in Amsterdam, and Caspar Gruber in Brunswick.[23] This task in particular provided Leibniz with ample opportunity to show off his skills as a smooth operator and gifted negotiator, and one of notable successes was to acquire the Dutch publicist Wicquefort's library in 1677, on which he probably drew to

---

18. Palumbo, 'Leibniz as Librarian', 614.

19. Palumbo, 616: 'The concept of the *bibliothèque de parade*, based on the ostentation of luxury and puissance, was always and radically extraneous to Leibniz's vision.'

20. See also Leibniz to Duke Johann Friedrich, January 1677, A I, 2:15–16.

21. Leibniz to Lorenz Hertel, 3 April 1705, A I, 24:508.

22. Antognazza, *Leibniz*, 80, 209; Leibniz to Hieronymus von Witzendorff, October 1678, A I, 2:94.

23. On the purchase of libraries, Leibniz to Johann Friedrich, 6/16 July 1678, A I, 2:56.

prepare *Caesarinus Furstenerius* and other political works.[24] In the case of the valuable collection of the recently deceased medical physician Martin Fogel, Leibniz managed to acquire 3,600 volumes for a much lower price by concealing his true interest in the library.[25]

Eventually and through his efforts, Leibniz could reassure the duke in early 1678 that the ducal library had become 'one of the choicest in Germany', from which the ruler would not fail to derive much glory.[26] Crucially, too, helping bolster the ducal library's prestige would also further cement Leibniz's own status as a scholar of distinction and allow him to further position himself, within the Republic of Letters and at court, as a purveyor of 'all sorts of knowledge, correspondences and curiosities'.[27]

Widening His Sphere of Influence

With the exception of a few weeks spent in Hamburg in the hope of securing the savant Joachim Jungius's library, mainly through the intermission of the latter's disciples, Heinrich Siver and Johann Vagetius, Leibniz spent all of the three years from 1676 to 1679 in Hanover and the Harz mountains.[28] There, he was determined to go beyond the duties of his librarianship—'I undertake somewhat more than the care of the Library requires, and than a Librarian is generally accustomed to doing'—and extend his influence by seeking, if unsuccessfully, to persuade the duke to create the new office of Director of the General Archives especially for him and to entrust him with administering cloisters and confiscated church property.[29]

As an adviser to the duke, Leibniz suggested reform in a variety of domains, such as the judiciary by reactivating his project for a new imperial Code, a project he had already initiated in Mainz and would ultimately intend for the Kaiser.[30] He also continued his work as a political publi-

---

24. Davillé, *Leibniz historien* (1909), 31.
25. On Fogel, see Marten and Piepenbring-Thomas, *Fogels Ordnungen* (2015).
26. Leibniz to Johann Friedrich, May 1678, A I, 2:55; 7 August 1678, A I, 2:67: 'However, the choice that His Royal Highness has made of this Library will be publicized universally; and everyone will think that he has chosen well.'
27. Leibniz to Johann Friedrich, January 1677, A I, 2:15.
28. On Leibniz and Jungius's papers, see Kangro, 'Joachim Jungius und Gottfried Wilhelm Leibniz' (1969); and Wahl, 'Leibniz' Beziehung nach Hamburg' (2017).
29. Leibniz to Johann Friedrich, January 1677, A I, 2:12; Autumn 1678, A I, 2:nos. 71, 72, 73; December 1678, A I, 2:110; Autumn 1678, A I, 2:no. 73.
30. Leibniz to Johann Friedrich, 3 December 1676, A I, 2:no. 2; December 1677, A I,

cist.³¹ He advised the duke and defended proposals on a wide range of issues, from internal administrative reform, which aimed at improving the efficiency of public administration and introducing social security allowances, to more consequential diplomatic matters and ongoing European conflicts.³² Leibniz drafted several proposals and memoranda towards government policy, typically drawing on historico-legal scholarship to promote the political interests of the House of Brunswick or the prospects of peace in May 1677.³³

In the summer of 1677, Leibniz began working on a proposal for the House of Brunswick's co-equalization of status with the electors of the Holy Roman Empire—its status having previously been limited to that of observer—by developing a new concept of sovereignty. Drafted under the pseudonym of Caesarinus Furstenerius, his tract *De jure suprematus ac legationis principium Germaniae* drew on the history of the Empire to argue the claims of German princes to participate directly in the electoral system and in international negotiations, while claiming that the princes' independence would not impair the cohesiveness of the Empire itself.³⁴ Abridged and translated into French in the form of a dialogue entitled *Entretien de Philarète et d'Eugène*,³⁵ the tract was distributed to all the delegations at the Nijmegen peace conference, which had been meeting since 1675 to try to resolve the war incurred by the belligerent policies of Louis XIV.

Despite his hope of becoming a high-flying political and diplomatic operator, Leibniz was never appointed to an official political post or entrusted with diplomatic duties. Nonetheless, he set out to transform Hanover into an important centre for learning and, in this effort, aspired to act as a *touche-à-tout*, an adviser 'without portfolio', an expert-in-residence on matters of learning, science, and technology in a way

---

2:no. 33; Leibniz for Ludolf Hugo, March 1678, A I, 2:no. 37. On Leibniz's activities as a jurist, see Berkowitz, *The Gift of Science* (2005); Armgardt, 'Leibniz as Legal Scholar' (2014); and Artosi and Sartor, 'Leibniz as Jurist' (2014).

31. On this role, see Beiderbeck, 'Leibniz's Political Vision' (2014), 370–71.

32. See Leibniz to Johann Friedrich, A I, 2:no. 70 (September 1678), AI, 2 nos. 71–74 (Autumn 1678), and A I, 2: no. 94 (December 1678); Schneiders, 'Sozietätspläne und Sozialutopie' (1975); Antognazza, *Leibniz*, 210.

33. Leibniz to Johann Friedrich, December 1678, A I, 2:110; 3 December 1676, A I, 2:2l; May 1677, A I, 2:no. 13.

34. Antognazza, *Leibniz*, 205. For more on Leibniz's work in this context, see the essays in Altwicker and Cheneval, *Rechts- and Staatsphilosophie* (2020); and Beiderbeck, 'Leibniz's Political Vision'.

35. Antognazza, *Leibniz*, 205.

unprecedented before and unparalleled after.³⁶ This would enable him to exert influence—if not by virtue of birth, then through his intellect—even if in an unofficial capacity. Thus, for instance, he documented various legal matters without having an official role in the legal system. In both Paris and London, Leibniz explained, he had had 'the opportunity to advise excellent people' in a broad array of fields, including 'in matters of religion and state' as well as 'in science, the arts and curiosities'.³⁷

A 'Walking Encyclopedia' at the Court of Hanover

No sooner had he arrived in Hanover than Leibniz undertook to set some of his designs in motion. A 'walking encyclopaedia' (*inventaire ambulant*), as he described himself in the long memorandum he intended for the duke shortly after his arrival,³⁸ Leibniz set out to establish himself at the court of Hanover as a recognized purveyor of information, schemes, and proposals, and more generally as a fount of ideas and information, in the hope of slowly but surely carving out a niche for himself.³⁹ He committed himself to sharing all 'useful or pleasant reflections or inventions', hoping, too, that the duke would entrust him with undertaking projects that would be useful to the public and advantageous to the state and more generally would advance the arts and sciences for the benefit of all humanity—projects that, he assured the duke, could be carried out 'cheaply and expediently', in a manner surpassing even the efforts of 'foreign Academies'.⁴⁰

In Hanover, Leibniz was housed in the librarian's quarters immediately adjacent to the duke's private apartments and dining rooms, and this may have contributed to the development of a personal relationship between the two, whereby the duke could informally consult the savant on a wide range of issues.⁴¹ That Leibniz was in the employ of a naturally 'curious' and 'enlightened prince', who expected to be kept up to date with the latest scientific developments, granted him a privileged access to the ruler— 'which no one else had', as he would state to Gallois, not without exaggeration—as well as the possibility to 'occasionally cultivate his first loves' and 'keep [himself] in the habits of his fellow savants'.⁴² The duke treated him

36. Rescher, *On Leibniz*, 130.
37. Leibniz to Johann Friedrich, February 1679, A I, 2: 121.
38. Leibniz to Johann Friedrich, January 1677, A I, 2: 12.
39. Rescher, *On Leibniz*, 130.
40. Leibniz to Johann Friedrich, February 1679, A I, 2:121.
41. Antognazza, *Leibniz*, 200.
42. Leibniz to Jean Gallois, September 1678, A III, 2:503; first half of December 1677, A III, 2:294.

with 'a bit more liberty' and gave him free rein to absent himself from his official duties 'however often this would seem to be necessary for the sake of other labours', that is, in order to 'pursue useful studies' and 'elaborate in spare hours, on what [he had] once invented'.[43]

From Hanover, Leibniz set out to cultivate an extended network of correspondents, some of whom were 'undoubtedly some of the most skilled in Europe'; and he kept himself abreast of new developments, curiosities, and controversies, including the melting of iron and the invention of 'perpetual light', as well as relayed 'all kinds of knowledge' to the duke.[44] Accordingly, 'hardly a week [would] go by without us learning something curious and pleasant in this way'.[45] He also commissioned the production of various scientific instruments intended for the duke, such as barometers and thermometers, and he informed the duke of the books he had received and shared with him scientific news, including of microscopes and astronomical experiments.[46] In this manner, Leibniz placed his position in the Republic of Letters in the service of his ruler and the pursuit of the public good—two aims that did not, however, always coincide, as we shall see.

While advocating new projects, Leibniz was eager to reassure the duke that he was concerned solely with the latter's best interests and the advancement of the public good: 'my designs only seek the satisfaction of my prince.'[47] He presented himself as the 'instrument' of the duke's glory and the welfare of mankind: 'The great design which I have and which encompasses all others consists in the advancement of the useful arts and sciences.'[48] Leibniz matured in a tough marketplace of ideas in which he analysed opportunities acutely and strategized accordingly. He was always on the lookout for new information and schemes and drew on the knowledge provided to him by his various informants in his determination to help transform Hanover through the promotion of invention, technology, and manufacture. He was convinced that the natural and theoretical sciences should be made to yield practical applications and

---

43. Leibniz to Conring, June 1678, A II, 1:633

44. Leibniz to Johann Friedrich, 6 January 1677, A I, 2:12–17; May 1678, A I, 2:53; 6 July 1678, A I, 2:57.

45. Leibniz to Johann Friedrich, 6 January 1677, A I, 2:16–17.

46. Leibniz to Johann Friedrich, 6 July 1678, A I, 2:57; 10 September 1678, A I, 2:71; November 1678, A I, 2:99. Leibniz reports that 'a craftsman [*Mechanicus*] from Delft has improved the microscope to such an extent that one can track every day the changes that occur daily in the growth of plants'.

47. Leibniz to Johann Friedrich, November 1678: A I, 2:99.

48. Leibniz to Johann Friedrich, Autumn 1678, A I, 2:88.

ensure the kingdom's economic prosperity, especially against the backdrop of European conflict.[49] Even seemingly improbable schemes, such as Wenzel von Rheinburg's and Johann Joachim Becher's gold-making processes in Vienna and Holland, respectively, should be given adequate consideration—even if this prompted a formal complaint against a charlatan to the city of Amsterdam on at least one occasion.[50]

Negotiations for the acquisition of trade secrets or manufacturing processes depended on trust between the parties and the conviction of the good faith of those with whom one was dealing, in which case some reward could be promised so long as the secret produced viable results.[51]

Leibniz assessed the validity of potential projects and recommended those whose 'solidity' he could attest to and believed would be of interest and benefit to the duke, especially those of a commercial nature. In this manner, he identified and brought to the duke's attention various commercial opportunities, such as the production of phosphorus and cast iron and the development of machines to clean ports and canals.[52] From the inception of his employment, he set out to acquire for a 'reasonable price' a device to 'prevent the spread of fires', 'Rabel's cure, 'the art of melting iron so that it [became] malleable', and 'a particular cart invention'.[53] He also extolled the merits of phosphorus, urging the duke to negotiate its acquisition before Brand sold it off to the first comer.[54] Other interests were the extraction of iron ore from the Harz mountains for industrial purposes; the use of machines for the mechanization of silk and wool manufactures; and even the potential curative power of various drugs such as wound water, antipyretic, and Moxa.[55]

In addition to commending to his patron new projects he deemed 'useful or agreeable', Leibniz promoted individuals whom he considered reliable and potentially useful. In a letter of May 1677, Leibniz suggested recruiting a certain Johann Crafft whose highly successful plan for wool manufactures

49. See introduction to A I, 2:xxxi–viii.
50. Leibniz to Johann Friedrich, end of December 1678, A I, 2:106; mid-June 1679, A I, 2:no. 139; Johann Friedrich to the City of Amsterdam, September 1679, A I, 2:no. 170.
51. Leibniz to Johann Friedrich, end of December 1678, A I, 2:106.
52. Leibniz to Johann Friedrich, May 1678, A I, 2:54; end of December 1678, A I, 2:175.
53. Leibniz to Johann Friedrich, January 1677, A I, 2:18.
54. Leibniz to Johann Friedrich, 6 July 1678, A I, 2:57: 'One of the most admirable discoveries that we have ever heard about and I have no doubt it will have consequences'; Leibniz to Johann Friedrich, 31 July 1678, A I, 2:64: 'Many people have assured me that there is something solid, and many reasons lead me to believe that Mr Brand truly possesses it; I believe that is worthwhile considering something so important.'
55. Introduction to A I, 2:xxxvii.

for Saxony and Erfurt would be of interest to the duke of Hanover as well as setting up a commercial partnership between the courts of Hanover and Saxony, from which further negotiations and collaborations could potentially develop.[56]

Throughout the correspondence, Leibniz was thus keen to recruit and promote new savants and inventors, seeking funding and protection for them in order to induce them to reveal their secrets and help them perfect their inventions, and he regularly suggested individuals he believed would be assets to the service of the duke and useful to the kingdom. One individual in particular Leibniz singled out as a potential asset to the Hanoverian court was the mathematician Tschirnhaus, and in early 1678 he recommended the latter's appointment at the Hanoverian court.[57] He told Tschirnhaus, 'I have always considered you to be one of the most deserving of esteemed people I know whose genius is for things that are genuinely beautiful and solid, and who is capable of realizing them. This is why . . . I took it upon myself to tell my master about you in terms befitting your merits . . . and if you were to come here . . . you would be received as a gentleman of the court.'[58]

Leibniz touted the surprising benefits that Tschirnhaus would accrue from being in the service of the duke, whose knowledge spanned nearly 'all sciences and curiosities'. Tschirnhaus would welcome joining the court not only on account of the duke, but also because of the presence of able ministers and such deeply erudite men as Molanus, Fleming, Steno, and Hugo. In addition to vaunting Tschirnhaus's 'able qualities', Leibniz was quick to point out the latter's lack of 'ambition' or 'interest in advancement', thus implying the latter's potential pliability.[59]

Leibniz's Own Projects and Specialness

Leibniz's stay in Paris had provided him with renewed impetus, enthusiasm, and expertise by bringing him into close contact with many of its scientific personalities and societies. His return to Germany, which he viewed as the opportunity finally to implement some of his schemes, began a period of innovation that would only end with the death of

56. Leibniz to Johann Friedrich, May 1677, A I, 2:23–24.
57. Leibniz to Johann Friedrich, beginning 1678, A I, 2:46.
58. Leibniz to Tschirnhaus, January/February 1678, A III, 2:340–41.
59. Leibniz to Tschirnhaus, 2:341. Leibniz had worked with Tschirnhaus in Paris. Ironically, the latter would plagiarize Leibniz's differential calculus in a publication that appeared in the *Acta eruditorum*, thus precipitating a ten-year hiatus in their relationship.

Johann Friedrich. Leibniz presented himself as a new kind of state counsellor and a superior kind of projector, 'capable of seeing the openings necessary to improve technology and polity at once' by converting 'the imperfectly understood and ill-distributed knowledge of artisans into practices essential to develop the state and its people'.[60] In Paris, Leibniz had busied himself with 'extraordinary tasks' which he had not pursued to completion, and in which the duke, 'whose sophisticated taste [did] not satisfy itself with common things', was bound to be interested. He offered 'to bring new inventions and curiosities from extraordinary cabinets' to the court of Hanover, 'having learned from workmen themselves many things, some curious, others important for commerce and manufacture'.[61]

While his Parisian experience had impressed on him the value of organized research and knowledge, it had also exercised a moderating influence on him: as he was able to witness the daily workings of academic societies and encountered firsthand the difficulties of implementing his own calculating machine, Leibniz scaled down his projects for societies to more realistic and modest scales, henceforth preferring more practically oriented and focused projects, such as his *Bedencken über die Seidenziehung*, on silk manufacture.[62]

Many of his projects now assumed narrower concerns, often in the shape of local ventures aimed at promoting commerce through manufacture. Some of these more concrete and targeted projects entailed ways of improving the efficiency of existing public administrative structures, as in the organization of archives or the development of water drainage systems. Leibniz even suggested creating an academy of commerce and languages, so that the young could acquire necessary linguistic skills, and he recommended the establishment of a bureau of information modelled after Théophraste Renaudot's Bureau d'Adresse from which people could find particular goods for sale or hire or indicate goods they needed.[63]

---

60. Jones, *Reckoning with Matter* (2016), 87.
61. Leibniz to Johann Friedrich, 21 January 1675, A I, 1:491–3.
62. See Roinila, 'G. W. Leibniz' (2009), 165–79; Totok, 'Leibniz als Wissenschaftsorganisator' (1966), 29.
63. Leibniz to Johann Friedrich, September 1678, A I, 2:110. Leibniz's idea for an information bureau was not original: precursors include projects by Sir Arthur Gorges and Sir Walter Cope in the 1610s for an office of general commerce, Théophraste Renaudot in 1630 for a *bureau d'addresse*, and Samuel Hartlib in the 1640s for an office of intelligence—this last one having inspired Wilhelm von Schröder's project of 1686 for an *Intelligenzwerk*. These information bureaus would play a crucial role in promoting informational and economic exchange. Anton Tantner regards these offices as a 'transitional phenomenon between feudal and 'capitalist conditions' (450). For more on this see, inter alia, Keller, 'A Political *Fiat Lux*' (2016); Tantner, 'Intelligence Offices in the Habsburg Monarchy' (2016);

By 1676 Leibniz had encountered two of Europe's most prominent academies, the Parisian Académie des Sciences and the London Royal Society, and he sought to apply the expertise embodied by them when he joined the court of Johann Friedrich, a prince who seemed amenable to his scholarly and scientific plans and whose concern for 'public welfare' Leibniz judged might play to his advantage. Still, as Leibniz had to navigate the realities of court life and win over his employer, it quickly dawned on him that he would have to contend with the political realities of his time and attune his musings to them. He no longer enjoyed the freedom to dream that his younger years had afforded him but now had to work within the parameters and strictures of a relatively rigid system geared above all towards efficiency, public utility, and financial return. Knowledge, above all, was to be rendered productive.

In addition to his official commissions, Leibniz therefore used his position as a counsellor to put forward various—generally unsolicited—proposals of his own, many of which, he claimed, would consolidate his position at court, and establish his 'brand' of knowledge by helping expand his employer's coffers and authority. These included, from 1678 onwards, his plan for more extensive work on the *Demonstrationes catholicae* and his recommendations for insurance against natural hazards.[64] His project for a calculating machine promised to be 'one of the most remarkable inventions of this time', capable of 'delivering men from the slavish toil of calculating'.[65]

Among the other 'chimera' (*chimères*) Leibniz conjured up to provide valuable sources of profit for the kingdom were a special form of wagon that could be used 'in peacetime for commerce and the traffic of travellers, and in war for cannons and convoys', a cipher-machine, mechanized silk production, improved watches, a number of pharmacological remedies, and a proposal for promoting the brewing of a local beer.[66]

Nipperdey, '"Intelligenz und "Staatsbrille,"' (2008); Solomon, *Public Welfare, Science, and Propaganda in Seventeenth Century France* (1972).

64. Leibniz to Johann Friedrich, October 1679, A I, 2:221. On the *Demonstrationes*, see Schepers, 'Demonstrationes Catholicae—Leibniz' (2011); and Dingel, 'Leibniz und seine Überlegungen' (2019). On Leibniz's pioneering of insurance schemes, see Zwierlein, *Prometheus Tamed* (2021).

65. Leibniz to Johann Friedrich, 26 March 1673, A I, 1:488; February 1679, A I, 2:125. On the calculating machine, see Jones, *Reckoning with Matter*; and Lenzen, 'Leibniz and the Calculus Ratiocinator' (2018).

66. Leibniz to Johann Friedrich, December 1678, A I, 2:111; February 1679, A I, 2:125; Autumn 1678, A I, 2:79.

A more significant proposal was for a new pumping system in the Harz mines in order to improve ore mining.[67] First broached by Leibniz in spring 1678 and culminating, after a long series of letters and memoranda, with the ratification of a treaty in October 1679,[68] this proposal encountered much resistance both in the mining authority and in the duke's entourage. Drawing on what Leibniz claimed to be his unique skills in mechanics and natural philosophy, the project was intended to solve, once and for all, the flooding of the Harz mines and to guarantee the kingdom's economic independence through harnessing the power of perpetual motion, thus extending 'man's dominion over nature'.[69]

In this manner, the forces of nature would yield many practical uses and benefits for the duke and his state. Furthermore, Leibniz reassured, the project would not inconvenience the duke or risk damaging his reputation. On the contrary, the duke would be able to extract from this 'almost inexhaustible treasure' many 'advantages', including increased revenue and glory for himself, in particular by attracting foreign investment.[70]

The cost of the project would be minimal but its 'effect . . . as great as possible.' Within this exchange Leibniz remained careful to couch profit-making in noble terms, transmuting the duke's self-interest into the pursuit of the greater good.[71] In this manner, Leibniz hoped to reconcile personal advancement with the satisfaction of the public good and the advancement of the sciences (as he saw it), along with other proposals for societies such as the Ordo Caritatis or Societas Theophilorum, Consilium de Scribenda Historia Naturali, and Consultatio de Naturae Cognitione.[72] This project would combine profitmaking with the promotion of progress: the income from the privilege promised by the duke for Leibniz's technical inventions for the Harz mines would, if successful, support the establishment of a foundation, an 'academy' that would help

---

67. On the Harz mine project, see Hecht and Gottschalk, 'The Technology of Mining' (2014).

68. Leibniz to Johann Friedrich, Autumn 1678, A I, 2:no. 73.

69. See Ash, 'Expertise and the Early Modern State' (2010), 17; and Wakefield, 'Leibniz and the Wind Machines' (2010). Also Leibniz to Johann Friedrich, February 1679, A I, 2:80, 121–25.

70. Leibniz to Johann Friedrich, Spring 1679, A I, 2:83; February 1679, A I, 2:130. 'I have no doubt that there is still much hidden wealth that we could already attain if we made use of all the industry we are capable of and all the strength that nature has provided us. Why should we leave to posterity what is in our current power, especially since the life of metal found in mines . . . diminishes when neglected.'

71. See Smith, *The Business of Alchemy* (1994), 9.

72. Leibniz to Johann Friedrich, September 1678, A I, 2:76–77; Consilium de Scribenda Historia Naturali, 1679, A IV, 3:nos. 132, 133.

implement—and 'immortalize beyond his death'—Leibniz's plan for a universal philosophical language.[73] This so-called universal characteristic would 'carry the forces of the mind as far as the microscope has pushed those of sight' and reduce all reasoning to calculation in such a manner that it would be possible to 'invent with ease' and establish truths 'in an irresistible fashion', providing an incomparable aid and acting as 'the table of things, the inventory of knowledge, and the judge of controversies'.[74]

On several occasions throughout his correspondence, Leibniz acknowledged the unorthodox and bold nature of some of his projects: 'But this is because even the most solid things can look strange, when the mind is distracted or when it is unprepared. This is why I had reason to fear that the Duke might not have tasted something so far removed from the thoughts and occupations He has every day.'[75] Once again, as he regularly did when assessing the projects and ventures that were presented to him, Leibniz provided for the possibility for the seemingly impossible, for that 'which ha[d] never entered the minds of men' and which 'no one would believe to be possible'. For this reason, preserving his mechanical secret in the Harz would be easy since 'no one could ever imagine that air could be harnessed for such a purpose'.[76]

Notwithstanding his proximity to the duke, Leibniz effectively remained in competition with other projectors, many of whom he was determined to cast as either useless or deceitful.[77] The impression he sought to convey

73. Leibniz to Johann Friedrich, 29 March 1679, A II, 1:702, 703: 'Your Highness may one day strengthen this beautiful plan that He has formed of an Assembly for the advancement of science, which by this means can be made perpetual, and pass to posterity. . . . But the real way to fix this thought and make it immortal will be through the foundation.' See Keller, *Knowledge and Public Interest* (2019), 277 and 282: 'Gottfried Wilhelm Leibniz's projects for the social recombination of knowledge and society in service to the state demonstrate how the advancement of political and epistemic empire drew ever closer.' On the universal characteristic, see Rossi, *Logic and the Art of Memory* (2006), chap. 8; and Mugnai, 'Ars Characteristica' (2014).

74. Leibniz to Johann Friedrich, February 1679, A I, 2:126; 29 March/8 April 1679, A II, 1:701.

75. Leibniz to Johann Friedrich, April 1679, A I, 2:167. The Hanoverian court was a smaller court, and Leibniz assumed the authority to address the duke directly on all kinds of subjects.

76. Leibniz to Johann Friedrich, February 1679, A I, 2:127.

77. This did not stop him from tracking their activities through the use of spies or by bribing their servants. Leibniz to Johann Friedrich, mid-September 1678, A I, 2:68: 'Dr Becher is very eager for this secret, which compels me to believe that it, namely to produce gold from Dutch sand, is worth nothing, or at least very doubtful.' On Leibniz's rivalry and careerism in relation to other projectors, see Schaffer, 'The Show That Never Ends: Perpetual Motion in the Early Eighteenth Century' (1995); and Whitmer, *The Halle Orphanage* (2015). On Leibniz, Becher, and Wihelm von Schröder, see Keller, 'Happiness

to the duke was that, if his projects seemed particularly unusual, it was because of his unique talents and knowledge. Thus he took pains to distinguish himself from the mass of (bad) projectors by reasserting the well-founded nature of his own projects:

> It has been some time since I promised His Royal Highness proposals which I believe to be advantageous and without incurring any burden to his subjects. I wavered a long time before reaching this resolution. For I know that all new proposals are naturally suspect, and that we usually have a poor opinion of those who venture to make them concerning matters of state, especially nowadays, when one finds everywhere adventurers and opinion givers, whose only aim is to catch something, oblivious to the success of their ventures, when they have their purpose, and never lacking in excuses and pretexts to blame someone else when they fail to succeed.[78]

Leibniz defended himself from possibly being branded a braggart or intriguer; his claims were not 'idle boasts' (*fanfaronades*) but the by-products of sheer intellectual prowess and the application of his 'restless mind', which he insisted had been admired by the 'greatest intellects in France'—even if he had not felt comfortable sharing such 'advantageous sentiments' since he 'was not so favourable' to himself'.[79] Leibniz had 'no other interest than that of acquiring honour by considerable discoveries in the arts and sciences and of obliging the public by useful works' under 'the protection of a great prince', who could 'penetrate to the bottom of things' and was inclined to 'favour beautiful things'.[80] He promoted 'only his industry and work' and could be faulted only for his modesty, a trope which he rehearsed on several occasions over the course of his correspondence:

> I therefore readily admit, and in good faith, that I do not have a good memory, nor a strong imagination, and even less this prominence of spirit and this pleasant disposition, which enables one to shine in society, but instead I have good will and determination, which is the only thing that depends on us, whereas the other talents are bestowed upon us by nature and strengthened by a few conjunctures of fortune. This desire that I had to produce things of considerable importance had

---

and Projects between London and Vienna' (2021). On Leibniz and von Schröder, see Keller, 'A Political *Fiat Lux*' (2016).

78. Leibniz to Johann Friedrich, Spring 1679, A I, 2:79.
79. Leibniz to Johann Friedrich, February 1679, A I, 2:122.
80. Leibniz to Johann Friedrich, 19 January 1675, A I, 2:274.

opened up unknown roads for me, and made me study an art which has not yet been fully cultivated by men.[81]

Leibniz's sincerity, in particular, distinguished him from the mass of 'bad' projectors.[82] Acting, he insisted, out of 'affection' rather than 'interest, or flattery', and abhorring 'the superfluous', he averred, 'I agree that there are many who would be more capable of this task than myself: but good intention coupled with mediocre talent is sometimes better than great strength of mind which is unaccompanied by sincerity.'[83] By actively peddling this image of himself, meticulously fashioning this 'disadvantage' into a virtue, Leibniz sought to single himself out for special treatment and exclusivity in certain realms, hoping that the Elector would entrust him—and ideally him alone—with the task of masterminding schemes of various natures, including those pertaining to theological reconciliation: 'What I desire in general is that His Highness grants me a little more trust regarding serious matters than he generally accords others.'[84]

Leibniz's letters thus reveal not only his deeply ambivalent relationship to projects and projectors, but also his determination to assert his exceptionality and become the duke's exclusive adviser, especially in matters of technology and industry.[85] He continually insisted that his proposals were not only 'very reasonable' but 'novel' and 'remarkable'[86]—just as he himself was, as he took trouble to point out. (By contrast, he characterized his rival Becher as a 'very odd man' [*tout à fait bizarre*] with whom he 'wanted nothing to do' and whom he would later dismiss as a 'man well known for his exaggerations mixed with black malice' to his new employer Ernst von Hessen Rheinfels.)[87]

---

81. Leibniz to Johann Friedrich, February 1679, A I, 2:122.

82. Of course, Leibniz's determination to distinguish himself from the other 'bad' projectors was hardly unique. In his *Fürstliche Schatz- und Rentkammer*, in fact, Wilhelm von Schröder extolled the right kind of cameralist—industrious, financially knowledgeable, and attentive to his master's interests—in his own bid to secure a position in the emperor's Hofkammer. See Wakefield, *Disordered Police State*; and Freudenberger, *State and Society in Early Modern Austria* (1994).

83. Leibniz to Johann Friedrich, Autumn 1678, A I, 2:87.

84. Leibniz to Johann Friedrich, 2:229: 'Fourth, if there is any semblance of success in the main part of this pious business, it will take some secret negotiation, and I hope that His Highness will use me.'

85. Leibniz to Johann Friedrich, 29 November 1678, A I, 2:87: 'This being granted, I said, it is useless to think of other inventions which are not worth our own, and all that remains is to make a firm resolution to work hard for the demonstration and execution of this design.'

86. Leibniz to Johann Friedrich, January 1677, A I, 2:9.

87. Leibniz to Johann Friedrich, mid-September, 1678, A I, 2:68–70; Leibniz to Ernst von Hessen Rheinfels, 24 March 1683, A I, 3:278.

The German mathematician now turned courtier had no small estimation of his own intellect, placing himself on par with such luminaries as Aristotle, Bacon, and Galileo. In fact, his gift for invention was such that it would soon eclipse all other methods by effecting irreversible change on men's conduct and reasoning.[88] Furthermore, his talent was not limited to mathematics, as had hitherto been thought, but extended to theological matters. Indeed, his primary meditations concerned theology, and he 'had only applied himself to mathematics and scholastics to perfect his mind and learn the art of invention and demonstration'—an art, he believed, he had taken as far as possible. In fact, only he held the key to the solution of certain intractable problems, as he had recently been able to demonstrate within the realm of mathematics:

> It is common knowledge that I was the only one able to solve certain arithmetical problems set forth by [the mathematician] Ozanam: as for geometry I was the first to find the secret of the tables of sines, and remedied this lack of practical geometry in such a way that it is now possible to easily solve the problems of trigonometry by rule without having recourse to books and tables; this is something of vital importance to travellers and engineers, who are not always certain to have their books on them, and one of the greatest advancements in practical geometry that we have witnessed in our time.[89]

In this regard, a letter Leibniz wrote to Johann Friedrich in the autumn of 1679 is particularly striking. In this letter, which effectively served as a job application, Leibniz, through the guise of a fictional third party, engaged in a lengthy and hyperbolic, albeit altogether rather conventional, self-description to vaunt his exceptional talents and signal his special value to the duke. The fictional author of the letter describes how in Paris he had met a 'man of religion' whose 'merit was universally acknowledged' and who had 'meditated at length on the controversies' without, however, ever 'falling into excess'. This man was truly remarkable in his 'ability to explicate a passage and to show its true meaning', which he did 'with force and a singular clarity', and he also displayed a 'perfect knowledge of what we call the humanities [*humanités*]'. In fact, he was able to move his readers 'as soon as he started composing verses, which happened to him only very rarely' but as if he had done nothing else in his life. His style was

---

88. Leibniz to Johann Friedrich, February 1679, A I, 2:122.

89. Leibniz to Johann Friedrich, 2:124. Leibniz claimed to have developed successful methods of making the trigonometrical table superfluous and was seeking here to take advantage of popular interest in Diophantine analysis.

'simple and natural', whether in Latin or in the vernacular, but also 'strong and incisive' when needed. Furthermore, this was a man unencumbered by any type of artifice and untainted by any form of deceit or dissimulation (*serré sans obscurité, agréable sans fard*): 'He disliked putting on appearances [*couleurs empruntez'*] and believed that the beauty of a speech should consist in the strength of its arguments. He was, for these reasons, master in the art of reasoning: everyone agreed with him.' As he immersed himself in the study of scientific controversies, Leibniz's curiosity had been aroused by new discoveries in mathematics and physics, and he had become determined to leave his mark in those fields too. A significant discovery in the realm of mathematics remained the 'most assured mark of a solid mind', for its value lay not in rhetorical eloquence, but in 'demonstrations whose strengths and flaws were visible'.[90] This had prompted Leibniz to abandon his studies and duties to reach Paris, 'which is the centre of all beautiful curiosities': it was there 'that he [had shown] what he could [do] . . . distinguish[ing] himself from all others', and had been 'acknowledged by the great scholars of Paris as one of the finest geometers, capable of making discoveries of consequence'. Indeed, 'no foreigner of his kind (and he was) was ever received more favourably by persons of merit'. Yet in spite of all these achievements, Leibniz's 'demeanour did not seem to promise anything extraordinary', and 'his ordinary conversations were rather weak'. In fact, he 'did not possess nor feign to possess the art of showing off his skills', so that his fictional interlocutor was 'surprised not to recognize in him the marks of that which had been reported to him' about Leibniz. The former had nonetheless been very quickly disabused of this most deceptive initial impression. Leibniz had proved such a remarkable savant in Paris that he had been offered a post there, which he had declined for the pleasure of serving the duke and in the hope of undertaking a world-changing mission with the latter's support:

> People in Paris were so convinced of my talents in these matters that the most illustrious members of the Royal Academy of Sciences testified publicly to Mons. Colbert, that the latter could not choose anyone other than myself to fill the place left vacant at the Academy by M. Roberval, who had been considered one of the first geometers of his time. . . . People thought I had already been appointed at the Academy when His Royal Highness called me to him. I did not hesitate as to what I had to do, even if I reluctantly left a city where I had already acquired so many friends and admirers; but I considered that it was

---

90. Leibniz to Johann Friedrich, Autumn 1679, A II, 1:761.

still quite another thing to receive the support and be listened to by a great Prince, who is enlightened and inclined to beautiful things as I had recognized immediately when I had the good fortune to greet His Royal Highness in Frankfurt. I believed that it would be possible to perform things which would otherwise be buried in an eternal oblivion, and which I could not dare hope for elsewhere under His Majesty's auspices.[91]

This would not be the last time that Leibniz exaggerated his position—an exaggeration here bordering on misrepresentation—in order to leverage a more important status and greater influence. Throughout his correspondence, in fact, Leibniz was frequently simultaneously self-deprecating and boastful. While false modesty was a common feature of the projector's persona,[92] this evident contradiction would have compromised Leibniz's credibility in some recipients' minds.

Secrecy and Control

For someone who enjoyed presenting himself as a fearless outsider who was not stirred by common opinion, Leibniz emerges from his correspondence as extremely insecure, especially when confronted with difficulty or when the beautiful projects he had conjured up failed to materialize as easily as he had expected. He was particularly anxious about acquiring the reputation of a 'projector'. In fact, in the case of the Harz project, such was his fear of being tainted by failure that he remained adamant that his own scheme to help pump water out, his compensation, and even his identity be divulged only once it had proved successful.[93] This secrecy would also foster the impression that the scheme was being carried out on the duke's direct orders and discourage others from trying to render the scheme 'odious' to the duke in the hope that he would give it up.[94] Somewhat oblivious to the practical difficulties and dismissive of the expertise required to implement a scheme of this magnitude, Leibniz in fact accused some of misunderstanding his project

---

91. Leibniz to Johann Friedrich, February 1679, A II, 1:684.

92. See Wahl, 'Die Gier nach Ruhm unter dem Mantel der Bescheidenheit' (2016); Abou-Nemeh, 'Daring to Conjecture in Seventeenth- and Eighteenth-Century Sciences' (2022).

93. Leibniz to Johann Friedrich, October 1679, A I, 2:221; also beginning of March 1679, A I, 2:134.

94. Leibniz to Johann Friedrich, October 1679, A I, 2:222; mid-October 1679, A I, 2:206.

and of deliberately obstructing it.⁹⁵ He pled with Johann Friedrich to ensure his 'safety' and help 'prevent plotting' against him, especially in the duke's absence.⁹⁶ He imagined conspiracies being fomented against him with the aim of 'corrupting' his workers or taking over his project.⁹⁷ He was consumed with a fear of scandal and sabotage, especially when the viability of the Harz project began to be questioned: 'I have already noticed a subtlety which one meditates to my prejudice.'

With his 'honour [finally] committed' once the matter had been made public, something 'which [he] had avoided as far as [he] could', in the autumn 1679 and amidst mounting difficulties, Leibniz pleaded with Johann Friedrich to support him while he tried to overcome these difficulties so that he might 'be spared unbearable confusion and shame' because it was no longer just about 'gaining money' (*lucro captando*), but 'avoiding damage' (*damno vitando*).⁹⁸ Leibniz had, however, omitted to mention that many were self-created. In his anxiety to secure for himself a unique and indispensable position at court as a mediator between the government and the mines, he had often denigrated the extensive practical experience of the Harz mine engineers.⁹⁹

Furthermore, in 1678 the Dutch engineer and court councillor Peter Hartzing had submitted a plan for draining the water from the mines by the use of pumps activated by wind-power, a plan that the Harz mining office had received with enthusiasm. Leibniz, however, had dismissed the project in a letter to the duke before making a very similar proposal— ratified by the duke on 25 October 1679—for draining a colliery by means of windmills (rather than water wheels powered by rainwater).¹⁰⁰ This submission had prompted accusations of plagiarism against Leibniz and created conflict between him and Hartzing, with the mining experts at the Bergamt of Clausthal overwhelmingly supporting Hartzing.

Much of Leibniz's correspondence regarding the Harz project consists of requests to be remunerated for his labour and to be protected in the event that age, illness, mistaken conjectures, or hostility from others—especially

---

95. See Antognazza, *Leibniz*, 335. Such an arrogant approach would not have endeared Leibniz towards mining practitioners, who were of course experts in the field.

96. Leibniz to Johann Friedrich, mid-October 1679, A I, 2:214.

97. Leibniz to Johann Friedrich, 2:214.

98. Leibniz to Johann Friedrich, 2:206.

99. See Ash, 'Expertise and the Early Modern State' (2010); and Wakefield, 'Leibniz and the Wind Machines' (2010). Leibniz made a point 'not to submit to the judgement or whim of these people who claim to be knowledgeable in mechanics'. A I, 2:207.

100. Leibniz to Johann Friedrich, 9 December 1678, A I, 2:no. 87; Antognazza, *Leibniz*, 227.

since his ideas were 'not to everyone's taste and not for the vulgar'—prevented him from working so successfully in the future. While remaining always careful to seek for himself 'a perpetual and hereditary advantage', Leibniz requested that the duke conceal the intent of the privilege for as long as possible.[101] When negotiating for what he felt was his reasonable due, Leibniz combined two strategies. On the one hand, he couched his argument in terms of the practices and parlance of the emergent financial and commercial world. On the other hand, he was careful to present himself as the subject of an absolutist ruler to whom he would always defer. Hence he was simultaneously an 'independent contractor' and a 'self-effacing courtier'.[102]

Interestingly, and perhaps not altogether surprisingly, Leibniz nonetheless requested that the duke compensate him before the validity of his scheme had been firmly ascertained or empirical evidence of its efficacy had been obtained.[103] In his own case, he suspended his principle that projects should be viewed with suspicion until examined carefully, for his projects represented an opportunity that would otherwise be lost if not seized immediately: several centuries might pass before anyone would dare to take up his ideas again, not least because these ideas did not come easily and were apt to be dismissed as chimerical by most people, who believed only what appeared before their eyes.[104] Growing increasingly desperate amidst mounting difficulties, Leibniz tried to render his propositions more attractive by offering to forfeit his fee altogether if he failed to realize his scheme within a certain amount of time.[105]

Apart from performing his official duties, then, Leibniz laboured hard to secure additional responsibilities and a tailor-made role at the court of Duke Johann Friedrich. He wanted the duke to know that he was no ordinary projector or courtier. Already in a lengthy memorandum of January 1677, drafted at the very beginning of his tenure, he had bemoaned the discrepancy between his fame and recognized ability and his rank and compensation. By offering him adequate financial compensation and enabling him to advance his schemes, the duke would 'derive satisfaction . . . as well

---

101. Leibniz to Johann Friedrich, Autumn 1678, A I, 2:90.

102. See Jones, *Reckoning with Matter*, 120.

103. Leibniz to Johann Friedrich, 29 March 1679, A II, 1:703: 'This is why it will not help to write a book about it: since I cannot provide samples of it yet because all things are connected and will not allow themselves to be detached from each other.'

104. Leibniz to Johann Friedrich, 29 March 1679, A II, 1:702.

105. Leibniz to Hieronymus von Witzendorff, mid-October 1679, A I, 2:217; Leibniz to Johann Friedrich, April 1679, A I, 2:167.

as glory now and for posterity, wherever the sciences are in esteem'.[106] In return for his loyal service, Leibniz expected 'public marks of esteem' and pleaded on several occasions for the equalization of his status as *Hofrat* with the duke's other councillors so as to secure his position at court, a distinction he finally obtained in December 1677.[107]

Leibniz's concerns with his reputation and financial security constitute a thread that runs throughout his correspondence with the duke and betray his sense of precariousness within the Hanoverian establishment. He was particularly attentive to maintaining and elevating—as much as was possible within such a highly hierarchical system—his status at court, regularly comparing his situation to that of others. Various slights he received, whether from the duke's personal doctor, Dr Jakob Kozebue, who remonstrated with Johann Friedrich that Leibniz had taken a place ahead of him in church and thus beyond his rank, or from a malicious French secretary who had failed to wait for Leibniz, only served to heighten his sense of insecurity.[108] Similarly, the attribution of credit was a particularly sensitive issue for Leibniz. He was keen to assert his intellectual authorship over his inventions, especially in connection with the Harz project: 'I have demonstrated everything, and experienced every single part of this invention which I can valiantly call original and mine.'[109] No doubt recalling Robert Hooke's unpleasant attempted recreation of his calculating machine in London a few years earlier, he expressed irritation with those he perceived as plagiarizing his ideas: 'I dislike the boasts of the people who take credit for what was explained to them in substance, when they [added only] . . . some trifle.'[110]

More broadly, Leibniz appears in his correspondence as someone obsessed with maintaining control over his situation at all times and ideally being always one step ahead of others. This obsession manifested itself, among other ways, in uncovering new information while preserving his own secrets and often in acting behind the scenes. Indeed, he was not averse to petty intrigue, including the theft of letters and the use of disinformation and dissimulation. One notable instance was the acquisition of the physician Hennig Brand—and his secrets—for the court of Hanover

---

106. Leibniz to Johann Friedrich, Autumn 1678.
107. Leibniz to Johann Friedrich, Autumn 1678, A I, 2:85; Autumn 1679, A I, 2:229: 'A coach would be useful to me. His Majesty has been generous to all others regarding coaches and horses.'
108. Johann Kahm to Leibniz, 21 June/1 July 1679, A I, 2:182–83.
109. Leibniz to Johann Friedrich, Autumn 1678, A I, 2:84.
110. Leibniz to Johann Friedrich, early August 1679, A I, 2:188.

against Joachim Becher's various attempts at recruiting him. Leibniz tried to develop a somewhat ambiguous role at the Hanover court, often advising on or seeking to influence policy in areas beyond his official remit, including in theological matters, for which Leibniz hoped the duke would 'use him in secret negotiations'.[111] Cultivating opacity in various dealings could yield many advantages, especially when developing stratagems to help bring about—nay, mastermind—the reconciliation of the churches.

In this regard, Leibniz's connection to the bishop and court favourite Jacques-Bénigne Bossuet (1627-1704) warrants a brief mention. Bossuet was concerned primarily with vindicating the literal meaning of sacred scripture and with defending the Catholic faith, for which purpose he had brought together apologists tasked with protecting the status of scripture from the doubts of rational biblical criticism.[112] Bossuet's *Exposition of the Catholic Doctrine* had impressed Leibniz, and from 1678 to 1680 he enthusiastically scoured Germany for Latin translations of Talmudic volumes and directed books and catalogues from his own network of scholars to Bossuet.

While Leibniz was careful to keep Bossuet provided with books sourced from across Germany, he seems to have had greater designs and conceived a larger role for himself, especially once the pope approved the French theologian's doctrine—which Leibniz himself qualified as 'the most moderate' he had seen, giving him hope for an eventual reconciliation of the churches.[113] Encouraged, Leibniz suggested forging a piece of writing in the spring of 1679 whose origin it would be impossible to trace, and which would be addressed by a Catholic to a Protestant with the aim of converting him.[114] Whether this stratagem, which Leibniz hoped would be approved by the pope and have an even greater impact than Bossuet's brief, was indeed implemented remains unknown, but the absence of further reference to it suggests that it was not.[115]

At times, as Leibniz put it, it was even necessary to conceal sound projects for the greater good: 'This writing remains buried in silence and

---

111. Leibniz to Johann Friedrich, Autumn 1679, A I, 2:229.

112. Preyat, *Le Petit concile de Bossuet* (2007), 58.

113. Leibniz to Johann Friedrich, June 1679, A I, 2:176-77, 181: 'And the affair will be easy since the Pope now passes for a good, enlightened, and equitable man: and he has shown his zeal by condemning several unreasonable propositions of false morality, and his moderation by approving the exhibition of faith published by the bishop of Condom [Bossuet], which is the most moderate that I have seen yet.'

114. Leibniz to Johann Friedrich, Autumn 1679, A I, 2:227-28.

115. This may indicate that the duke did not fully trust Leibniz, despite the latter's clear conviction that the duke supported him.

secrecy.'¹¹⁶ His plan for a portable watch, intended as an improvement on Huygens's, is one example: 'I have not yet revealed this thought to others, but have preserved it for a worthy recipient' and he was reluctant to communicate 'the essence of [t]his invention'.¹¹⁷ There lay, it seemed, an unbridgeable gap between him and the rest of the court.

Between Artifice and Sincerity

One form of this modus operandi was Leibniz's deployment of rhetoric that, in certain respects, went beyond the epistolary conventions of the time and served again to emphasize his exceptionality. Preterition (or apophasis) was Leibniz's rhetorical device of choice throughout his correspondence with the duke.¹¹⁸ It is a rhetorical figure of speech in which the speaker or writer draws attention to something by professing to pass over it, revealing by obfuscating. The implicit is rendered explicit by the pretence of not making it so. More broadly, by resorting to this seemingly oxymoronic device, the writer blurs the distinction between the implicit and the explicit. Crucially—and what concerns us here—the peculiarity and force of this device lie in its staging and exposition of that very pretence. Leibniz regularly deployed the device when submitting requests to the duke, as in his long memorandum of 6 January 1677 with its financial demands.¹¹⁹ His most characteristic use of preterition, however, was to advertise his exceptionality while professing his modesty.

On a later occasion in the autumn of 1678, Leibniz delivered a lengthy panegyric of the duke and his numerous qualities before concluding that it would actually be 'useless and even ill-advised' to do so.¹²⁰ While claiming that he did 'not say all these things to justify [his own] conduct', he must have been trying to do precisely that. In general, Leibniz's deployment of this device can be seen as emblematic of an impatient and ironic attitude towards court conventions and protocols. The court society in which he moved was strictly codified and relied on the perpetuation of a naturalized artifice through tacit social consensus. Without openly defying these conventions, Leibniz nonetheless distanced himself from them by adopting their rhetoric in exaggerated or paradoxical ways, thereby creating the

---

116. Leibniz to Johann Friedrich, Autumn 1678, A I, 2:89. Also A I, 2:82: 'The things of greatest importance must remain in silence.'
117. Leibniz to Johann Friedrich, February 1679, A I, 2:125.
118. See Henkemans, 'La préterition' (2009).
119. Leibniz to Johann Friedrich, 6 January 1677, A I, 2:12.
120. Leibniz to Johann Friedrich, Autumn 1678, A I, 2:81.

impression of a disjunction within the text. In a letter from the middle of June 1679, for instance, Leibniz suggested taking up Counsellor Block's position immediately after expressing regret at the latter's demise.[121] By proceeding in this manner, Leibniz exposed both the fictionality of pro forma expressions of regret and the genuine motivation behind them. His focus on personal advancement and his own projects often led him to disregard proper decorum, especially when he felt that time was of the essence. In his dogged pursuit of truth—and of his own cause—he adhered only half-heartedly to conventions from which he probably considered himself absolved to a great extent; he was, after all, no ordinary courtier.

Precisely his failure to conform to social norms more fully—which is to say, more successfully insincerely—made Leibniz's rhetoric appear gauche and disingenuous. His persistence, indeed pushiness, especially when experiencing difficulties, was bound to make him appear an ambiguous figure at the very least. Exposing artifice for what it was and yet still resorting to it cast Leibniz as a social projector in spite of himself. Ironically, then, Leibniz's reluctance to properly fashion himself as a courtier, owing to his commitments to sincerity and truth deriving from his pursuit of a rationally guided society, contributed to his appearing unreliable and slippery.

In what proved a difficult balancing act, Leibniz deliberately placed himself simultaneously within and without the court system. He acknowledged the discrepancy between court life, with its strict hierarchy and rigid codes, and his own projects, many of which, he recognized, were rather far-fetched. His letter to Johann Friedrich of April 1679 is particularly noteworthy in this regard. In it, Leibniz struck an unusually candid note, explaining that truth does not always square with everyday matters and can thus strike one as unconventional and bizarre:

> It is not that they are far removed from the truth—for after all the design of perfecting the human understanding is so important and so certain that one could not speak of it in terms significant or confident enough—but on account of the fact that even the most solid things can appear strange, when the mind is distracted or unprepared. I feared somewhat that Your Highness had previously not encountered something so far removed from his daily preoccupations and activities.... My experience has been that most people do not suffer to concern themselves with certain matters that are not to their taste, however valuable those matters might be.[122]

---

121. Leibniz to Johann Friedrich, mid-June 1679, A I, 2:178.
122. Leibniz to Johann Friedrich, April 1679, A I, 2:166.

This tension was even reflected at a linguistic level. While most of the letters from the correspondence were drafted in French, some were written in German. The latter tend to address administrative and financial issues. Whereas the letters written in German, with their formalisms and circumvolutions, seem to have been written in Leibniz's official capacity as a counsellor, the letters in French, in which he expressed himself and discussed his projects more freely and took more liberties, gave voice to the audacious and ambitious thinker. When writing in French, he felt emboldened to confide in someone he believed, or at least wanted to believe, understood and shared his vision of progress and was prepared to support his loftier plans. This distinction between the two languages illustrates the tension between Leibniz's official capacity within a highly hierarchized German administration and the niche he sought to carve out for himself in the wake of the intellectual and scientific excitement he had experienced in Paris. As Tim Hochstrasser has accurately observed, Leibniz wanted success at court, but only on his terms: 'While he desired the favour of rulers so as to enable him to enjoy academic freedom and diplomatic responsibility, he was not unwilling to fulfil the minimum requirements of court service. He hoped that the status he enjoyed in the "Republic of Letters" would translate automatically into equivalent recognition in court society, disregarding the subservience to the hierarchy which other courtiers regarded as routine.'[123]

Leibniz's rapport with the duke is symptomatic of the tension between court-life (or 'artifice') and intellectual life ('sincerity'), and of how the latter, in his relationship with the duke, ultimately prevailed. To be sure, Leibniz's tone toward the duke was always, as one would expect, deferential. He always concluded his letters by reassuring the duke of his devotion and his deference to the will of the duke, whom he knew would act justly and always in Leibniz's best interest. Above all, Leibniz was keen to present himself as a sincere and self-effacing subject who placed the duke's happiness and glory above everything else and whose fate happily depended on the latter's goodwill: 'Your Royal Highness knows that I am wholeheartedly devoted to whatever pleases Him: and that He only has to turn me down to make me almost as happy. But not yet knowing His decision on this matter, I flatter myself that He could perhaps give me this place.'[124] In keeping with his official position, too, Leibniz spared no effort flattering and grooming the duke:

---

123. Hochstrasser, 'G. W. von Leibniz' (1998), 7–8.
124. Leibniz to Johann Friedrich, mid-June 1679, A I, 2:178.

All these reasons would no doubt have silenced me, if I did not have the honour of serving a Prince like His Majesty. Because He does not fear new proposals, like those who never dare to be generous; and the charms of novelty do not bind Him to anything rash. He has however undertaken some very great and extraordinary things since the happy accession to His regency, which in this manner will be rendered forever memorable. Moreover, since He is so naturally inclined and displays a curiosity found in great souls, He neglects nothing but pays no attention to people or external circumstances which are irrelevant to the matter at hand but do not fail to make great impressions on weak minds. It is also no small thing that one can be assured of His confidence and discretion. His Majesty knows how to keeps secrets . . . and He possesses insights which dispense Him from having to consult others except when He wishes to unburden Himself. For (and this is the main thing) He possesses all the promptness and foresight necessary to delve into any matter, as difficult or embarrassing it may be, as long as He is made aware of its importance.[125]

Nonetheless, Leibniz's rapport with the duke exceeded that between a subject and his patron. Leibniz perceived the duke as more than a simple means to an end. His correspondence with the duke is wide-ranging and regular, and his letters bear out a certain degree of intimacy, or at least the desire for such intimacy with the duke. The duke had conceived a 'beautiful design . . . for the advancement of the sciences', and unlike most, and in particular 'the vulgar', could appreciate Leibniz's vision—or so Leibniz hoped.[126] Throughout the correspondence, Leibniz seems to view himself, and persisted in representing himself, as the intellectual and scientific extension of the duke, the latter's arm or 'instrument', and conversely the duke as his agent, in what he wanted to believe were their shared goals. Behaving towards the duke in a manner that far exceeded his official position, Leibniz often used the personal pronoun 'we' (*nous*) when discussing his projects. Frequently resorting to the formula, 'I take the liberty of . . .', he repeatedly sought to act above his station and would suggest what his employer should write and do according to the canons of natural justice or enlightened behaviour (which generally happened to correspond to Leibniz's own aims).[127] Occasionally Leibniz even dictated to his employer

---

125. Leibniz to Johann Friedrich, Autumn 1678, A I, 2:80.
126. Leibniz to Johann Friedrich, February 1679, A II, 1:680.
127. Leibniz to Johann Friedrich, mid-October 1679, A I, 2:213: 'I am taking the liberty of submitting to His Highness a proposal which strikes me as very reasonable.'

instructions for the speedy execution of a project or the terms of a contract: 'He will be kind enough to declare that this right will belong to me, as inventor and entrepreneur, with all obligations attached and under the same conditions.'[128] Though always careful to affirm their conformity to royal intent, Leibniz often presented his arguments as self-evident, so self-evident in fact that only 'blindness' could prevent one from recognizing their validity.[129] Should irrefutability not be sufficient, however, Leibniz would present the duke with an array of arguments or objections emanating from different perspectives and from which the duke could then select. This strategy would at least create the illusion of offering the duke maximal freedom of choice.

Leibniz encouraged the duke to exercise his authority in various ways. On some occasions he characterized the materialization of his great projects as being purely contingent on the duke's approval and 'will'. On other occasions, he tried to pressure the duke into using his influence to maintain secrecy—as in the case of the Harz project—or sought the duke's protection when his plans failed to materialize. More broadly, Leibniz hoped that others would form a good opinion of him by virtue of his mere association with the duke.

Consequently, Leibniz's relation to the duke perhaps evinces more complexity and subtlety than the 'self-fashioning' described by such historians as Mario Biagioli.[130] The model of self-fashioning illuminates the paradox at the heart of the legitimation process of scientific claims, especially those at odds with a dominant conceptual paradigm. The solution for scientific practitioners lacking in high social status by birth lay in entering into patronage relationships with society's powerful. This patron-client exchange was a polite fiction concealing mutual interests; it was based on mutual obligation, even though the exchange was generally presented as voluntary and disinterested by both parties and couched in the language of courtesy.[131] One rather curious but effective strategy consisted in effacing oneself behind one's patron and transferring authorship of the discovery to the latter.[132] Galileo, for instance, presented his 1610 discovery of the satellites of Jupiter as 'Medicean Stars', that is, as emblems of the Medici dynasty rather than scientific discoveries.[133] As Biagioli has

---

128. Leibniz to Johann Friedrich, February 1679, A I, 2:132.
129. Leibniz to Johann Friedrich, mid-October 1679, A I, 2:213.
130. See Biagioli, *Galileo, Courtier* (1993); Meyer, 'Les courtiers du savoir' (2010).
131. Kettering, 'Gift-giving and Patronage' (1988), 131–51.
132. Biagioli, 'Etiquette, Interdependence' (1996), 193–238.
133. Biagioli, 'Galilée bricoleur' (1992), 85–105.

demonstrated, Galileo's courtly role was integral to his science.[134] The legitimation of his scientific discoveries lay in his skilful self-fashioning in the Medici court and his ability to forge for himself a new socioprofessional identity as court philosopher rather than as a mere mathematician, mathematics at the time being subordinated to philosophy and theology.

Leibniz's case, however, can be characterised adequately neither merely in terms of 'self-fashioning' nor in terms of a straightforwardly instrumental relationship. A passionate and original thinker, Leibniz had to contend with epistemological and social paradigms that were at odds with his, and he did not always negotiate the discrepancy between them well. The correspondence reveals a reluctant courtier who was ready to abide by the conventions of the court—but only up to a certain point. Ultimately, Leibniz was more interested in his own visions and brand of knowledge, which he hoped he would, with the support of the duke, gradually impose at court. Ironically, sincerity and projection: Leibniz in this case appeared as a projector precisely because of his commitment to sincerity.[135]

Whereas Leibniz wrote numerous letters to the duke, the latter is known to have sent only six documents to him, most of which were decrees, in particular concerning the Brand and Harz projects. The duke's opinion of Leibniz is thus difficult to gauge; to what extent he shared his employee's commitment and enthusiasm for great intellectual and scientific projects remains uncertain. This much can be said: as far as the duke was concerned, Leibniz's projects were viable and ultimately successful only insofar as practical applications resulted from them and they were financially profitable. Only rarely, then, did the duke's and Leibniz's visions actually converge. Though Leibniz himself was not discouraged

---

134. Biagioli, *Galileo, Courtier*, 2.

135. That said, there are significant parallels in Galileo's and Leibniz's individual trajectories. In Leibniz's case too, being a savant implied being a courtier at a time when boundaries between polite and courtly societies were still porous. Both discerned which roles were valued at court and strategized accordingly, and both found out that these strategies ran up against certain limits once they no longer coincided with their patrons' priorities. By building his new socioprofessional identify for himself, Galileo hoped to gain acceptance for his work on motion, but his attempts to legitimize Copernicanism and mathematical physics were not well received. Still, while Leibniz devoted much effort to fashioning himself as a state adviser and counsellor over the course of the period under purview at a time when the court required that science yield commercial and technical applications, he remained his own man in the sincere belief that his discoveries, inventions, and projects should benefit and contribute to the state and, ultimately, mankind. Leibniz abided first and foremost by this strong commitment to progress and the advancement of the common good, which he assumed (or wanted to assume?) Johann Friedrich shared.

from promoting projects that he recognized might seem outlandish to most, only his technical proposals would ultimately be put into practice.[136]

Leibniz's correspondence with Johann Friedrich thus reveals not an isolated demiurgic genius, but an extraordinarily human character driven above all by ambition and vision. Leibniz clearly angled for a special position and had singled himself out for special treatment from the outset of the relationship, a risky strategy that was bound to antagonize many. In fact, he had been in no hurry to take up his new appointment. While he had originally received confirmation of it in January 1676, he finally arrived in Hanover only in December, after several complaints from Johann Carl Kahm, the payment of 200 thalers to Leibniz, and even the threat of cancelling his appointment. Leibniz would eventually ascribe his failure to comply with Kahm's instructions and incredible tardiness to ill-health and difficulty of transport.[137] At the court of Hanover, he was constantly acting above his station, testing the norms and conventions of the court, drafting unsolicited petitions, requesting money, and pushing for a higher rank and ever-greater responsibilities. Worse, he risked undermining himself by acquiring the reputation of a mere projector, since many of the projects he promoted failed even as he sought greater recognition for himself. Leibniz was already, and intent on remaining, a misfit.

136. Introduction to A I, 2:xxxviii.
137. Antognazza, *Leibniz*, 174–75.

CHAPTER TEN

Leibniz with His Fellow Projectors

LEIBNIZ CULTIVATED CLOSE RELATIONSHIPS with a number of projectors over the course of the 1670s, and most of these relationships were coloured by a strong ambivalence. He followed from afar various schemes, especially those relating to the separation of gold and silver, even inviting the alchemist Jakob Vierort for a public demonstration of one of his transmutations at court.[1] Some of these projects—including the mercantile project-maker Martin Elers's schemes to set up a new town near Harburg for émigré Huguenots in late 1680 and even a Brandenburg-African company that would provide Africans as slaves, farm labourers, or soldiers—were judged impractical and potentially embarrassing.[2]

Within the context of a new natural philosophy which had given rise to a flurry of novelties and propositions, the question of how to assess experimental findings that often lacked rigorous verification or 'solidity' was posing itself with increasing urgency.[3] From the tide of 'beautiful propositions' emanating from all corners and saturating the epistemological horizon, only a handful could actually be verified or realized, and it was 'most reasonable that we should mistrust these things, and not believe anything before seeing them'.[4]

These projectors engaged in a wide range of ventures, many fanciful and foolish, including the pursuit of 'the *Philosopher's stone*, perpetual

---

1. Leibniz to Vierort, 12 November 1681, A III, 3:61.
2. Crafft to Leibniz, 24 February 1682, A III, 3:273, quoted in O'Hara, *Leibniz's Correspondence* (2024), 167–68.
3. See Duchesneau, 'Leibniz et les hypothèses' (1982), 223–38.
4. Leibniz to Johann Friedrich, 15 October 1679, A I, 2:219.

motion ... malleable glass or incombustible oil', and even 'an Elixir of immortality'. In a striking and previously largely unexamined letter to Carcavy drafted several years earlier in his bid to obtain a privilege, Leibniz—undoubtedly writing self-referentially—had elaborated in greater detail on that particular breed of inventors: 'harmful to oneself, useless to all ... but of greater use for the state in the future than one would assume if only it were known. They alternate between insanity and wisdom. And because there is no great mind without this mixture, at least the *appearance* of stupidity has prevailed among them, or, to say it philosophically, has acquired its residence.' That 'type of men', he continued, for the main part, did 'not breathe or speak anything but inventions, experiments, new ideas, and not always in vain' for there was often 'a lot of gold hidden in this manure, if one [had] the patience to look for it', including 'the secrets of art and nature, argumentation, metallurgy' as well as 'emended medicine'. And yet many of these projects were all too often squandered or fell into oblivion on account of the neglect of their authors, who 'wither[ed] away to the last bit with ruined faculties, lost reputation, neglected, despised, poor, embittered by their misfortunes and, as they believe, an ungrateful world', and took 'what was in the state's interest not to perish' to the grave with them since 'nobody cared for them ... listened ... [or bought] what they did not want to give out for free, after it was created with great labour'. Men who devoted their lives to the pursuit of 'neglected and vanishing things' and the creation of excellent ventures by exploring the productive potential of art were often left at the mercy of circumstances that lay beyond their control:

> I know those who have used up ten—who would believe it?—even twenty full years wandering throughout Europe in order to compile hidden and curious discoveries; and yet these good men, once they returned home, had hardly enough to sustain their lives; not on account of a flaw within their venture, of which they have many and excellent ones, but [*sc.* by a flaw] of character ... of someone who is rolling that Sisyphean rock in the vain hope of finding something, neglecting everything else.[5]

Still, although many schemes were intended to 'deceive', while others yielded disappointing or no results, thus rendering projectors even more 'suspect', some—such as Leibniz's own—did produce the most unexpected

---

5. Leibniz to Carcavy, early November 1671, A II, 1: 287–90. See also Keller, *Knowledge and the Public Interest*, 279–280.

and marvellous results, and they should not all be dismissed out of hand.[6] What was 'despised' in one instance could well be 'deemed as the highest invention or secret' in another.[7] Poor appearances could conceal a great treasure: 'And without doubt, for the most part, such Phaethon-like ventures go astray, but some stumble upon heroic medicines . . . or machines useful for life. . . . Some of these experiments are delivered in a way that they are useless in themselves, but combined with others that are hidden in another part of this dung heap, they can in the end yield a huge gain.'

Under proper supervision, these inventors' enthusiasm and abilities could be harnessed to a properly organized programme of scientific research whereby the mechanical and practical arts could produce useful and even profitable inventions.[8] It was important, above all, among the merchant arts and sciences to 'subtly transfer the raw material of others to oneself, and to cultivate it by one's own artifices' and, if necessary, to establish a 'kind of trade' with 'those . . . outstanding minds . . . who drive forward such excellent and far-reaching things'.[9] As he would write to Crafft, 'in tinkering, a blind person [was] as good as a sighted one'.[10]

At the same time, these projects and propositions were not to be taken at face value, but examined and subjected to a combination of 'reflection' and 'experimentation'.[11] Against the prevailing 'insolence of besmirching whatever onto the documents', Leibniz therefore 'listened to everything but believed nothing' until it had been adequately put to the test 'independently of others' beliefs'.[12] In this manner, more popular errors would be avoided.[13] Leibniz himself took stock of this stance for his own research

---

6. By the same token, Leibniz did not completely rule out the possibility of transmuting base metal into gold. Ariew, 'Leibniz on the Unicorn' (1998), 276–77, describes Leibniz 'as a sober, cautious interpreter, a skeptic one might say, but one who is prepared to concede the possibility of many strange phenomena'.

7. 'The arts of Englishmen are known to me and I know in what manner so many famous experiments have reached them; [I know] how long they [these experiments] were despised among the ranks of vile German artisans, and how now that they have reached the grandees of the Royal Society, they are deemed the highest inventions or secrets.'

8. Leibniz, *Preface to Nizolius*, 1670, A V, 1. 2:413; to Carcavy, November (?) 1671, A I, 1:182; *Bedenken von Aufrichtung einer Akademie*, 1671(?), A IV, 1:543–52; Leibniz to Tentzel, 29 July/August 1692, A I, 8:366–67.

9. Leibniz to Carcavy, early November 1671, A II, 1:288

10. Leibniz to Crafft, 5 November 1691, A III, 5:195. Leibniz would echo the same opinion several years later in a letter to Justel: 'It does occur that out of a thousand men there will be one who will find something rare and extraordinary.'

11. For more on Leibniz's combination of 'conjecture' with experiment, see Wahl, '"Im tunckeln ist ein blinder"' (2015), 225–59.

12. Leibniz to Conring, 8 February 1671, A II, 1:131.

13. Justel to Leibniz, 30 July 1677, A I, 2:285–86.

and speculations—it was imperative not to rush into new discoveries but to apply oneself diligently to their demonstration before 'divul[ging]' in an untimely fashion', and to 'refrain from expounding [his] dreams in mathematics and mechanics until . . . [he could] convey them by some result and sample'.[14] It was, in fact, precisely by means of rigorous demonstration and theoretical knowledge that 'the certain would be separated from the uncertain' and that 'demi-savant[s] puffed up with an imaginary knowledge project[ed] machines and buildings that could not succeed' would be exposed.[15] Ingenuity in itself was insufficient but needed to be verified through experience and the deployment of hypotheses: both were inextricably connected.[16] It was necessary to join reasoning to observation and to extend the realm of demonstrative knowledge to the world of natural philosophy and experience by rendering probabilistic calculations demonstrative.[17] Leibniz mocked scholars who peddled 'empty thoughts and remote speculations' that did not yield 'anything useful', but he equally criticized those who were overly preoccupied with practical details and, as a result, could 'hardly meditate on things'.[18]

Johann Daniel Crafft and the Phosphorus Affair

Though Leibniz naturally sought to promote his own projects, it appears that he was more successful as a purveyor of others', especially at a time of German economic reconstruction.[19] The German inventor and projector Johann Daniel Crafft (1624–97) emerges, in this context, as a privileged interlocutor in Leibniz's correspondence. In Crafft he seems to have found a kindred spirit. For Crafft shared Leibniz's entrepreneurial drive, energy, and outlook, and both were imbued with the belief that scientific discoveries should yield tangible schemes of improvement for the public good.[20] Leibniz and Crafft had known each other since Leibniz's stay in

---

14. Leibniz to Conring, 8 February 1671, A II, 1:131.
15. Leibniz, 'Recommendation pour instituer la science générale', 1686, A VI, 4:712.
16. See Wahl, '"Im tunckeln ist ein blinder"', 251–52. As Wahl astutely points out, Leibniz emphasizes the role of chance in discoveries, notably in the field of chemistry. See also Abou-Nemeh, 'Daring to Conjecture in Seventeenth-and Eighteenth-Century Sciences' (2022) on the role of conjecture in the advancement of the sciences.
17. Leibniz to Huygens, 29 December 1691, A III, 5:241; Leibniz to Conring, 3/13 January 1678, A II, 1:582.
18. Leibniz, 'Plan zu einer deutschliebenden Genossenschaft' (c. 1691–95), A IV, 6:791, quoted in Wahl, '"Im tunckeln ist ein blinder"', 252–53.
19. P. Ritter, quoted in Moll, 'Von Erhard Weigel' (1982), 57.
20. For more on this, see Slack, *From Reformation to Improvement* (2014).

Mainz, and they would continue to exchange numerous letters touching on commercial, technological, and scientific ventures until Crafft's death in 1697. Both were constantly seeking new ventures, and Crafft in particular had a remarkable ability to conceive technical and commercial projects. While he failed ultimately to leave much of a mark—and hence his legacy contrasts markedly with Leibniz's—Crafft was no less an impressive figure in his way. Crafft has been called a 'visionary of the early economic enlightenment'.[21] He was equally committed to Leibniz's cherished dictum of *Theoria cum Praxi* and spent much of his life seeking practical applications of projects.[22] Under the patronage of his Elector, Crafft would turn Saxony into a testing ground for most of his theories and schemes in a bid to introduce manufacturing into what had, according to him, remained an essentially feudal—and outdated—trading system dominated by guilds.[23]

Though little is known of Crafft, a few elements can be gleaned from his correspondence with Leibniz and the curriculum vitae in the shape of a short autobiography that he penned when applying for employment at the imperial court in Vienna. Born in 1624, he studied medicine, botany, and chemistry before gaining a medical degree and finding employment in the Harz mountains as a doctor at the Brunswick-Luneburg mining office in Zellerfeld, where he also devoted himself to the study of metallurgy.[24] He later gave up his position to travel throughout Europe and even North America, eventually returning with an excellent knowledge of chemistry as well as of manufacturing and commerce.[25] Newly appointed as a commercial adviser to the Elector of Mainz, who sent him on various assignments throughout Europe to collect useful information and propositions, Crafft put his hand to the production of all sorts of small commodities and manufactured products, including glass and metalwork, the refinement of wine, and the production of wool and silk.[26] At the Mainz court he also distinguished himself through his efforts to promote manufacturing in the Erfurt region, notably through glass and steel manufacture, a project that was cut short, however, by the Elector's untimely death.[27]

---

21. Forberger, 'Johann Daniel Crafft' (1964), 63.
22. As he would write to Leibniz, 'My opinion is strengthened daily in more and more that everything is possible and laid in nature' (A III, 3:489).
23. A III, 3:489.
24. Forberger, 'Johann Daniel Crafft', 70.
25. Peters, 'Leibniz als Chemiker' (1916), 105.
26. Forberger, 'Johann Daniel Crafft', 71.
27. Peters, 'Leibniz als Chemiker', 87.

After the death of the Elector of Mainz in February 1673, Crafft approached the Elector of Saxony with proposals for the introduction of various industries, including the cultivation of mulberry trees and the extraction of silk in Dresden—a project he realized in the following years, setting up Saxony's first silk factory in Leipzig in 1674.[28] In 1675 he was appointed commercial adviser to the Saxon court, and he set in motion, with Elector Johann Georg II's support, the preparatory work for the construction of a wool factory in the newly erected town of Ostra near Dresden. This manufacture subsequently failed after facing much resistance from the mercantile community.[29] Crafft later travelled to England and Holland to collect information and hire workers that would be able to carry out his designs.[30]

Crafft was held in equally high esteem among the various electors who employed him, the historian H. Saring eventually describing him as a 'gifted and ingenious personality' whose life's work found itself cut short by a tragic destiny.[31] Leibniz vaunted Crafft's merits to his own employers on several occasions, presenting a summary of the latter's most important research to the duke,[32] and even trying, unsuccessfully, to have him brought to the Hanoverian court in 1693. In a May 1677 letter to Duke Johann Friedrich, Leibniz described Crafft as enjoying unusual success— and enhanced 'credit'—in implementing his projects in Saxony, at a time when such ideas were mainly ignored.[33] Crafft was a man who could not only 'make proposals but also execute them' and, as an unselfish man of good character, let down only by his credulity.[34] In letters to Tschirnhaus and Johann Friedrich, Leibniz emphasized Crafft's scientific and practical abilities, especially in matters of trade and commerce, and his knowledge of dyes and mining, even recommending Crafft's innovative wool manufacture to the duke.[35] Crafft was not only aware of the technical innovations of his time, but he was a great inventor, grasping early on the value of practical implementation.[36] Both Leibniz and Crafft were devoted to the

---

28. Forberger, 'Johann Daniel Crafft', 71.
29. Forberger, 67.
30. Peters, 'Leibniz als Chemiker', 106.
31. Saring, 'Crafft, Daniel' (1957), 387.
32. Antognazza, *Leibniz* (2009), 285; Leibniz to Crafft, end of July 1677, A III, 2:201.
33. Leibniz to Johann Friedrich, May 1677, A I, 2:23–24.
34. Leibniz to Johann Friedrich, 2:23.
35. Leibniz to Johann Friedrich, 1678, A I, 2:64; Leibniz to Tschirnhaus, 30 January 1693, A III, 5:487; Leibniz to Johann Friedrich, May 1677, A I, 2:24; Antognazza, *Leibniz*, 51.
36. Forberger, 'Johann Daniel Crafft', 63.

pursuit of the public good,[37] and over the years, they would collaborate on a number of schemes, including the extraction of ore, the manufacture of dye, the reform of coinage, and plans for a sugar distillery company. In mid-1680 they even jointly devised a mercantile policy for the emperor, recommending the establishment of manufactories across the territory (together with the introduction of import tariffs on French goods) with the double aim of enriching the empire and weakening French trade—a move Crafft later advised be adopted by all German princes at the diplomatic conference on reunifications, meeting at Frankfurt.[38]

The extensive correspondence between Leibniz and Crafft, spanning a wide array of topics, reveals a close and mutually supportive relationship in which Crafft for his part offered to look after Leibniz's interests and help him in any way he could.[39] Crafft eagerly reported on his own progress and activities to Leibniz, his successes, and the various challenges he faced, including during his trip to England in 1677.[40] He also divulged to Leibniz how he had been invited to join a scheme that produced gold from silver and had been granted a privilege by the King of France for the production of steel—a process coveted by Johann Friedrich.[41] Crafft requested that the latter provide Leibniz leave to travel to Amsterdam.[42]

Resourceful and talented in numerous fields, Crafft conjured up schemes for a cheap and improved way of manufacturing silk, for constructing a watermill inexpensively, and even—what he hoped would prove profitable for him—for a mysterious special chair from which 'a queen [would] rejoice with great benefit, since the hard and troublesome work can be done altogether unwittingly by other people'.[43]

---

37. Crafft to Leibniz, 25 July 1678, A III, 2:475.

38. Leibniz hoped that domestic goods, like silk and wool products, would eventually be exported out of Germany. O'Hara, *Leibniz's Correspondence* (2024), 166–67.

39. Crafft to Leibniz, 14/24 March 1673, A I, 1:413; Crafft to Leibniz, 6/16 May 1672, A I, 1:406: Crafft hoped to 'lend Leibniz a hand whenever he [could]'.

40. Crafft to Leibniz, 25 November 1672, A I, 1:406: 'In sum, my affairs are so well, according to all human appearances, that I hope to experience an inner peace and satisfaction before my end'; Leibniz to Crafft, mid-August 1677, A III, 2:210.

41. Crafft to Leibniz, 14/24 March 1673, A I, 1:414; Johann Christoph Crafft to Leibniz, 5 October 1675, A I, 1:426.

42. Crafft to Leibniz, 7 December 1677, A III, 2:289.

43. Crafft to Leibniz, 19 June 1677, A III, 2:163: 'I possess wonderful and compendious speculations regarding the introduction of manufactured goods which are quite infallible. I can now accomplish with a hundred what I could before with a thousand, in such a manner that I could reliably serve a prince, who would grasp their usefulness to his country, at minimal cost'; Crafft to Leibniz, 20 December 1672, A I, 1:408; Christoph Pratisius to Leibniz, 29 June 1677, A III, 2:465; Crafft to Leibniz, 6/16 May 1672, A I, 1:40fe6.

He was also a fount of information, reporting to Leibniz on book auctions and the outbreak of plague in Austria—even seeking to formulate a remedy against it—and bringing potential proposals and inventions, such as imitation porcelain produced in Normandy, to Leibniz's attention.[44] The two exchanged information about, among other things, the benefits presented by porcelain, experiments on longitude, a process to improve metal, a new liqueur, news from the court in Vienna, and Leibniz's operations in the Harz mountains. For his part, Leibniz congratulated Crafft on the status of his factories in Dresden, reported on his correspondence with Georg Schuller, and asked for more details about the process whereby lead was processed with mercury and sulfur.[45] They also compared notes on the progress of fellow projectors, such as Becher, and on their impressions of courtiers, such as the cameralist Philipp Wilhelm von Hornigk, whom Crafft described as honest and reliable.[46]

Leibniz found in Crafft an asset of great value who was keen to serve him and could be instrumentalized in a variety of different ways, as in forming a commercial partnership between Hanover and Saxony for wool manufacture in the late spring of 1677. Later, in 1680, Leibniz would rely on Crafft's mediation in his scheme to obtain a position as *Reichshofrat* at the imperial court.[47] But while Crafft seems to have been forthcoming with his information and remained eager to collaborate with Leibniz on his schemes and ventures, strengthening what he saw as a mutually beneficial partnership, their collaboration clearly ran up against limits as Leibniz occasionally chose to withhold some information from Crafft, such as his trip to Paris.[48] Upon hearing that his 'trusted friend' had reached Paris in 1672, Crafft suggested travelling himself to Paris, under the pretext of having been invited by Leibniz,[49] and with the hope of joining Leibniz in the construction of Crafft's new smelting furnaces, which Leibniz had previously promoted at the Académie for the French state.

44. Georg Hermann Schuller to Leibniz, 24 November/4 December 1677, A III, 2:287; Crafft to Leibniz, 9/19 September 1679, A III, 2:869: 'The reports from Vienna are bad . . . the contagion threatens to spread to other parts of Austria, even at court'; Crafft to Leibniz, 2:862; Crafft to Leibniz, beginning December 1678, A III, 2:544.

45. Leibniz to Crafft, early October 1678, A III, 2:508.

46. Crafft to Leibniz, A I, 1:415: 'Dr Cardilucius . . . is nothing to look at. . . . His medical reflections are fabulations. . . . In my judgement his work is bad'; Leibniz to Crafft, October 1679, A III, 2:878; Crafft to Leibniz, early December 16788, A III, 2:545.

47. See Smith, *The Business of Alchemy* (1994), 187.

48. Crafft to Leibniz, 4 July 1674, A I, 1:421: 'Wherever you may have the luck and opportunity to find yourself, we may perhaps be able to create some mutual benefit . . . for I can serve you better than anyone.'

49. Crafft to Leibniz, 20 December 1672, A I, 1:407.

It seems, however, that Leibniz was intent on 'enjoying the effects' of this invention at court on his own, especially at a time when he was himself hard at work on his calculating machine.[50]

Later in Hanover, when the duke requested his advice, Leibniz responded tepidly to Crafft's scheme for iron processing and manufacture.[51] Crafft sought the support of the Hanoverian court for a silk-winding venture in Haarlem, hoping that Leibniz would be permitted to visit the Netherlands, but to no avail.[52] Leibniz, evidently disinclined to share the limelight with anyone else or to serve as a conduit between Crafft and experimental philosophers in Paris, eventually advised Crafft to abandon his plans concerning the Hanoverian court.[53]

Leibniz was happy to advertise Crafft's merits—as in connection with his scheme in late 1678 for producing silk stockings at one-tenth of the usual cost—but only as long as this did not risk compromising Leibniz's own standing with the duke.[54] So while he was happy to take advantage of Crafft's help and information, their relationship could be rather one-sided, with Leibniz always seeking to maintain the upper hand. Although Crafft appears to have demonstrated unreserved devotion to Leibniz, Leibniz was more cautious and perhaps calculating, for example in response to Crafft's demand for payment (which Leibniz may have regarded as excessive) from the Hanoverian court for his dyeing technique.[55] Accordingly, Leibniz was willing to help Crafft only up to a certain point—and from a safe distance. He must have greeted the prospect of inviting Crafft to Hanover with some ambivalence, to say the least, for so brilliant a projector as Crafft was bound to detract, however unwittingly, from Leibniz's own lustre and perhaps destabilize the position he was trying so assiduously to create for himself in Hanover.

Crafft's unreserved zeal and his ingenuousness may have contributed to the numerous difficulties that plagued his career, and he often shared with Leibniz his frustration at the lack of funds necessary to implement

---

50. Moll, 'Deus sive harmonia universalis' (1982), 57.
51. Leibniz to Crafft, end of June 1677, A, III, 2:165.
52. Crafft to Leibniz, end of December 1677, A III, 2:303.
53. Leibniz to Crafft, mid-July 1677, A III, 2:190.
54. Leibniz to Gallois, 19 December 1678, A III, 2:571: 'There is another *curieux* versed all his life in the realm of manufactures and commerce, who wrote to me after having found and executed the art of making silk in such a convenient new way, that he assures to be able to produce ten for the price of one. I do not know if Mons. Colbert would give this some consideration at the present moment.'
55. Leibniz to Crafft, end of July 1677, A III, 2:165.

his designs.⁵⁶ These difficulties seem to have culminated in mid-1679, when he deplored the bad state of his chemical factories and his powerlessness, as well as the hostility of the guilds and his lack of trustworthy collaborators, since 'this profession generally seems to have the common misfortune that those who are otherwise honest are led astray when the project has a purpose'.⁵⁷ Crafft's financial difficulties were compounded by ill-health, in particular gout, for which Leibniz, who suffered from the disease himself, recommended a milk-based diet.

In exchange for the regular communication of valuable information and of reliable results conducive to the public good, Leibniz would continue supporting Crafft financially to the end of the latter's life, actively championing his friend, regularly lavishing him with praise, and emphasizing his knowledge, abilities, creativity, and integrity.⁵⁸ Still, as Crafft increasingly failed to hold up his end of the bargain, to Leibniz's growing irritation and rebuke, Leibniz complained about his greed and impaired judgement, which led Crafft to invest in dubious schemes. Some of his projects, including one for the production of brandy, left much to be desired on account either of their content or their association with shady characters.⁵⁹

Still, one technical project on which Leibniz and Crafft did collaborate was the production of phosphorus.⁶⁰ The history of its discovery in the 1670s involved several protagonists, including Hennig Brand, a self-styled physician from Hamburg, who isolated the element in 1669; the projector Joachim Becher; Crafft; and last but not least Leibniz, who in his correspondence at the time and in his *Historia inventionis Phosphori* of 1710 reported on the unfolding of events and later defended Brand against Crafft's and Kunckel's claims to authorship.⁶¹

This episode, rich in intrigue, is perhaps symptomatic of a particular kind of epistemological and social fluidity that ultimately left Brand largely cheated out of his discovery. A junior army officer during the Thirty Years' War, Brand had gleaned some chemical and technical knowledge that he

---

56. Crafft to Leibniz, 20 December 1672, A I, 1:407: 'I see no end to my travels. . . . I do not want to return home as I have not finished my task . . . but everything remains paralyzed due to a lack of money.'

57. Crafft to Leibniz, 6/16 June 1679, A III, 2:760.

58. Forberger, 'Johann Daniel Crafft', 70.

59. O'Hara, *Leibniz's Correspondence*, 568–70.

60. For more on the discovery of the phosphorus, see Krafft, 'Phosphorus' (1969). On attempts to instrumentalize the natural sciences for technical purposes, see Keller, *Knowledge and the Public Interest* (2015).

61. See *Misc. Berl.*

now deployed to extract large fees as an alchemist. Leibniz's impression of Brand was mixed:

> Dr. Brand does not have the capacity to judge what he is able to achieve, nor to make claims for himself. Not that he has often evoked imaginary or vain things—but, just like everyone, he also has his own character. Namely: he easily lets himself be patronised, has little power of judgement and leads an irregular lifestyle, but he is quick to act and very skilful in his work.... I often noticed him making a big racket about trifles, but he does not make much of things that warrant it. He looks for great mysteries and pipe dreams but he neglects the small discoveries he makes which could improve his life.... He is a man to launch twenty attempts in a week.[62]

Like many of his contemporaries, Brand had pursued alchemy, and he discovered phosphorus in 1669 after exposing the residue remaining in the distillation and crystallization of urine to very high temperatures in a tightly closed retort.[63] In 1676 he apprised the chemist and alchemist Johann Kunckel of this luminous substance, and Kunckel, sensing a wonderful commercial opportunity, in turn wrote enthusiastically to his colleague Crafft, who at the time was serving as trade adviser (*Handelsrat*) to the court of Saxony and living in Dresden. Negotiations with Brand began shortly after, Crafft quickly betraying Kunckel by turning directly to Brand behind Kunckel's back and buying the process and Brand's stock of phosphorus samples for two hundred Reichsthaler. Although Brand refused to share the process with him, Kunckel eventually succeeded in replicating it later that year and would, on this basis, portray the discovery of phosphorus as a largely personal achievement, first in his *Oeffentliche Zuschrifft von dem Phosphoro mirabili und dessen leuchtenden Wunder-Pilulen* (Public communication of the phosphorus mirabilis and its wonderful luminous pellets, 1678) and later the 'Historie vom Phosphoro und Rubin' in his posthumously published *Laboratorium Chymicum* (1716).

Crafft, for his part, set out to demonstrate this 'eternal fire' (*ignis perpetuus*) at various electoral courts, vaunting its possible alchemical powers and commercial potential and offering it up for sale. Leibniz drew Johann Friedrich's attention to this new chemical element as well as its possible commercial applications, and in the spring of 1677 Crafft stopped in

---

62. Quoted in Prinzler, 'Aus der Geschichte' (1993), 4.
63. Emsley, *Shocking History of Phosphorus* (2000), 405. Brand was one of a number of many investigators searching for and producing phosphorus of various kinds in the period. See Lawrence Principe, *The Transmutations of Chymistry* (2020).

Hanover, where he demonstrated phosphorus and its peculiar properties before the duke and other members of the court, including his old friend Leibniz. In mid-July Leibniz reported on this phenomenon to his friend l'Abbé de la Rocque in a letter entitled 'Le phosophore de Mr Krafft ou liqueur et terre seiche de sa composition qui jettent continuellent de grands eclats de lumiere'. During his demonstration in Hanover, Leibniz reported, Crafft had presented two vials.[64] One contained a 'liqueur that glows at night much like glowworms . . . which at night produces rather beautiful effects': 'If one rubs the face, the hands, and the clothes with the liquid, it lights them all up . . . [and] the clothes are not ruined by it.'[65] The other vial contained the same matter but in a dry form that seemed to sparkle 'without having been exposed to sunlight'. One large piece of this phosphorus would undoubtedly suffice to light up an entire room, according to Leibniz, even if it remained difficult to produce one of considerable size.[66] In September 1677 Leibniz marvelled at phosphorus's intense luminosity to the duc de Chevreuse, Colbert's son-in-law: 'I was recently presented with a truly extraordinary species of phosphorus or glowing matter. We know that Lapis Boniensis, exposed to strong light, receives it and preserves it in the dark for some time, but this is altogether something else. Phosphorus does not require to be exposed to any outside light since it preserves it in itself.'[67]

In both letters to de la Rocque and Chevreuse, Leibniz seemed to suggest that the perennial search for 'perpetual light' (*lumière perpétuelle*) may have come to an end, for this matter's incandescence seemed boundless inasmuch as it could 'last at least several years'.[68] In one particular chunk sent to Oldenburg in London and examined through a piece of glass, it was even possible to discern 'various images of the sun, each one bigger than the next'.[69] Phosphorus in itself remained a mysterious substance whose 'considerable qualities would perhaps one day be elucidated'. It was surely animated by a 'true flame only visible in the dark but otherwise not strong enough for touch to discern it.'[70] Neither its toxicity nor its great self-ignitability were then known, even though Brand already suspected these, as he later reported to Leibniz.[71]

64. Letter from Hansen and Justel for De La Rocque, mid-July 1677, A III, 2:191.
65. From *Journal des Sçavans*, 2 August 1677, 244–46, quoted in Smith, *The Business of Alchemy* (1994), 248.
66. Leibniz to Chevreuse, mid-July 1677, A III, 2:192.
67. Leibniz to Chevreuse, September 1677, A III, 2:232.
68. Leibniz to Chevreuse, 2:232; mid-July 1677, A III, 2:191.
69. Leibniz to Chevreuse, 2:192.
70. Leibniz to Chevreuse, September 1677, A III, 2:232.
71. Brand to Leibniz, 30 April 1679, A III, 2:731: 'When I had the fire in my hand . . .

Such a marvellous invention would yield the most fantastical (and entertaining) applications, including 'fire fountains' and 'fire numbers and figures' capable of illuminating the night sky; but its dissemination and the divulgation of its process were also a matter of general interest.[72] The discovery of this 'perpetual light' had, however, come at considerable expense, and its secret was now preciously guarded 'as a matter of consequence'. Still, a reward, by a monarch for instance, might entice the inventor to reveal the method, which many had sought to know, of producing what he had invented.[73] Such a ruler could in fact be none other than Duke Johann Friedrich, who, having been convinced by Leibniz of the importance of phosphorus, ordered his official immediately to begin negotiating with Brand for its acquisition.

A formal contract was drawn up and signed on 14 July 1678 between Brand and Leibniz, who acted on behalf of the duke. Brand committed himself to communicate the composition of phosphorus along with other curiosities he was privy to and to 'travel to Hanover when called upon, cultivate a diligent correspondence ... and work toward the perfection of his fire'. Leibniz, on the court's behalf, promised to pay Brand 10 thalers each month and an additional 120 thalers per year as a commission. Further payments would be made upon the communication of 'other curiosities' that Brand may come to know—a reference, in all likelihood, to Brand's claim that he could transmute silver into gold.[74] Leibniz's haste in acquiring the process was also a way of preempting the projector Johann Joachim Becher's rivalry in acquiring the process for his own employer, the Duke of Mecklenburg-Gustrow.

Even after the conclusion of the contract, Leibniz repeatedly expressed his concerns to the duke that Becher would still attempt to lure Brand into the service of his employer.[75] With a little bit of money, as he knew firsthand, Brand could be 'induced to anything'. To prevent any possible attempt to influence Brand, Leibniz resorted to all the means at his disposal, not only offering Brand more money and promising him to head a large-scale production, but also commissioning the theft of a cache of letters from Becher's apartment.[76] In November 1678 Becher left Hamburg

---

[it] lit up God help me ... [leaving] my skin burn[t] ... [and] my children screaming, so terrible it was to look at'.

72. Leibniz to Chevreuse, September 1677, A III, 2:233; mid-July 1677, A III, 2:192.
73. Leibniz to Chevreuse, September 1677, A III, 2:233.
74. Leibniz to Brand, 14 July 1678, A III, 2:474.
75. Leibniz to Johann Friedrich, 31 July/10 August 1678, A I, 2:63: 'The only problem is that Brand, because he is in bad straits, will sell it for a small amount to the first bidder.'
76. Leibniz to Johann Friedrich, 2:63-68.

for Holland, where he attempted unsuccessfully to sell his project of an 'eternal mine' (that is, the programmed extraction of gold from sand dunes) to the Dutch government. For the slight he had suffered at the hands of Leibniz, Becher would try to exact revenge through ridicule, satirizing Leibniz's ideas for the possible improvements to wagons by spreading the rumour, in his *Foolish Wisdom and Wise Foolishness* (*Närrische Weißheit und weise Narrheit*) of 1682, that Leibniz wanted to build a special type of carriage capable of covering the Amsterdam-Hanover route in six hours.[77]

After the contract was concluded, Brand, now 'well ensnared',[78] came to Hanover, where he worked on producing a larger amount of phosphorus over the course of five weeks. During that time, Leibniz learned the exact manufacturing process, so that he could now replicate it.[79] In the late summer of 1679, Brand returned to Hanover to produce a larger quantity of phosphorus, and for nearly eight weeks he worked with his stepson on extracting between 1 and 1.5 kilos of phosphorous from more than a hundred tons of urine collected from army latrines. Unbeknownst to Brand, and despite the various promises that had been made to him, Leibniz was all this time betraying his trust. Even prior to the conclusion of the contract, Leibniz had set out in confidential letters to the duke how to get hold of Brand's secrets—ideally including the method for producing gold—while persuading him to work for the court for the least amount of money and in the most efficient manner.[80]

An overworked and underpaid Brand, however, did not remain in Hanover long, even breaking off his stay to renegotiate the terms of his contract and demand more money. Already after his first trip to Hanover, he had complained bitterly to Leibniz and Crafft about his low pay, not least because he had a large family to provide for, and had threatened to accept Becher's generous offer, despite Crafft's requests that he avoid Becher. In desperation, Brand tried unsuccessfully to blackmail Crafft and threatened to take drastic measures, especially as he knew that Crafft

77. Becher, *Närrische Weißheit* (1682), 147: 'This Leibniz is known by his works, [he is] a very learned man ... [who] has written his own philosophy and other things. But I do not know who set him on this post-wagon, from which he does not want to dismount, even though he has sat on it for several years, and sees that it does not depart.' Cf. A I, 2:76, 126.

78. Leibniz to Johann Friedrich, mid-September 1678, A I, 2:68.

79. Leibniz to Johann Friedrich, 2:69: 'The urine of soldiers standing in the camp was collected in pens, and when there was a sufficient supply, Brand came to us and performed the display outside the city. Everything that he did himself, I did with my people in other laboratories.'

80. Prinzler, 'Aus der Geschichte', 10.

'could help [him]' but chose 'not to'. If Crafft failed to behave 'honestly' towards Brand, he would regret his actions and a 'sombre horizon would lie ahead' he warned. Leibniz placidly appeased Brand by referring to the terms of the contract and promising further payments. At the end of 1679, however, the Duke of Brunswick died and Brand returned, unpaid, to Hamburg. He later endeavoured to dig up a treasure in Schipbeek, but this endeavour too ended in failure, and he was even beaten up by the disappointed soldiers and spectators before vanishing from the horizon. He probably died in 1692.[81]

Crafft, meanwhile, had entrusted a sample of phosphorus to Robert Boyle, who had long been fascinated with the substance and would manage to replicate its production before going on to lay the foundations of phosphorus chemistry (which he detailed in his 1680 publication *The Aerial Noctiluca or New Phenomena and a Process of a Factual Self-shining Substance*) and to produce the first commercial phosphorus.

Holding up a Mirror to Himself: Leibniz and Johann Joachim Becher

Inquiries about and references to Johann Joachim Becher (1635–82) pepper Leibniz's correspondence in such a manner that it would be remiss not to examine his relationship, or more accurately his fascination, with Becher. A pioneering cameralist, Becher was also an alchemist, a fellow courtier and projector.[82] Ten years older than Leibniz, he too had been a member of the Mainz court and had had Boineburg and Schönborn as patrons. Aside from Herbert Breger's and Pamela Smith's studies, which constitute excellent forays into the matter, little attention has been devoted to the relationship between Becher and Leibniz.[83] Yet from 1669 to 1679 the figure of Becher looms large in Leibniz's correspondence, appearing dozens of times in letters to a wide range of recipients. Leibniz followed Becher's career with great interest and not without a certain amount of admiration and envy.[84] While they certainly hailed from

81. Prinzler, 10.
82. Freudenberger, *State and Society in Early Modern Austria* (1994), 141–42. For more on Becher, see Lorber, *Theatrum Naturae & Artis* (2017); Smith, *The Business of Alchemy* (1994).
83. See Breger, 'Becher, Leibniz' (2016); and the valuable juxtaposition of Becher and Leibniz in Smith, *The Business of Alchemy*.
84. Bredekamp, *Die Fenster der Monade* (2004), 147, notes that Leibniz perceived Becher as a problematic personality but nonetheless admired his ingenuity and courage.

different backgrounds, and Becher's vision remained more overtly commercial than Leibniz's, the two were more similar than either would have liked to concede.[85]

A polymath, Becher published on a wide array of topics, including politics, economics, chemistry, alchemy, moral philosophy, ethics, and even social utopia. Like Leibniz, he had moved in and out of different worlds, scientific, artisanal, and commercial, in his attempt to fuse theory with practice. From 1657 onward he was engaged as a court physician, mathematician, and commercial adviser in Mainz, Bavaria, and Vienna, respectively, while travelling widely throughout Europe.[86] Crucially, both Becher and Leibniz believed in the power of philosophers to make knowledge yield practical applications and direct material progress, helping transmute raw materials into new sources of revenue for their patrons and thereby providing a solution to the economic difficulties that had befallen the German territories in the wake of the Thirty Years' War.[87]

Expert at mediating between the courtly and commercial worlds and exploiting the credit he had previously gained as an alchemist and physician, Becher established himself as a trusted economic adviser.[88] Both Leibniz and Becher—whose realms of expertise were complementary—set out to capture the knowledge and methods of artisans and bring them within the sphere of the state, in this manner insinuating themselves as irreplaceable intermediaries between artisans and the court.[89]

Alchemy in particular became for Becher the perfect metaphor for the regenerative power of art in the material world, 'the vehicle by which he spoke at the court of production and material growth' and the 'model of civic negotia and manufacture', and he enhanced his credit by transmuting lead into silver at the Viennese court.[90] Both, too, used their scientific expertise to land court appointments from which they sought to carry out their programmes.

---

85. Commenting on their relationship, the German historian Wilhelm Dilthey, 'Leibniz und sein Zeitalter' (1926), 24–25, writes, 'No one among Leibniz's contemporaries took the new ideal of a universal culture, in which this great century lived, more deeply to heart than Johann Joachim Becher.... Political and scientific arrangements finally coincide for him, as in the great utopias of all times. His ideas and plans went as far as Leibniz's ever did. And he too devoted his life to their realization.'

86. See Lazardig, '"Masque der Possibilität"' (2006); and Teich, 'Interdisciplinarity' (1988).

87. Cf. Smith, *The Business of Alchemy*, 156, 9, 265.

88. Cf. Smith, 6.

89. Cf. Smith.

90. Smith, 241, 244. This occurred in a context in which cameralism and court alchemy seem to have been 'coproducing' each other. See Keller, '"A Political *Fiat Lux*"' (2016).

Leibniz was intrigued by Becher, whose career he closely followed and sought to emulate in several respects, even on occasion competing for the services of the same artisans. Like Leibniz, Becher strove to create a space for himself and his ideas within the rigid structures and codes of courtly society and to project his own plan onto the world around him.[91] At the court of Mainz, he had, as the Elector's personal physician, refined his knowledge of the natural and technical sciences and begun working at the intersection of natural philosophy, economic policy and mercantilism at a time when monarchical absolutism and the mercantile system were merging.[92] As one of the leading German political economists of his time, he was guided by what seemed to be a genuine concern for the welfare of people and the flourishing of the peasant and artisan classes, and he pioneered a set of audacious economic policies, such as a policy for repopulation in the wake of the Thirty Years' War, proposals for fair taxation including on previously untaxed commercial products, a 'luxury tax', and an appropriate salary pension fund—many of which may have influenced Leibniz, and which brought Becher into conflict with the guilds on more than one occasion.[93]

Becher's activities were wide-ranging: in 1669, for instance, an enthusiastic Prince Casimir of Hanau commissioned him to found a colony in Guyana for the West India Company. 'Regenerative' commerce would act like a 'magic wand' through which it would be possible to produce 'everything divine', as Leibniz commented, not without a mix of sarcasm and incredulity.[94]

Back in Munich, Becher opened a large alchemical laboratory, supported by the Bavarian government, before entering the service of Emperor Leopold I in 1670 as economic and trade adviser. More generally, he proposed economic policies for the principalities that constituted the Holy Roman Empire as well as measures against the import of French goods into the Free Imperial Cities, and he oversaw the establishment of an agency for the promotion of commerce and industry.[95]

In Vienna in late 1672 he founded a *Kunst- und Werkhaus*, which he envisaged as a learning and production centre where he hoped to train

---

91. Smith, 10.

92. See Hassinger, *Johann Joachim Becher, 1635–1682* (1951) and 'Becher, Johann Joachim' (1953).

93. Hassinger, *Johann Joachim Becher*, 123; Nipperdey, "'Intelligenz' und 'Staatsbrille'" (2008).

94. Smith, *The Business of Alchemy*, 171.

95. Freudenberger, *State and Society in Early Modern Austria*, 141.

vagrants and orphans in various crafts, including the arts and trades of porcelain, the weaving of silk and wool, the preparation of chemical products (including medicines and dyes), the manufacture of glass, and many other practical processes. The factory building eventually burned down during the Turkish war in 1683 and was never rebuilt.[96]

Leibniz's engagement with Becher, who had left the Mainz court only a few years before Leibniz's own service, started early and not uncritically. Already in his *Dissertatio de arte combinatoria* of 1666, Leibniz had rejected Becher's project for a universal language as impractical—something Becher would later dismiss in his own *Psychosophia* by stating that Leibniz had 'neither read nor understood' his work—and had read his *Methodus Didactica* of 1668 with a mixture of approval and criticism.[97] On the other hand, he was lastlingly inspired by and sought to improve on Becher's 'Theatre of Nature', which attempted to link words with physical objects.[98]

In 1669 Leibniz had mentioned Becher in the preface of his *Societas Philadelphica*, which resembled Becher's own project for a philosophical community.[99] Consulting various experts on the subject, Leibniz sought their opinion of Becher and inquired after the success of the manufactures he had set up in Bavaria.[100] His first, gently mocking reference to Becher in his correspondence appears in a letter of September 1669 to Jakob Thomasius, his former teacher, where he describes Becher as 'a clever but meddlesome man' (*homo ingeniosus, sed Πολυπραγμων*).[101] Thomasius had preferred for his part to reserve judgement until after 'he had seen [Becher] and had had space to mull things over',[102] but Leibniz's curiosity had been piqued by the 'fantasy' Becher had conjured up for the Count of Hanau. His own attitude toward Becher was markedly ambivalent. To his friend Johann Graevius, Leibniz described Becher, whom he had just met in Mainz, as 'a man thoroughly polished in every kind of select inquisitiveness of exquisite learning' whose work should be followed closely.[103] While numbering among the 'very few men who instructed Germany' and 'from

---

96. Smith, *The Business of Alchemy*, 190–1.
97. Leibniz, 'De arte combinatoria', 1666, A VI, 1:201; Becher, *Psychosophia* (1678), 382; Bredekamp, *Die Fenster der Monade*, 43.
98. Bredekamp, 43.
99. Leibniz, 'Societas Philadelphica', 1669, A IV, 1:552.
100. Leibniz to Oldenburg, 28 September 1670, A II, 1:102.
101. Leibniz to Jakob Thomasius, September 1669, A II, 1:41.
102. Jakob Thomasius to Leibniz, 22 November/2 December 1669, A II, 1:43.
103. Leibniz to Graevius, 7 June 1671, A I, 1:154.

whom great things might be expected', Becher also 'attracted resentment everywhere'.¹⁰⁴

Another manifestation of Leibniz's preoccupation was his attempt to probe the validity of Becher's method, laid out in the first book of his *Subterranea Physica*, for extracting metal from limestone by which 'nature as well as art had been able to make one reign pass into the next'.¹⁰⁵ While Leibniz seemed won over to some extent by Becher's new method, Oldenburg reserved his judgement pending closer examination.¹⁰⁶

While it was imperative to obtain as much information as possible on Becher's plans and activities, Leibniz was intent on keeping him in the dark about his own. Becher was not to be trusted, especially when it came to Leibniz's prospects at the Viennese court for the position of *Reichshofrat*.¹⁰⁷ Leibniz's public disdain was actually more nuanced in private, and it is clear that he perceived Becher as a potential rival whose efforts needed containing. On one occasion at least, Leibniz is likely to have sought to dismiss Becher's desire to present some suggestions on trade and manufacture at the Hanoverian court on the recommendation of the Count Palatine of Sulzbach.¹⁰⁸

Leibniz kept abreast of Becher's whereabouts and speculated as to his plans and possible chances of success, though at times he had to concede an absence of knowledge. Crafft regularly updated Leibniz with news about Becher, with whom he shared quarters for a period, and his activities—generally at Leibniz's request.¹⁰⁹ When it came to Becher, nothing was unimportant, and Crafft would act as a key informant on his activities, confiding to Leibniz that Becher had scammed a monk in the process of making tincture,¹¹⁰ or disclosing the latter's project for obtaining a million Reichsthaler credit for the emperor as early as 1671:

> I have some news about Dr. Becher's affairs in Holland.... He produced a million Reichsthaler for His Majesty, in such a way that His

---

104. Leibniz to Oldenburg, 9 May 1671, A II, I:169.
105. Leibniz to Otto Tachenius, 4 May 1671, A II, 1:101.
106. Oldenburg to Leibniz, 8 October 1671, A II, 1:256.
107. Leibniz to Lincker, 7 June 1671, A I, 1:392: 'I request that you recommend me to Hocher ... [but] Becher and Lambeck ... should avoid knowing this. Please write if they do anything new and remarkable. Crafft may be in the know only if warned not to share this.' See also Crafft to Leibniz, 24 May/2 June 1677, A I, 2:272.
108. Writing notably that 'one does not rush into such things here, especially since the time is not right either' (A I, 2:319).
109. Crafft to Elers, 20 November 1677, A III, 2:271: Crafft reported to Martin Elers that he 'lived in close friendship with Becher'.
110. Crafft to Leibniz, 20 December 1672, A I, 1:407.

Majesty must pay 5 percent interest for forty years, after which time the capital will be his.... I am still of the opinion that this million will be coined in the kingdom [in America] of the Count of Hanau. Or perhaps the need for money or his greed drove [Becher] to contrive such a work so he would be sent to Hamburg and Cologne... because he knows that besides a new deceit, he will receive a comfortable allowance and respect from both places in carrying out his duties.[111]

To be sure, this reporting on Becher throughout the 1670s did not prevent Crafft from collaborating with him on a number of projects, including an instrument in Haarlem that wound spools of wool, or admitting that he sought to turn to his own advantage the good credit Becher enjoyed at the imperial court.[112] Crafft even discerned opportunities for himself in Becher's Viennese workshop:

> From Dresden I travel to Vienna to sort out the steel and glass furnaces erected by Dr Becher along with other things, from which several thousand yearly thalers are to be expected. Above all else, Dr Becher is in very good standing with His Majesty, and, because of certain uncommon suggestions that will bring in millions, so well loved and in good credit that my planned journey will prove most beneficial to me.... Aside from this... another good friend... has invited me to [code: produce gold] from [code: silver], with [code: great profit]. I have reason to believe that the process is genuine and will send more about it my very learned Sir.[113]

Overall, although his correspondents' opinion of Becher was overwhelmingly negative—with the notable exception of the Jesuit priest Adam Kochanski—Leibniz remained remarkably balanced in his appraisal of Becher, doubtful but open-minded, even if he perceived Becher as rather gullible and superficial.[114] Leibniz's Paris trip, however, marked an interruption in his preoccupation with Becher, and in later years the 'ingenious' inventor lost some of his lustre in Leibniz's eyes, especially as the latter increasingly grew to consider the former a potential rival for the post of *Reichshofsrat* at the Viennese court.

---

111. Crafft to Leibniz, 3/13 October 1671, A I, 1:224–25.
112. Smith, *The Business of Alchemy*, 186; Crafft to Leibniz, 14/24 March 1673, A I, 1:414.
113. Crafft to Leibniz, 1:413–14.
114. Breger, 'Becher, Leibniz', 35.

Becher struck Leibniz as an ingenious man, albeit one who often promoted others' ideas uncritically. He displayed 'occasional cockiness, wastefulness, vanity' and had a knack for antagonizing many on account of either his boastful character and arrogance or the failure of some of his projects—including a nonfunctioning machine at the Palatinate court.[115] Leibniz had heard 'contradictory judgements' about Becher: 'those who were aware of his widely known shenanigans did not attach any value to him', while 'others held him as a great philosopher'.[116]

Still, Leibniz's interest in Becher continued unabated, even after the latter had left the Viennese court. Before getting wind of Brand's secret and subsequently trying to recruit him, Becher had left for the Netherlands, where, with Crafft and others, he formed in February 1678 a company to operate the instrument he had invented to wind spools of raw silk.[117] In the same year, Becher undertook a new venture involving the extraction of gold from the seashore near Scheveningen. This project attracted attention and fuelled much speculation from his contemporaries, who then relayed the information back to Leibniz. Before he could even carry out a trial, however, Becher left for Hamburg, where he came into direct competition with Leibniz for Brand's knowledge (and for Brand himself) at the end of July 1678.[118] In mid-June 1679, after learning that Becher's recent process for extracting gold from sand had succeeded in a small-scale trial, Leibniz wrote to Duke Johann Friedrich in the hope of being able to replicate Becher's process:

> I draw my Lord's attention to another matter: a man whom I think I already mentioned to His Majesty has offered to draw gold from sand. He advances that through this method there will be almost as much gain, all costs included, as in the mines of Hungary: his claims were disputed but I now learn from several letters that an experiment has succeeded in such a manner to give hope for greater successes. If this is indeed the case I will reveal to His Majesty something I dared not yet raise since I doubted its veracity; it is that I know this person's secret at least as it was when he submitted his proposal last year to the gentlemen of the Estates; if he has discovered something new since then I do not know nor do I believe it. One of his friends, who also happens

---

115. *Misc. Berl.*, 105; Roscher *Die österreichische Nationalökonomik unter Kaiser Leopold I* (1864), 38–59.
116. Prinzler, 'Aus der Geschichte', 9.
117. Smith, *The Business of Alchemy*, 190.
118. Smith, 190.

to be mine, recopied it word for word and what he has copied does not amount to a mere recipe or process but a detailed account of experiments already carried out, as meticulously recorded by this person over the course of his work in order to memorize the materials, proportions and procedures. This has encouraged me to try and attempt the experiment one day especially since His Majesty disposes of a great advantage since his country offers the elements most necessary for this operation.[119]

Leibniz even discussed Becher with his mentor Christiaan Huygens in Paris in September 1679, both parties expressing their cautious scepticism:

> You will no doubt have heard about Mr. Becher's endeavour in Holland to extract gold from sand. Some people seem to have been won over to this idea. . . . Mr. Becher says he is also in negotiation with the French. I would be grateful to know if you have heard anything about this in Paris. On my part I doubt this venture's success for I believe to know more or less what his experiment consists of. There is a trace of gold: but I do not know if there is enough to produce anything.[120]

Having himself inquired about Becher's experiments and ascertained from his brother-in-law Philips Doublet that even his projects' most enthusiastic supporters were starting to lose faith, Huygens wrote back a few weeks later: 'I would like to know if there is indeed gold hidden in the sand of our dunes, as implausible as this strikes me, irrespective of what chemists and even our Mr. du Clos. advance.'[121] After reading his *Trifolium Hollandicum*,[122] Leibniz had become aware of one very evident flaw in Becher's process for extracting gold from sand, as he reported in his next letter to Huygens in December 1679. Not only did Becher's essays fail to establish the validity of his proposition, namely, that gold could indeed be extracted out of sand by melting silver with sand, but they also failed to show that the operation could be repeated with the same silver. Failing this, 'all the silver of Europe would have to pass through his furnace before he could gain the million per year that he promises'.[123]

---

119. Leibniz to Johann Friedrich, mid-June 1679, A I, 2:176–78.
120. Leibniz to Christiaan Huygens, 8 September 1679, A III, 2:849.
121. Huygens to Leibniz, 22 November 1679, A III, 2:888. See Doublet to Huygens, 5 October 1679, in Huygens, *Oeuvres complètes* (1888–1955), 8:231–34, esp. 233.
122. Becher, *Trifolium Becherianum Hollandicum* (1679).
123. Christiaan Huygens to Leibniz, end of November/early December 1679, A III, 2:902.

Another source of information on Becher was Christian Philipp, the Elector of Saxony's ambassador in Hamburg, who had played an instrumental role in helping to recruit Brand, the 'real inventor of phosphorus' who had yet to achieve any public recognition comparable to Crafft's or even Becher's.[124] Philipp was particularly unforgiving toward Becher, denigrating his projects (and in particular his venture for mining gold from sea sand) as mere 'chimeras' (*chimères*), and his *Psychosophie*—a copy of which Philipp had provided Leibniz at his request—as the 'most confused book in the world'.[125] According to Philipp, the 'Illustrious' Mr Becher's writings were full of 'nonsense' and 'boasts' (*rodomontades*), and his words 'subject to caution'.[126] Whereas Leibniz took some interest in Becher's ventures and publications, including his *Psychosophie* and his *Politischer Diskurs*, and even praised his medical knowledge,[127] Philipp remained thoroughly unimpressed and did not pull any punches, noting that he would only ever agree to being Becher's patient if he had grown tired of life:

> A leopard doesn't change its spots: all Becher's ventures are so many paradoxes.... His quarrel against reason shows that he doesn't have much of it.... I would not want in the world to fall into his hands for anything, if I am ill, except if I had grown tired of living: what he states about his own pills on page 303 smacks of pure charlatanism: I believe that the money that he received to extract gold from sand, is not the first in his life that he has received in this manner.[128]

Surprisingly perhaps, Leibniz still did not dismiss Becher, instead asking Philipp to relay back to him further details about Becher's proposals, which 'would no doubt not fail to be arduous'.[129] Philipp in turn continued to express his doubts as to the success of Becher's latest gold-extraction venture, noting that it remained inconclusive.[130] He even questioned the authenticity of Becher's claims, observing tartly that it 'would be impossible

124. Christian Philipp to Leibniz, 16/26 November 1678, A I, 2:387. On Leibniz's defence of Brand's claim to have discovered phosphorus, see the section 'Johann Daniel Crafft and the Affair of the Phosphorus' earlier in this chapter.
125. Christian Philipp to Leibniz, 2:387.
126. Christian Philipp to Leibniz, 26 October/5 November 1678, A I, 2:378.
127. Leibniz to Christian Philipp, 22 April/2 May 1679, A I, 2:469: 'I would like to know in what state lies the affair of the melting of sands (*fonte des sables*). Because after receiving the money for it, he cannot fail to complete it without damaging his reputation.'
128. Christian Philipp, 26 April/6 May 1679, A I, 2:472.
129. Leibniz to Philipp, 30 November/10 December 1678, A I, 2:385.
130. 'I am informed from Holland that Mr. Becher has tested his proposal on sand, and that many are optimistic as to its success. For my part, I find it hard to believe that any good can come out of it.' Christian Philipp to Leibniz, 24 September/4 October 1679,

for him to return all the money amassed through these kinds of promises'.[131] After Becher had once again fallen out of view, it was Christian Philipp who informed Leibniz in early 1680 that 'the affairs of Mr. Becher in Holland ha[d] ended like all his previous undertakings, and they are yet to see a single effect of his grand promises'.[132] A few months later, he could further inform Leibniz that Becher, 'notwithstanding the cunning spirit of the Dutch, ha[d] succeeded in extracting a good sum of money' before disappearing and leaving his investors seething. He had resurfaced in England, where he had 'found his dupes as always' but had begun to 'lose all his credit' after the failure of many of his projects.

Years later, Leibniz remembered Becher as an 'ingenuous man and knowledgeable in German writings, but who wrote too much, and discoursed on known and unknown topics with equal ease'.[133] And yet Leibniz was more similar to Becher than he would probably have liked to admit. For both sought to validate their claims to superior expertise through the material production of various instruments and the realization of various technical schemes, many of which actually failed. Crucially, both were liminal individuals with an interest in natural philosophy who sought to instrumentalize a state of pervasive doubt and epistemological uncertainty to explore the 'realm of possibilities', at a time when a more experimental approach prevailed over any static concept of 'truth'.[134]

In this context, only experience and successful reduction to practice showed that the seemingly impossible could be achieved, and the gap between theory and practice often remained irrepressible. Becher's volume *Foolish Wisdom and Wise Foolishness* paired a list of projects that seemed foolish or impossible but nevertheless succeeded in practice with others that appeared rational but failed. These included Russian-Chinese trade, Henry IV's introduction of silk production in France, machines for hosiery, a perpetual mobile, lighter-than-air bullets, and even the use of

---

A I, 2:519: 'You will undoubtedly have read in the gazettes that Mr. Becher demonstrated his transmutation into gold [*chrysopoeia*]; some say it is successful, but others disagree.'

131. Christian Philipp to Leibniz, 26 April/6 May 1679, A I, 2:472; 17 May 1679, A I, 2:478: 'Mr. Glauber wrote a long time ago of how to extract gold from sand, and a man who is versed in these sorts of things, told me that surely there is gold in almost all kinds of sand and pebbles, but that the costs incurred were higher than the profit yielded, especially since the damage is much greater when the experiment is carried out on a large quantity than on only a little sand: it is therefore possible that the test, which Mr. Becher has done, will be disproved on a large scale.'

132. Christian Philipp to Leibniz, 26 April/6 May 1679, A I, 2:472.

133. *Misc. Ber.*, 102–8.

134. See Smith, *Business of Alchemy* (1994); and Keller, *Knowledge and Public Interest* (2015).

elephants in agriculture. This pairing, in which Becher listed six of his own failed inventions, served to emphasize the hazy limits of possibility and capture the fundamental uncertainty and ambiguity of project-making. Far from mocking his fellow projectors' projects, Becher's book was an account of the complexity of conditions necessary to bring about a project, ranging from the lack of patent law to court intrigues. The implementation of technical and economic projects was often contingent on external and unpredictable conditions, and even a well-conceived project was unlikely to be brought into practice due to '*wunderliche Conjuncturen*'.[135]

Placing themselves on a fault-line, Leibniz and Becher blurred the distinction between fact and fiction, sincerity and artifice, cultivating an interesting confusion of genres that contributed to the poetic of ambiguity characteristic of their age. And although Leibniz increasingly disparaged and mocked Becher over the years, his preoccupation with the projector betrayed the parallels between them. As Herbert Breger has pointed out, Becher fascinated his contemporaries in personifying the reigning atmosphere of both relentless curiosity and suspicion, intrigue and rumour, and above all of unexpected surprises but also dashed hopes.[136] In obsessively tracking Becher's every movement across Europe, Leibniz was also holding up a mirror to himself, trying to clarify his own projects and self-perception. By means of another perhaps, he sought to gain a better understanding of himself.

Leibniz's Double Standards

While Leibniz felt that his own projects should be accepted at face value, he regarded his peers' ventures rather differently. For Leibniz, most projectors were charlatans who were not to be trusted, and he remained particularly anxious not to be exploited or taken advantage of by them. His correspondence is particularly instructive in this regard, testifying to radically different perspectives and understandings, for Leibniz's fellow projectors rarely shared his view that he was treating them fairly. He repeatedly showed himself prepared to drive a hard bargain with them, holding their projects to a higher standard than he applied to his own.

One potential project was that advanced by Georg Schuller, Spinoza's physician and confidant, whom Leibniz had met through the mediation of Tschirnhaus during his stay in Amsterdam in 1676 en route to Hanover.

---

135. Becher, *Närrische Weißheit*, 179.
136. See Breger, 'Becher, Leibniz', 35.

After returning to Germany, Leibniz would consult Schuller for information relating to the preparation of Spinoza's *Ethica and Opera Posthuma* as well as about procuring books from Amsterdam.[137] They also discussed a wide range of topics, in particular relating to metallurgy and its transmutations and Becher's latest activities.[138] Moreover, Schuller helped disseminate Leibniz's *Entretien de Philarète et d'Eugène* among the envoys at the peace congress in Nijmegen and offered to collect and send information on various individuals' activities, such as Peter Hartzing.[139]

In mid-1678 Schuller offered to sell a process for gold transmutation that he assured could be completed in three weeks and would provide enough revenue 'for a lifetime'.[140] Leibniz, however, remained sceptical on account of Schuller's vagueness (as he reported to Crafft) and requested more details.[141] He seems to have remained adamant on insisting for the detailed description of the process, including the dosage and number of experiments involved, as well as the quantity produced and wasted, before proceeding to any payment.[142] Despite Schuller's mounting frustration and protestations that the process had successfully been carried out 'without any magic or deceit', both Leibniz and Crafft seem to have lost confidence in its validity.[143] Considering the lack of details and funds, as well as the dubious prospective profitability of the venture, Leibniz finally declined participation in preference for Brand's process.[144]

Another projector to find himself on the receiving end of Leibniz's implacable negotiating technique was the French engineer Noel Douceur, who had invented a method for increasing the malleability of cast iron, a procedure with obvious military benefits, as Leibniz was quick to discern.[145] Leibniz thus set out to acquire this secret on behalf of Duke Johann Friedrich, initiating a lengthy negotiation that was conducted

---

137. Leibniz to Schuller, 19 November 1677, A III, 2:267. See Totaro, 'Manuscript of Spinoza's Ethics' (2013), 465–76.

138. Schuller to Leibniz, 19/29 March 1678, A III, 2:150.

139. Leibniz to Schuller, 18 October 1677, A III, 2:249; Schuller to Leibniz, 27 August / 6 September 1678, A III, 2:487.

140. Schuller to Leibniz, 15 October 1678, A III, 2:522.

141. Leibniz to Crafft, 10 November 1678, A III, 2:535; Leibniz to Schuller, mid-November 1678: A III, 2:538; Leibniz to Schuller, beginning of November 1678, A III, 2:532.

142. Leibniz to Schuller, 26 November 1678, A III, 2:552.

143. Schuller to Leibniz, 21 January 1679, A III, 2:542–613; Crafft to Leibniz, beginning of December 1678: A III, 2:542.

144. Leibniz to Schuller, mid-January 1679, A III, 2:585, 598; Leibniz to Crafft, mid-July 1677, A III, 2:190.

145. Leibniz to Mariotte, 18 March 1678, A III, 2:358.

partly in Paris by Edme Mariotte and Brosseau, who waited for Leibniz to assess Douceur's claim so that they could proceed with the rest of the payment.[146] Testing, however, proved close to impossible in Germany, and although Douceur had communicated his secret, the rest of the payment was not forthcoming, prompting the French inventor, who was well aware of the value of this discovery, to request it repeatedly: 'I have sufficient hope to believe that you will have enough kindness, to reward me, if by means of this secret, I offer you the opportunity to produce considerable revenue outside France by obtaining a privilege from the prince, similar to the one that I hope to obtain soon from the King of France.'[147] Douceur assured Leibniz of the number and quality of 'people of honour' with whom he had shared the knowledge of his process or who had witnessed it first-hand, and he complained that he was 'greatly astonished' at the way he had been treated by Leibniz and at some of the 'impossible' requirements the latter had set him, including asking for puncture-proof guns.[148]

His secret had been tested very successfully at the Académie and would succeed when Leibniz commanded it.[149] Failing this, Douceur even promised a refund.[150] Mariotte, who later supported Douceur's claims in this venture, would also be able to verify these claims.[151]

Despairing at Leibniz's seeming failure to live up to his word, Douceur eventually threatened to sell his secret—which he had consented to give up for a 'modest present' for Mariotte's sake—to other bidders.[152] Venting his frustration with the numerous difficulties that had plagued the negotiation, Douceur rebuked Leibniz for his failure to have had the process tested after Douceur himself had complied with all the terms.[153] Leibniz, it seems, did not take kindly to Douceur's plea not 'to refuse [him] what was his right',[154] and he berated Douceur for his ingratitude:

146. Brosseau to Leibniz, 28 April 1679, A I, 2:466. See also introduction to A III, 2:xxix.
147. Douceur to Leibniz, 27 March 1679, A III, 2:675–80; 25 February 1679, A III, 2:633.
148. Douceur to Leibniz, mid-December 1679, A III, 2:912.
149. Douceur to Leibniz, 2 December 1678, A I, 2:547; Douceur to Leibniz, mid-December 1679, A III, 2:912.
150. Douceur to Leibniz, 25 February 1679, A III, 2:636.
151. Mariotte to Leibniz, May 1681, A III, 3:434: 'In the matter of Mr Douceur, I believe he is entitled to complain'; Douceur to Leibniz, 2 December 1678, A I, 2:547.
152. Douceur to Leibniz, mid-December 1679, A III, 2:912; 2 December 1678, A I, 2:546.
153. Douceur to Leibniz, second half of November 1679, A III, 2:891.
154. Douceur to Leibniz, 2:891.

I received your letter, but I must admit that I am not too satisfied with your ways of writing. When one has done what I've done for you, one expects more to be thanked than to be quarrelled with. It is not I who owes you the money that remains but up to my master to pay you when He finds you have satisfied his requirements or finds that your secret is valid. I am under no obligation to make enquiries for you at court, nor to undertake trips or incur expenses to test it.... Finally I am very tempted not to get further involved in this matter in any way. For it provides me neither profit nor honour, and you do not even acknowledge my efforts.[155]

Perhaps concerned that he would not appear as a man of his word, and after Ernst August, Johann Friedrich's successor, refused to pay the outstanding sum for what he considered to be a superfluous expenditure, Leibniz eventually relented, though the full amount—coming partly from his own funds—would be paid only by 1685.[156]

A final example consists in Leibniz's handling of Brand's repeated requests for money. Increasingly irked by them, Leibniz had asked Christian Philipp to intervene in order to reduce Brand's claims to 'something reasonable', now portraying Brand, whom he had previously praised, as someone who made 'extravagant requests' and advanced things to 'which one should pay no attention'.[157] Leibniz did not make light of Brand's ingratitude, as a 'man who failed to appreciate all the good done for him': for promoting and helping him in exchange of the execution of phosphorus, Brand, who presently found himself in severe financial straits, had had the temerity to request more money. Leibniz remained inflexible to Brand's predicament, blaming his wife (who 'gobbled up everything he earned') and deriding 'these people's ridiculous pretensions'. Brand had failed to respond after being reprimanded by Leibniz, and Leibniz, concerned that this would reflect poorly on himself in the duke's eyes or that he would lose face with the latter if he pressed the duke for money, asked Philipp to intercede and try to 'reason' with that 'venerable fool' (*venerable sot*) so that he would live up to his promises.[158] The alternating

---

155. Leibniz to Noel Douceur, mid-April/May 1681, A III, 3:399.

156. Introduction to A III, 3:xlii. On not appearing as a man of his word, see Mariotte to Leibniz, 29 November 1681, A III, 3:520. While it remains unknown whether Douceur had already approached others to sell the idea, it was claimed that Prince Rupert of the Rhine was already in possession of a similar process. O'Hara, *Leibniz's Correspondence*, 9.

157. Leibniz to Christian Philipp, 17 December 1678, A I, 2:395.

158. Leibniz to Christian Philipp, 2:394.

portrayals of Brand as a naïve victim who had fallen prey to the promises of impostors or simply an imbecile, and as charlatan himself shows the sheer instability and ambiguity at the heart of the concept of projector.[159]

And yet, even as he treated Brand like an unreliable and ungrateful child led by his emotions,[160] Leibniz praised him as an incredible worker who found ingenious solutions, and he would make a point in crediting him with the invention of phosphorus and granting him what he felt was Brand's due. His was thus a fundamentally ambivalent approach to his fellow projectors: torn between seeking to treat them well and exploiting them, Leibniz admired projectors for their discoveries while lamenting their unreliability, lack of 'firmness of character', and pursuit of glory and money by practising 'dubious arts' over the common good.[161] Even his opinion of Becher seems to have remained unsettled till the end.[162]

Leibniz's correspondents invariably had their own side of the story, Brand, for example, praising Becher as 'honest' (*ein ehrlicher Mann*) while disparaging Leibniz as 'fickle' (*ein unbestendiger mensch*) and 'similar to a clown'—and the truth probably lies somewhere in between.[163] The project would come to nothing with funding running out, Brand leaving permanently after having fallen ill, Ernst August, duke Johann Friedrich's successor, showing no interest in it, and Leibniz now engaged in new projects. While it is difficult to assess the fairness of the situations, it is, however, undeniable that in his negotiations Leibniz held his correspondents to a more exacting standard than he applied to his own ventures, and his correspondence reveals that he repeatedly drove to desperation the other projectors with whom he dealt.

159. Wahl, '"Im tunckeln ist ein blinder"', 233; Christian Philipp to Leibniz, 10/20 April 1680, A II, 3:386: 'Mr Brand is a man without manners and who boasts of having knowledge of many things he does not: it is true that he finds his dupes, but these are stupid people who are even more ignorant than him.'
160. 'He is a man who receives very badly the services that are rendered to him.' Christian Philipp to Leibniz, A I, 2:393.
161. *Misc. Berl.*, 95.
162. *Misc. Berl.*, 105.
163. Brand to Crafft, 26 November 1678, A III, 2:553.

CHAPTER ELEVEN

Establishing a European Network from Hanover

AFTER HIS DEPARTURE from Paris and over the course of the next three years, Leibniz cultivated an extensive correspondence with savants, theologians, diplomats, and others across the whole of Europe. Finding himself isolated in Hanover, he actively sought to mobilize his epistolary network to maintain a presence beyond Johann Friedrich's court by continuing many of the relationships he had forged in Paris. Having failed to secure a position in France's state-backed learning apparatus, Leibniz now discerned in the Republic of Letters an alternative path to advancing the cause of science, animated by savants driven less by the pursuit of glory but sincerely devoted to the search for truth and willing to 'open themselves up to others' in what amounted to a communally impulsed vision of science, a practicable model of communal learning and scholarship.[1] To this aim, he maintained contact with some of his old Parisian acquaintances, including Henri Justel, the natural philosopher Edme Mariotte, Christiaan Huygens, Jean Paul de la Roque (the editor of the *Journal des Sçavans*), Simon Foucher, Nicolas Malebranche (whose *Conversations Chrestiennes* Leibniz had received from the Countess Palatine Elisabeth), Pierre-Daniel Huet, and Jean Gallois. To this impressive number of correspondents he would add, while in Hanover, Georg Hermann Schuller, the Cartesian and former Rinteln professor Arnold Eckhard, and the philosopher Niels Stensen, who now held the office of Vicar Apostolic for

---

1. Leibniz, 'Recommendation pour instituer la science générale', April–May 1686, A VI, 4: 692–93, quoted in Laerke, 'Ignorantia inflat' (2013), 27. Laerke notes, 'Leibniz thus envisaged an intellectual community where each member made a modest contribution to the collective production of something useful and directed towards the common good' (20).

the northern provinces of Europe after having given up the study of natural sciences.² The correspondents also included Bishop Ferdinand von Paderborn, the legal scholar Vincenz Placcius, Johann Eisenhardt (with whom Leibniz debated about credibility of historiography), and the doctor Johann Sigismund Elsholz (to whom Leibniz sent his programme for the organization of natural research).³ Leibniz's correspondence with four individuals in particular stands out in significance: Christophe Brosseau, Adolf Hansen, Christian Philipp, and Hermann Conring.

Building the Ducal Library: Leibniz with Brosseau

The diplomat Christophe Brosseau was the Hanoverian court's resident at the French court. His correspondence with Leibniz, which spanned thirty years, stands out as a 'remarkable manifestation of humanity and civilisation in the turbulent reign of Louis XIV'.⁴ Brosseau's predecessor, the French politician and diplomat Louis de Verjus (1629–1709), had succeeded in convincing Duke Johann Friedrich to enter a contractual alliance with France and later recommended Brosseau to the post of Resident in Paris.⁵ Leibniz found in Brosseau a reliable helper and friend with whom he felt a certain kinship. Both had lost their father at a young age and had studied law before joining princely courts. Both, too, owed their successes to generous patrons, Louis de Verjus in Brosseau's case, Boineburg in Leibniz's.⁶ Johann Friedrich approved of Brosseau, who was devoted to his employer and understood his expectations and aims well, as the diplomat's frequent letters to Leibniz (after the latter's departure from Paris) attest.⁷ For this reason, while entrusting Hansen with his more personal and philosophical tasks and projects, Leibniz generally relied on Brosseau for the execution of his courtly ones.

Brosseau was more than happy to oblige: he played an instrumental role, for instance, in Leibniz's assembling the ducal library, and most of their correspondence bears on the purchase of books and the logistics of shipping them to Hanover via Hamburg. He also sent Leibniz booksellers'

---

2. Leibniz regretted that such studies were now alien to Stensen: Leibniz to Conring, 3/13 January 1678, A II, 1:579.

3. With Oldenburg's death in 1677, Leibniz lost his agent in London and would rely instead on the *Philosophical Transactions* to keep abreast of scientific news in England.

4. Jurgens and Orzschig, 'Korrespondenten von G. W. Leibniz' (1984), 102.

5. Jurgens and Orzschig, 103.

6. Jurgens and Orzschig, 103.

7. At least thirty letters survive from 1676 to 1679. The complete list is available on *Die Leibniz-Connection* (2020).

catalogues, suggesting that an inventory of the books already owned by the library be compiled and then supplemented with the best books that were missing on each subject.[8] Some books were too big to be sent by overland post and required shipping.[9] Despite his efforts, Brosseau was at times unable to procure all the requested books, either because a book had not yet entered circulation in France or because of censorship.[10] Brosseau also made payments on Leibniz's behalf, delivered some of his letters, and checked on Hansen's progress, especially when Leibniz felt the latter to be slacking in his duties or to be insufficiently forthcoming about the execution of Leibniz's arithmetical machine.[11] Another topic of the correspondence was the acquisition of scientific instruments for the Hanoverian court, a process with which Brosseau offered to assist, at one point ordering new instruments to be produced to replace equipment deemed inadequate for the duke.[12]

Leibniz's homme à tout faire: *Adolf Hansen*

After Leibniz's departure from Paris, Friedrich Adolf Hansen (1652–1711) became one of his most trusted agents there, ensuring Leibniz's continued presence by proxy. Little is known about him except that he was from Holstein, served as Swedish councillor in Greifswald, and would later act as master of ceremonies and court tutor (*Hofmeister*) to Count von Koenigsmarck in Oxford. In Paris, Hansen mainly looked after three young Danish nobles.[13] Hansen's correspondence with Leibniz charts the former's hyperactivity in seeking to mobilize Leibniz's Parisian circle: constantly running

---

8. Brosseau to Leibniz, 3 July 1679, A I, 1:488 'This bookseller, as you will see from this same letter, must provide me every week with a notebook of Spanish, Italian, Latin, French as well as other books, of which he has long since made a very curious collection; I will also send them to you so that you can choose the ones that will best please his Majesty'; Brosseau to Leibniz, 10 March 1679, A I, 2:435.

9. Leibniz to Brosseau, 13 March 1679, A I, 2:440.

10. Brosseau to Leibniz, A I, 2:28; April 1679, A I, 2:466.

11. Brosseau to Leibniz, 26 March 1677, A I, 2:261: 'I paid Mr Cassini the twelve écus that you owed him'; Hansen to Leibniz, 19 June 1677, A I, 2:276: 'Mr. Brosseau promised to take care of our correspondence.'

12. Brosseau to Leibniz, 13 March 1679, A I, 2:440: 'There is an English instrument maker called Butterfield who resides on the fauxbourg St Germain.... He makes microscopes of the latest manufacture, pedometers the size of a small pocket watch; new universal astronomical quadrans, which mark the minutes; pantographs, gyroscopes, and more new instruments. So, Sir, you could buy them for His Royal Highness as long as the price is not excessive and they are not too big. The worker should add a description of their use'; Brosseau to Leibniz, 28 April 1679, A I, 2:466.

13. Hansen to Leibniz, 9 November 1676, A I, 2:237.

errands, carrying parcels, delivering letters to and collecting responses from various *savants* in Paris, making appointments, reminding people to write to Leibniz, and sometimes even helping rekindle correspondences that had gone cold (although he did not succeed in Jean Gallois's case).[14] Much of Hansen's activity in Paris consisted of identifying various scholars' whereabouts and (often unsuccessfully) trying to chase them up.[15] One letter in particular gives a sense of the trials and tribulations involved in getting some of Leibniz's mail delivered:

> You would not believe how touched I am by the answer M. Huet gave you: I have been looking for him several times, sometimes alone, sometimes with some of my friends; and the first one who took me there was a French Gentleman, a great friend of Msle. le Fevre. Then afterwards I spoke about this letter to M. Justel and to M. L'Abbé de la Rocque, the latter promising to let me know when M. Huet was at home. Not being informed, I took the letter to court on New Year's Day, and I asked for news of M. Huet from M. Milet, who told me that he was in Paris. On my return I went to look for him, and not finding him, I sent my valet there several times. In the end, seeing that there was no way to meet him, I put the letter into M. L's hands.[16]

Hansen was equally important to Leibniz in supplying him with French scholarly journals (such as the *Journal des Sçavans*) and other volumes (such as le Chevalier de Mère's *Petit Traité des Agréments*), and helping to acquire books from France for the ducal library.[17] Crucially, Hansen was an efficient and determined informant who kept Leibniz abreast of

---

14. Hansen to Leibniz, 10 July 1679, A I, 2:494; 17 April 1679, A I, 2:463.

15. Hansen to Leibniz, 17 April 1679, A I, 2:463: 'I went looking for M. l'Abbé Gallois on Tuesday and not finding him I returned to his house last Friday, but he had gone to court with M. Colbert, I am leaving in an hour to look for him for the third time, and if I am fortunate to meet him, I will be able to learn from him everything you want to know'; 'One rarely finds M. L'Abbé Mariotte at home; after several trips I had made to find him, I was finally forced to leave your letters with his servant.'

16. Hansen to Leibniz, 10 July 1679, A I, 2:494.

17. Christian Philipp to Leibniz, 28 September 1678, A, I, 2:37; Hansen to Leibniz, 19 June 1677, A I, 2:276; 21 August 1679, A I, 2:511: 'I bought the conical sections from M. de la Hire for M. Collins, as you ordered, but I could not locate Viviani's resolution of M. Comiers' problems. Mr. Leonard a l'Escu de Venis has promised me to get more copies of it along with other books from that country, and I have asked Mr. Amelot to buy a copy for me, which I intend for you'; Hansen to Leibniz, A I, 2:383: 'Now that peace is established with Holland and there is no other danger to be feared on the seas than that of storms, I can easily send you in future whatever you ask of me for the ducal library, for ships regularly leave the port of Rouen for that of Hamburg.'

scientific developments, including news of publications in physics.[18] He remained on the lookout for information, often relaying to Leibniz accounts of 'practical commodities' (*commoditez de la vie*) and various spectacles and unusual sights, including the natural philosopher Comiers's talking head (*tête parlante*), and even an inventory of all items found in an Amsterdam cabinet of curiosities, including a table that 'changes colour according to its perspective', a giant globe, and a collection of skeletons belonging to various animals and other 'monsters'.[19]

Hansen also kept Leibniz abreast of French military and diplomatic developments and the latest Parisian cultural productions—on one occasion sending him a critique of *Phèdre*,[20] a tragedy whose author, Racine, Leibniz particularly admired:

> As for *La Galanterie* [i.e., Paris] things have been as usual; there are more bad books than good ones. There have been two exceptions, however, one of which is Racine, who is well known to you, and the other Pradon, whose Pyrame and Thisbe opened to great acclaim at the Hotel de Bourgogne. These two poets . . . have worked on the same topic, namely, Phedre and Hyppolite, but Pradon seems to prevail over Racine, although the latter has his play performed at the palace with the best actors. . . . We will soon have the Opera in Paris: the subject is Io, or Isis, whose story is told, as you know, in the first book of the *Metamorphoses*. The author however . . . departs from Ovid's thought. Reviews for this production have been mixed.[21]

The information Hansen sent Leibniz ranged from the mundane and the trivial, such as Le Brun's presentation at the Académie of how to 'represent passions', to the remarkable and even the cryptic.[22] On one occa-

---

18. Hansen to Leibniz, 28 February 1678, A I, 2:321: 'The composition of artificial stones as hard as marble is still in the hands of Mr. Comiers, I hope to have it soon and send it to you without fail'; 'I will try to introduce myself in Mr. Lanker's home in the hope of obtaining some advantage, and if I learn anything from him, I will let you know'; Hansen to Leibniz, A I, 2:323: 'As soon as I will be in Berlin I will try to introduce you to those you mention in your letter and will let you know either by letter or in person, next time I have the honour of seeing you'; Hansen to Leibniz, 7 August 1679, A I, 2:506: 'M. l'Abbé Mariotte tells me that he has answered you, he shows me many beautiful experiments, and the three parts of his *Physics*, his *Treatise on Colours* is progressing well, he has in his hands a treatise on salts composed by M. du Clos which he must examine before publication.'

19. Hansen to Leibniz, 4 June 1677, A I, 2:275; September (?) 1676, A, I, 1:458–59.

20. Hansen to Leibniz, 30 April 1677, A I, 2:270.

21. Hansen to Leibniz, 1/12 February 1677, A I, 2:244.

22. Hansen to Leibniz, 17 April 1679, A, I, 2:463: 'An important piece of news has started to spread; if it is true it cannot remain hidden any longer.'

sion, for instance, he wrote about the wife of the Commissaire Clavesaint, who was publicly hanged and burned after having had her husband murdered.[23] On another, Hansen, clearly rather mesmerized by the whole experience, devoted a whole paragraph to the description of an English fire-eater and his various ploys to perform his act to delight his audience.[24]

Hansen laboured diligently to perform the tasks Leibniz assigned him. His devotion and loyalty to Leibniz were perhaps nowhere more tested than in the case of Ollivier,[25] the master-artisan tasked with the realization of Leibniz's arithmetical machine in Paris. Throughout the correspondence Ollivier was a constant source of vexation. Before leaving for Hanover, Leibniz had entrusted Hansen with overseeing the final execution of the machine. The correspondence between Hansen and Leibniz up to Hansen's own departure from Paris recounts Ollivier's perpetual delays, excuses, requests for money, and indefinite postponement of deadlines. Hansen sought repeatedly to press Ollivier to finish the task he committed himself to undertaking while also maintaining its secrecy.[26] Yet Ollivier proved either unlocatable or already committed to the Jesuit priest Nicolas d'Harouis, who wished to see his own machines completed.[27] While Hansen persistently couched his exhortations in the language of a gentlemanly code of conduct, relying on Ollivier's sense of 'honour', Ollivier for his part deployed the language and negotiating skills of the world of mercantilism.[28] Much of the correspondence thus charts the gradual and very noticeable breakdown in the relationship between Hansen and Ollivier, with impatience gradually giving way to exasperation. Hansen regularly vents his frustration and growing annoyance at Ollivier's failure to honour

23. Hansen to Leibniz, 1 March 1677, A I, 1:253.

24. Hansen to Leibniz, 2/12 February 1677, A I, 2:243: 'There is here at the St Germain fair an Englishman who has found a secret to prevent the activity and action of fire: he can, without burning himself, nor even altering the skin of his lips or that of his tongue which is quite delicate, eat lit sulfur, and crush some in the shape of an egg, which he then places in an iron spoon ... which he then leaves to melt on the fire, and even ignite, and then swallow this drug as he would a broth: then a large piece of hot coal is placed in his mouth and a piece of raw flesh on it which remains there until it is cooked. Once it is cooked, he crunches it with the charcoal. Once this is done, a steel bar, which has been heated over the fire, is brought to him with tweezers, and placed between his teeth which he holds until it has cooled off.'

25. Fore more on this, see Jones, *Reckoning with Matter* (2016).

26. Hansen to Leibniz, 26 February 1677, A I, 2:250: 'I will urge him as much as I can, to hurry, and I will not forget to remind him to keep the matter secret.'

27. Hansen to Leibniz, 2:250: 'I looked for Mr. Ollivier three times in his house without having had the good fortune of finding him there'; Hansen to Leibniz, 5/15 March 1677, A I, 2:259.

28. See Jones, *Reckoning with Matter*, 76–80.

his promises to Leibniz. According to Hansen, Ollivier was 'lazier than ever' and constantly failed to fulfil his duties.[29] Hansen finally exploded, urging Leibniz to send him a copy of Ollivier's contract and invoking the possible need to enforce more strenuous measures in the face of Ollivier's obstinacy:

> M. Ollivier is really doing my head in [*me donne martel en tête*], having brought me neither brief, nor letter, nor machine, again this morning I sent my valet to his house who found him in bed, and for all answer he let me know that he was not working on our task and that he would send me a letter at some point: if you have made a contract with him, I beg you Monsieur to send me a copy, because if he does not want to live up to his duties we must think of other remedies. I assure you that this lazy idler mortifies me greatly since I cannot satisfy you as I wish with all my soul.[30]

Hansen remained steadfastly committed to 'coming to grips with this man', an expression he uses several times in the correspondence. Branding Ollivier 'lazy', 'idle', 'dishonest', and finally 'strange', Hansen grew increasingly anxious as the prospect of his departure for Germany via Rouen started to loom, and he requested that Leibniz send him assistance.[31] Leibniz would try to have Ollivier move to Hanover to no avail. In any event, Leibniz's calculating machine would be completed only forty years later.

In his devotion, Hansen often appeared—or presented himself—as Leibniz's surrogate. In Paris he acted zealously as Leibniz's representative and sought to relay information to him, offering, for instance, to send him—should Justel fail to do so—the process by which to 'grow metals in the shape of philosophical trees'.[32] One such occasion involved exploiting the inventor Comiers's vanity to glean what could be useful to Leibniz and thence 'better strike our goal'—note the plural pronoun.[33] Hansen also brought certain proposals or potential assets to Leibniz's attention, as when he promoted a young scientist and 'erudite' called Marcel who had designed an invention 'to relieve the memory in history and geography'.[34] Sometimes even he acted on his own initiative to seek to recruit assets for

29. Hansen to Leibniz, 9 January 1679, A I, 2:406.
30. Hansen to Leibniz, 7 November 1678, A I, 2:380.
31. Hansen to Leibniz, 7 August 1679, A I, 2:506.
32. Hansen to Leibniz, 4 October 1677, A I, 2:295.
33. Hansen to Leibniz, 16 August 1677, A I, 2:290.
34. Hansen to Leibniz, 14 November 1678, A I, 2:382.

Leibniz's network. On 31 October 1678, for instance, Hansen addressed to a certain Francois Regnauld:

> The eagerness with which Monsieur de Leibniz seeks the knowledge of the most beautiful minds of Europe makes him desire yours with a very ardent passion.... He asked me to assure you of the esteem he has for your profound knowledge and your rare merits, and would be very happy to one day maintain a correspondence with a man as enlightened as you. But as I cannot yet have the honour to pay you my respects, I believed that you would not find it amiss if I took the liberty to send you these things in advance in the hope of giving you complete satisfaction when I send you news of this scientist who, as you may have learned, is universally accomplished in mathematics and the other sciences, and who at least is on par with all the greatest scholars we have in Germany. You Sir, who are an excellent connoisseur, know how to distinguish the solid from the falsely brilliant, and do justice to everyone.[35]

Hansen's enthusiasm can be ascribed to his desire to reframe his status from something akin to an agent to that of a fellow valued member of the Republic of Letters.[36] Even though they were social equals, Leibniz and Hansen seem to have had a complex yet unequal relationship governed by a tacit arrangement whereby Hansen offered his services and information in exchange for Leibniz's recognition.[37] Norms of polite correspondence notwithstanding, Hansen sought to establish a more equal intellectual and scientific footing with Leibniz frequently deploying the rhetoric of learned epistolarity. If there was an obligation, it was that binding two '*savans*' governed by honour, courtesy, and mutual assistance.[38]

On the basis of this reciprocity, Hansen suggested a full-fledged exchange of information on several occasions, requesting that Leibniz share his designs with him, notify him of German publications about

---

35. Hansen to Francois Regnauld, 31 October 1678, A I, 2:521.

36. Hansen to Leibniz, 16 August 1677, A I, 2:290: 'The debt I owe you is greater than I can express because it is you, Sir, who provide me every day with new knowledge that I esteem more than all the inventions of Hermes.'

37. Hansen to Leibniz, 2/12 February 1677, A I, 2:242.

38. Hansen to Leibniz, 16 August 1677, A I, 2:290: 'I would be grateful Sir if you could send me from time to time what you learn from the correspondence you cultivate with foreign countries, because I am certain that there are hardly any scholars who do not know how advantageous it is for them to have the honour of speaking with you: you make them long to see their journals embellished with your curiosities, which they hope to obtain after the letter you addressed to the Abbé de La Rocque has encouraged them to believe that you will continue communicating curious inventions both in physics and in other sciences. All the *curieux* are convinced of their solidity, and await them with great impatience.'

contemporary Europe, and forward him the responses Leibniz received.[39] On one occasion, when handling a letter containing 'very beautiful remarks' intended for Justel, Hansen proposed to Leibniz a subterfuge whereby the latter would leave one end of his own letters open to be read by Hansen, who would then simply seal the edge with wax so that the intended recipient would never be the wiser. Through such a ploy Hansen would be kept informed directly of Leibniz's answers to his various correspondents, and make Hansen himself the 'happiest of all men in the world'.[40] Although it remains doubtful that Leibniz acceded to Hansen's request to let him operate unseen in his shadow, it is strikingly audacious.

While remaining keen to depict himself throughout the correspondence as a diligent student worthy of and ever grateful for his tutor's intellectual supervision, Hansen thus sought to reframe their relationship. The Parisian intellectual and scientific scene had not failed to dazzle Hansen, who expressed delight at all the 'beautiful things' he had 'learned [at Gallois's house]'.[41]

He clearly aspired to belong more fully to that world, and accordingly sought to exploit Leibniz's connections to widen his own circle of

39. Hansen to Leibniz, 8 March 1677, A I, 2:255: 'I received your letter dated of the 21st of the past month last Monday, and I hope that my response to your previous letter will reassure you that I took care of your commissions. I thank you from the bottom of my heart for the offer you are making me, and as you will not object to be taken at your word, I beg you most humbly to let me know from time to time what is printed in Germany regarding the present state of Europe, and to send me the catalog of Leipisic, and of Francfort when the season permits: please send me a message as well concerning the curious college in Germany, and its leader'; Hansen to Leibniz, 17 April 1679, A I, 2:463; 1 March 1677, A I, 2:254: 'You will oblige me very much by sending me from time to time what goes on in matters of letters and affairs, of which you have a full knowledge, I promise you that ingratitude will never be my vice'; Hansen to Leibniz, 4 October 1677, A I, 2:295.

40. Hansen to Leibniz, 30 July 1677, A I, 2:282–83: 'A curious man like yourself knows the obligation that I have toward you after having again received from you a letter so ample and filled with very beautiful remarks: having read it, I delivered it immediately resealed to Mr. Justel, who was so charmed with it that he said with an open heart that your correspondence was the dearest thing to him in the world, having thereby acquired a favourable opportunity to exchange with the most enlightened man in Europe, and without seeking to incense you Sir, I can assure you that all good people share this feeling, which indeed only do you justice. I will be the happiest man on earth if you could continue to inform me from time to time . . . to know what you are sharing with your friends: you have only to seal your letters, as long as you are kind enough to leave one end open, so that once I will have read the letter and placed it back in its envelope, all that will be left to do is to seal the letter with some wax and and no one will notice.'

41. Hansen to Leibniz, 17 April 1679, A I, 2:463: 'I dare say that I have found few companies in Paris from where I left with more contentment, nor that I left with more regret; so many beautiful things are learned there.'

acquaintances. Pressing to be instructed particularly in mathematics, Hansen repeatedly sought Leibniz's guidance and opinions, and reported on his own progress. He also proposed 'study trips' for himself, notably to England, where he would be able to hone his mathematical skills.[42] Leibniz for his part seems to have welcomed—and taken advantage of—Hansen's enthusiasm and assistance, but not in these years to have ever truly acknowledged Hansen's intellectual parity.

Leibniz with Christian Philipp

Much of Leibniz's correspondence occurred against a backdrop of European political conflicts. War appeared as a disruptive force that threatened the progress of science and the ideal of the free circulation of ideas promoted by the Republic of Letters.[43] From Sweden the Danish nobleman Christian Albrecht Walter commented on his country's attacks on Denmark under the influence of the French and deplored that he found himself 'in a country where the terrible noise of weapons [left] no room for muses', ironizing that 'everyone [was] talking about peace' all the while 'getting ready for a vigorous campaign'.[44] War had slowed down the progress of the *Ad Usum Delphini* series and had severely disrupted intellectual life, according to Henri Justel, to the extent that he 'did not have the heart to pick up anything'.[45] The French savant, however, hoped that such a sad state of affairs might 'dispose the minds of interested parties to its resolution.'[46]

One correspondent in particular on whom Leibniz relied to verify or refute political or military rumours and to discuss European geopolitics more broadly over the years 1676–82 was Christian Philipp. In spite

---

42. Hansen to Leibniz, 26 September 1678, A I, 2:367: 'Huet seems to count among your friends; please use me if you have anything you need to talk to him about'; Hansen to Leibniz, 4 June 1677, A I, 2:275; 31 October 1678, A I, 2:375; 10 July 1679, A I, 2:495: 'I still need three or four months for the work I have undertaken, and after that I will get started with the mathematical sciences'; Hansen to Leibniz, 24 October 1678, A I, 2:372: 'I have not yet read the books you were kind enough to mention to me, but I am preparing myself for it by reviewing the beautiful philosophy that I would like to master better. I still need at least three or four months for the work that I have begun, and then afterwards I will start working on mathematical sciences, of which I nevertheless examine some aspects from time to time.'

43. Leibniz to De L'Hospital, 20/30 July 1696, A III, 7:41; see Borowski, 'Republic of Letters' (2021).

44. Christian Albrecht Walter to Leibniz, 17 February 1679, A I, 2:637.

45. Justel to Leibniz, 22 July 1678, A I, 2:354; 30 July 1677, A I, 2:287–88.

46. Justel to Leibniz, 17 April 1677, A I, 2:267.

of their many differences, Leibniz and Philipp had much in common. Both were originally from Leipzig and had benefited from the support of eminent intellectuals, Oldenburg for Leibniz and Samuel Pufendorf for Philipp. Pufendorf had taken Philipp under his wing most probably through the mediation of Philipp's father in the late 1650s in Jena.[47] Master and protégé would remain in close contact over the course of their lives, Philipp continuing to serve Pufendorf diligently who in turn would recommend him for the position of Saxon envoy to Hamburg.[48] The similarities between Philipp and Leibniz, however, ran even deeper. During his time in Jena and before going on to pursue a course in jurisprudence, Philipp had stayed with Erhard Weigel, whose lectures Leibniz would attend in Jena.[49] Though Philipp would in retrospect express a more reserved view than Leibniz of Weigel,[50] he was, like Leibniz (who was his junior by seven years) exposed to Weigel's influence and intellectual openness and was instilled by him with a love of chemistry, the natural sciences, and the urge to derive practical applications from them. After his studies, Philipp had worked as a teacher before moving for a short period to Paris, where he lived in 1668–69 with Esaias Pufendorf, Samuel's brother, who at the time served as the representative of the Swedish crown. There Philipp was introduced to the Parisian scientific and intellectual scene where, as he would later report to Leibniz, he was made aware of the contradictions inherent in Cartesian geometry.

Philipp had known of Leibniz before corresponding with him, as a letter written by Christian Habbeus from March 1676, when Leibniz was still in Paris, indicates. The first surviving piece of correspondence is a letter from Leibniz to Philipp dated 22 July 1678, during Leibniz's trip to Hamburg, where Philipp was then stationed and where both attended a discussion on Cartesian philosophy held at Esaias Pufendorf's house and most likely met.[51] In November 1679, prompted by the attacks of the inveterate German Cartesian Johann Rabel, Philipp wrote to Leibniz requesting a summary of the discussion that had then taken place, and the latter's arguments as to how Cartesianism led to atheism.[52] This letter and Leibniz's subsequent response in January 1680 were two of the rare occasions

47. Döring, 'Korrespondenten von G. W. Leibniz' (1989), 104.
48. Delivering, for instance, a parcel to Adam Rechenberg in Lepizig: Pufendorf to Philipp, 27 May 1676: Meyer, *Samuel Pufendorf* (1894), 15.
49. Döring, 'Korrespondenten von G. W. Leibniz', 105.
50. Philipp to Leibniz, 16/26 March 1681, A I, 3:467; 21 April (1 May) 1682, A I, 3:534.
51. Leibniz to Philipp, 22 July/1 August 1678, A I, 2:357. Döring, 'Korrespondenten von G. W. Leibniz', 116.
52. Philipp to Leibniz, 22 November/2 December 1679, A I, 2:524.

when the two discussed philosophy.[53] After his stay in Paris and another brief stint in Sweden, Philipp finally took up the position of the Saxon court's representative in Hamburg, at the time a commercial, financial, and political hotspot in a context of mounting political friction. Against the backdrop of a fledgling diplomatic alliance between France and Sweden, he was tasked with collecting information and reporting back on local political and military developments and more generally the 'special passion' (*sonder passion*) that had developed in the northern European theatre of war.[54]

By writing to Philipp, Leibniz decided to keep alive his connection with someone who, on account of his familiarity with political and diplomatic activity in Hamburg and beyond, might prove to be a most useful contact. His letters to Philipp often take the form of short paragraphs, each addressing a different issue in a matter-of-fact way: the emphasis seems to have been on efficiently exchanging information, in particular on diplomatic and political issues, rather than nurturing a friendship. In Hamburg, Christian Philipp had honed his skills as a political analyst and deployed his excellent knowledge of political events of the time, much of which he would share with Leibniz, regularly supplying him with the latest diplomatic developments, especially regarding the much-anticipated agreement of peace across Europe.[55] This included keeping track of troop movements, sieges, the transit of diplomats through the region, and state of intergovernmental negotiations, as well as assessing the validity of rumours, some of which indeed he dismissed as 'fanciful'.[56] In exchange for this, Philipp himself inquired about the movement of Hanoverian troops and requested diplomatic information from Leibniz.[57]

The exchange of political news formed a large part of the content of the exchanged letters. Throughout his correspondence with Leibniz, Philipp regularly outlined the danger presented by a wealthy and bellicose France and its ambitions for European expansion, which, according to him, would entail the further weakening of an already frail and disunited Germany, many of

---

53. Leibniz to Philipp, end of January 1680, A I, 3:345–48.

54. Döring, 'Korrespondenten von G. W. Leibniz', 112–13.

55. Christian Philipp to Leibniz, 28 January/7 February 1679, A I, 2:417: writing that 'they awaited the conclusion of peace with impatience'.

56. Leibniz to Philipp, 14 March 1679, A I, 2:445: 'I still doubt all the rumours that are circulating about the marriage of the Dauphin and also those about the king of Spain.'

57. Christian Philipp to Leibniz, 26 March/5 April 1679, A I, 2:456; 13/23 August 1679, A I, 2:513: 'Mr de Rebenac is presently here with orders from the king of France to reconcile the Hamburgers with the [French-allied] elector of Brandenburg.'

whose leaders had already pledged allegiance to France.[58] Despite his fleeting hope for an improvement in relations between France and the German states upon the accession of Johann Georg III in Saxony in 1680, Philipp's outlook remained overwhelmingly bleak: 'I believe that all of Germany is mired in fatal blindness.'[59] Responding to Leibniz's grave misgivings about France—whose 'mind', fuelled by 'hatred', was now set only on 'Germany's ruin', and especially its 'division and impoverishment'—Philipp depicted the French as warmongering, harbouring 'wicked news' (*des méchantes nouvelles*), and revelling in European division: 'if France is fortunate enough to see division in the house of Austria, she can say that she is on the eve of her greatest power.'[60] He went on to venture that peace would already have been concluded had it not been for the petulant French who wanted 'peace to be received from them as a grace'. Through French expansionism, war would only persist, especially since it did not contravene French trade.[61]

Philipp hoped for the conclusion of the peace treaty but often despaired at the trivial nature of certain obstacles placed in its path, such as one prince's reluctance to take up the position of general, or a duke's lack of diplomacy in times of great necessity.[62] As he wrote in one letter, 'We take offence at trifles and neglect the essential.'[63] Still, on 1 February 1679 he reported the news of the Peace of Nijmegen, concluded earlier that month, and later himself requested news from Leibniz concerning the progress in negotiations and in the growing tensions between Sweden and Denmark:[64] 'The peace between the Emperor, France and Sweden was concluded in Nijmegen on the 6th of this month ... and the French project of April 5/15, 1678, was its foundation: accordingly, the Swedes will be compensated,

---

58. Christian Philipp to Leibniz, 12/22 November 1679, A I, 2:522; 22 January/1 February 1679, A I, 2:414; 16/26 November 1678, A I, 2:387; also 20/30 January 1682, A I, 3:515.

59. Christian Philipp to Leibniz, 20/30 August 1681, A I, 3: 498; also 20/30 January 1682, A I, 3:515: 'I believe that all of Germany is fatally blinded.'

60. Leibniz to Philipp, 12 November 1678, A I, 2:386; Philipp to Leibniz, 8/18 January 1679, A I, 2:411.

61. Philipp to Leibniz, 17/27 January 1679, A I, 2:412.

62. Philipp to Leibniz, 28 January/7 February 1679, A I, 2:417: 'This Prince is urged to accept the title of General of the Allied Armies on the Rhine Coast, but he refuses to do so until it is more apparent that everything necessary for him to carry out such a duty with dignity is in place. It is causing a stir in Regensburg that the Duke of Lorraine has summoned the neighbouring estates in rather strong terms to contribute to the upkeep of the garrison in Philipsburg.'

63. Philipp to Leibniz, 1/11 February 1679, A I, 2:421.

64. Philipp to Leibniz, 26 February/8 March 1679, A I, 2:432: 'I am told that M. le Comte de Revenac is coming and going in Zell, and that the negotiations are in very advanced stages. You are undoubtedly better informed than I and I beg you to share with me what you will come to know about it'; Philipp to Leibniz, 13/23 August 1679, A I, 2:514.

and the Emperor will remain neutral: it is hoped that the affairs of the North will calm themselves without French intervention; this would be desirable, so that the great scandal that French arbitration would cause be avoided.'[65]

Philipp remained in Hamburg until May 1681, when, his increasingly financially untenable position was terminated, as the Peace of Nijmegen in February 1679, followed by the peace treaties of St Germain and Lund, made a permanent presence in Hamburg unnecessary.[66] After being recalled, Philipp experienced several months of uncertainty until, in a quirk of fate similar to Leibniz's, he was appointed court librarian in Dresden. This proved to be a daunting task since no inventory of the library holdings had been carried out since 1595 and the library had been left in a state of relative neglect, prompting Philipp to request the more experienced librarian's advice regarding the best way to organize a library.[67]

An Impasse in Germany: Leibniz with Hermann Conring

Increasingly faced with the prospect of remaining permanently in Germany, Leibniz set out to cultivate a relationship with one of its most eminent scholars, Hermann Conring (1606–81).

With his groundbreaking work *De origine iuris Germanici* (1643), Conring, intent on bolstering the Protestant territorial princes' claims of autonomy against those of Hapsburg imperialism, had set out to trace the history of laws in force in the German Reich in order to debunk the mythical belief that there existed a line of continuity from the ancient Roman Caesars to contemporary Holy Roman emperors.[68] Throughout his life Conring remained a favourite in European royal circles, including with the duke of

---

65. Philipp to Leibniz, 1/11 February 1679, A I, 2:421. In the treaty signed in February 1679 between France and Sweden on the one side and the Holy Roman Empire on the other—part of the larger negotiations to end the Franco-Dutch War—Sweden regained German territory it had lost to the Prussian Army, while France gained Freiburg im Breisgau and conceded Philippsburg to the empire. No agreement was reached on the status of Alsace-Lorraine, which France continued to occupy. See Blanning, *Pursuit of Glory* (2007), 534, 539.

66. Döring, 'Korrespondenten von G. W. Leibniz', 114, 115.

67. Philipp to Leibniz, 21 April/1 May 1682, A I, 3:535.

68. According to Georg Schmidt, *Geschichte des Alten Reiches* (1999), Conring sought to reframe the imperial idea into a more complementary framework as the cooperation between the emperor and the princes. For more on Conring, see Stolleis, *Hermann Conring* (1983).

Brunswick, the Swedish king, as well as other princes who valued his legal skills and sought his advice, for which they rewarded him with pensions and appointments.

Indeed, Conring even found himself pensioned by the French state in relation to Colbert's attempt to install his state Republic of Letters, when Jean Chapelain, who had lauded the French statesman to Conring, suggested that the latter write a panegyrical history for the king and search German archives for documents that could be deployed in supporting French claims against the Hapsburgs.[69]

It is interesting, in this regard, to consider the similarities between Conring's and Leibniz's trajectories. Both were devoted to ending confessional disputes and the reform of German politics, and to the promotion of the scientific study of the past to this effect.[70] Both were imbued with the same values and driven by similar concerns, namely, encouraging the political unity of the German state by resolving political, religious, legal, and other controversies in a rational way in the wake of the Thirty Years' War. And like Conring before him, Leibniz, despite his undeniable talents, was regularly confronted with the limits of his position, meaning that he could never take an active part in the politics of this time, despite his efforts, but had to resign himself to the role of an adviser with limited actual influence. Conring would enjoy a long-standing and remarkably fruitful collaboration with the French court over several years, defending France's right to the Spanish Netherlands in 1667 and Louis's right to be considered a candidate in the imperial elections in 1668 and 1671, composing an advisory statement on the Triple Alliance in 1670, and even writing a memorandum outlining a strategy for the French king to take control of the Mediterranean.[71]

In *Consilium de maris mediterranei dominio et commerciis regi christianissimo vindicandis*, Conring—unlike Leibniz—carefully omitted questions of European politics or security in order to embrace Colbert's commercial goals. The plan presented France as Europe's natural leader, the only power capable of saving Europe from the relentless pursuit of profit. France's gradual monopolization of the Mediterranean trade could be achieved through friendly negotiations with the Gate, which would also help

69. Collas, *Jean Chapelain*, 432; Soll, *Information Master* (2009), 101. See Chapelain's praise of Conring to Colbert, 2 September 1666, in Collas, 432.

70. Fasolt, *Limits of History* (2004), 86. Like Leibniz, Conring was not opposed to having recourse to occasional little subterfuges to help mend the religious rift, using a pseudonym to compose the work on religious peace *Pro pace perpetua* in 1648.

71. Fasolt, 85. See also Fasolt, *Past Tense* (2014).

open the trade route to India and the Arabian Gulf. In another parallel with Leibniz, by the time Conring's project was conveyed to Paris, it had already been rendered superfluous by the resumption of French negotiations with the Ottoman Empire. In any event, Conring's proposal would have made Colbert and Pomponne disinclined to entertain a similar one shortly after.[72]

Boineburg actively recommended his protégé to Conring, his former professor at the Lutheran University of Helmstedt, sending him a copy of the *Nova Methodus* and writing to him on 26 April 1668 about the young man: 'He is a twenty-two-year-old doctor in law: very learned, a good philosopher, committed, able and prompt in speculative reasoning. . . . He is certainly a man of great knowledge, of disciplined judgement, and of great capacity for work.' That summer Boineburg sent Conring three copies of the newly published *Ratio Corporis Juris Reconcinnandi*, paving the way for the stimulating epistolary exchange on judicial and historical questions between Leibniz and the famous jurist.[73] When, two years later, Conring asked the baron for further information on his protégé, Boineburg sketched a second, still more detailed and appreciative intellectual profile of the young Leibniz:

> He is a young man, twenty-four years old, from Leipzig, a doctor of law, and a more learned one could scarcely be imagined. He understands the whole of philosophy thoroughly and is a productive thinker in both the old and the new. He is furnished above all with the ability to write. He is a mathematician, hard-working and ardent, who knows and loves natural philosophy, medicine, and mechanics. Of independent judgement in religion, he is a member of your [Lutheran] confession. He has mastered not only the philosophy of law but also, remarkably, legal practice[.][74]

Conring thus entered into communication with an evidently highly accomplished and promising young man who seemed as devoted as he was to the reconciliation of the churches. Their correspondence spanned a wide range of topics, including law and jurisprudence, theology, natural philosophy, and even medicine.[75] While Conring's *New Discourse* had sought, through historical analysis, to delegitimize the claims of the

---

72. See Wiedeburg, *Der Junge Leibniz*, 417–22.

73. Gruber, *Commercii Epistolici Leibnitiani* (1745), 1208–9. Leibniz's first letter to Conring dates from 23 January 1670, and the last one from February 1679. The nineteen extant letters exchanged between them are published in A II, I.

74. Boineburg to Conring, 22 April 1670, in Gruber, 1286–87.

75. See Mulvaney, 'Early Development' (1968).

Roman emperor, which he believed had caused the Thirty Years War, and defend the rights and sovereignty of the state of Germany over the last remaining vestiges of the empire, Leibniz was instead eyeing the broader reorganization of Europe and the strengthening of allied states, ideally under the guardianship of the Reich.[76] Leibniz kept Conring abreast of recent scientific discoveries as well as new publications he had received.[77] In return he requested news and sought Conring's opinion on matters of natural philosophy and medicine, and occasionally offered his own speculations for comment, especially in the realm of natural philosophy, for Conring was one whose 'ingenious mind extended to almost all kinds of studies', and Leibniz considered him deserving of reverence from 'all connoisseurs of true learning'.[78]

Conring's contributions to public science were invaluable, according to Leibniz, and the latter urged him on several occasions to collect and publish his 'manifold observations' and 'praiseworthy findings' for the greatest public good and that of the Republic of Letters.[79] He was eager to receive his approval and to present himself as emulating Conring—perhaps in the hope of appearing as the great master's intellectual heir and succeeding him as one of Germany's foremost savants.[80]

Merging genuine admiration and self-fashioning in his effort to win Conring over, Leibniz described himself as a young man 'led not by ambition but by the desire to improve' and 'eager to learn' from Conring before becoming a more public figure himself.[81] Such a prospect did not seem too far-fetched, especially because Conring seemed genuinely impressed by Leibniz; and while he was less effusive in the earlier years of their correspondence, he admired Leibniz's ambition and encouraged him to persist, even if he disagreed with Leibniz on certain points. In fact, Leibniz's letters conveyed 'a fiery and tireless talent, which [did] not limit itself within

76. Fasolt, *Limits of History*, 125, 182.
77. Leibniz to Conring, 8 February 1671, A II, 1:133; April 1677, A II, 1:504.
78. Leibniz to Conring, April 1677, A II, 1:504: 'Please indicate to me, when you have time, if Justel has written down anything new which concerns the Republic of Letters'; Leibniz to Conring, 3 January 1678, A II, 1:579; early May 1671, A II, 1:152: 'That you may be able to judge precisely theories of motion, and [*sc.* judge] my results in the Hypothesis for unravelling the phenomena of nature, I send you my—whims'; Leibniz to Conring, 19/29 June 1677, A II, 1:551: 'We know that the vastness of your ingenious mind extends to almost all kinds of topics, including religion, medicine or physics. Nothing you have to say is mundane'; Leibniz to Conring, April 1677, A II, 1:504.
79. Leibniz to Conring, 19 June 1677, A II, 1:551; 3 January 1678, A II, 1:584; 3 January 1678, A II, 1:584; 8 February 1671, A II, 1:132.
80. Leibniz to Conring, 8 February 1671, A II, 1:132.
81. Leibniz to Conring, 1:132.

the confines of any discipline'. What is more, Conring sensed that Leibniz's studies did 'not content themselves with bare attempts, but that already then they [had] made progress in disciplines of all kinds, and this at a very early age'.[82]

Leibniz's move to Hanover encouraged Conring to cultivate the young savant more intensely: his new position as head librarian at the court of Hanover was especially promising, and more particularly his proximity to Duke Johann Friedrich could prove beneficial, as Conring did not scruple to point out: 'Yet it will restore me not little, if the Excellent Prince will accompany me with pristine grace: that you procure this for me, I ask you again and again.'[83] For his part, Leibniz was determined to render himself useful to Conring, offering to act as his assistant and even volunteering his assessment of the current diplomatic and military situation in Europe.[84] In return, he sought to engage Conring primarily in exchanges of a scientific nature, and he used these as a sounding board against which to develop his own ideas—albeit necessarily within limits.

Over time, however, a distinct ambivalence towards Leibniz, punctuated by a measure of reproach, had become palpable, and Leibniz's requests to Conring met with increasing cageyness. The correspondence eventually ran its course, especially as, in declining health, Conring preferred to focus his attentions on more 'useful' things.[85]

Leibniz's idealistic inclinations, often far removed from practical needs or even bordering on delusion, may well have irritated Conring.[86] And apart from acting as an intermediary, passing on letters and information, locating books in the ducal library, or helping promote Conring's works to Duke Johann Friedrich, Leibniz was, as Conring evidently came to realize, of limited use to him.[87] Having established a solid reputation throughout Europe, Conring was in any event now much less dependent on personal

---

82. Conring to Leibniz, 8 February 1671, A II, 1:130.

83. Leibniz to Conring, 13/23 July 1677, A II, 1:554; Conring to Leibniz, 18/28 May 1677, A II, 1:505.

84. Leibniz to Conring, 29 June/9 July 1677, A II, 1:552.

85. Conring to Leibniz, 13/23 July 1677, A II, 1:554: 'I will not deny that I have observed many things in natural philosophy and medicine beyond those which I have publicized in my publications: but old age and a frailer health call me to much more useful things for human life.'

86. See Leibniz to Conring, 19 March 1678, A II, 1:597–607. In his last letter to Conring before the latter decided to end their epistolary relationship, Leibniz characterised his tetragonism as that which mathematicians 'ought to desire' rather than what 'they usually [did]'.

87. Conring to Leibniz, 8/18 May 1678, A II, 1:62.

patronage.⁸⁸ Despite his encouragement to Leibniz to persevere in his research, then, he must have felt increasingly overwhelmed by and resentful of the young man's claims on his own time and expertise. By the spring of 1678, unable to conceal his frustration any further, Conring remarked that the variety and amount of Leibniz's activities were such that there would be 'no time to spare for their treatment', not to mention that Leibniz's propositions struck him not only as 'new' but also, disappointingly, as 'very far from the truth'.⁸⁹

Finally, much to Leibniz's dismay, Conring drew the correspondence to a close by congratulating Leibniz for his new position in Hanover, one that would 'direct [his] mind towards thoughts useful to the Republic, once he had left his paradoxical ideas and idle speculations behind' and allow him to apply himself to 'practical and useful matters'.⁹⁰

88. Fasolt, *Limits of History*, 152.
89. Conring to Leibniz, 26 February/8 March 1678, A II, 1:594.
90. Conring to Leibniz, 8 May 1678, A II, 1:620. Leibniz would draft two letters seeking to defend himself from some of Conring's criticisms by once again expounding his various discoveries which contributed to raising science rather than revelling in paradoxical activities of little use. Perhaps resigned to Conring's uncompromising stance, however, Leibniz seems never to have sent these letters.

CHAPTER TWELVE

Setting up a Scientific and Intellectual Network

LEIBNIZ WAS PARTICULARLY PRODUCTIVE in the early years in Hanover as, reaping the benefits of his years in Paris, he established a network of contacts who served his needs as a librarian, courtier, and intellectual fascinated by what was occurring all over Europe. He frequently asked his correspondents to send him letters and parcels, and he inquired about or requested new books as well as booksellers' catalogues.[1] In return, he himself sent them books and journals and offered a range of services, which we shall examine below.[2] In his correspondence, Leibniz combined professional duty with personal interest, thus enabling him to remain invested and active in the main intellectual and scientific debates of his time. Within this setting, too, he viewed others and their replies as 'living mirrors' on which he projected his temporary intellectual self-understanding, in this manner maintaining a constant correspondence with himself.[3]

From Hanover, for instance, Leibniz pursued his historical research, responding to Eisenhart, a law professor at Helmstedt and a former pupil of Conring who had sent him a dissertation on *fides historica*, by setting out principles of the critique of testimony. Upon receiving an edition of

---

1. Hansen to Leibniz, 13 August 1677, A I, 2:289; Philipp to Leibniz, 1/11 February 1679, A I, 2:421: 'It's been a while since I learned anything new from Paris or London: but that's partly my fault. Here are the books that Mr. Bulow brought me from England, I beg you to send me this catalogue.'

2. In this respect, Leibniz was abiding by one of the key principles of the Republic of Letters: *do ut des*, i.e., reciprocal exchange.

3. Ramati, 'Harmony at a Distance' (1996), 443.

de moribus Germaniae from Conring, Leibniz also sought to probe the origin of modern names and ancient peoples across Europe and to sketch out a Christian philosophy of history.[4] Leibniz's interest in history was perhaps, as we have already seen, brought to the fore in his discussions with Huet. He also sought to oversee, albeit with great difficulty and through the tireless efforts of Soudry, Mariotte, and in particular Hansen, the efforts to bring his calculating machine to fruition. During this time, too, Leibniz refined his work on movement and collision, composing his dialogue *Pacidius Philalethi* and his work *De corporum concursu* (1678), as well as adumbrating his law of conservation of force and his project for the creation of an encyclopaedia of all knowledge building on his *ars inveniendi*.

Descartes and Cartesianism continued to constitute a major thread of discussion over this period with a number of correspondents, including Fabri, Berthet, Eckhard, Princess Elizabeth, and Christian Philipp, especially as far as physical questions such as the problem of movement and laws of collision were concerned. After having been able, with Tschirnhaus, to inspect Descartes's papers in Paris, Leibniz, upon arriving in Hanover, renewed his critical engagement with Cartesianism, which he dismissed as 'the antechamber of true philosophy' and as 'paving the way to atheism'.[5]

Encouraged by his work on motion, Leibniz was careful to distinguish his own conception of hypothesis, which adhered to a strictly hypothetico-deductive procedure, from the explanatory models of Descartes's physics. Descartes, for all his merits, had failed to substantiate his hypotheses through sufficient experimentation. While he had figured out 'many things . . . that had not occurred to anyone among the ancients', often reasoned 'splendidly' and 'perused everything so ingeniously and deeply',[6] Descartes had erected fragile hypothetical constructions that threatened to collapse at any moment for their lack of empirical support and verification. While they were indeed 'marvellously ingenious', Descartes's hypotheses were thus, as they stood, often 'uncertain and sterile' and could be easily superseded, according to Leibniz. Extending his criticism of Cartesian physics as a whole, Leibniz judged Cartesian methodology to be highly deficient:[7]

4. Davillé, *Leibniz historien* (1909), 35–6.
5. Christian Philipp to Leibniz, 22 November 1679, A II, 1:766.
6. Leibniz to Conring, beginning May 1671, A II, 1:153; 3/13 January 1678, A II, 1:581.
7. It is worth noting, however, that as Leibniz was beginning to formulate his own philosophical system, he would have been keen to minimize Descartes's significance.

Descartes can undoubtedly be compared to the greatest men of antiquity, both in his power of judgement and in his success in conjecture . . . but he was limited in the number of disciplines he mastered. For, immersed in mathematical and metaphysical contemplations, and later involved in controversies with the most learned men of his age, as usually happens to those who rather ambitiously strive to be considered originators of a school—he lost a great part of his time needed for more useful things, and could only buttress his physics by experiments commonly known. His pamphlets on meteors, on the passions of the soul, and on man show how much he would have been able to achieve if he had lived longer. . . . His physical system is undoubtedly ingenious, but little explored, because it can only be regarded as a hypothesis, as can many others.[8]

With Craanen and Tschirnhaus, Leibniz criticized what he thought was Descartes's deficient proof of God as *Ens perfectissimum*, attempting in his correspondence with Arnold Eckhard to create his own logically flawless proof of God. He also criticized the voluntarist concept of God and objected to the dangerous idea that matter could gradually take on all possible forms.[9]

Many of Leibniz's correspondences were built on the mathematical discoveries that he had made in Paris. Topics of discussion included the search for general equations of all differentiable or integral curves, the invention of the inverse tangent method, the investigation of transcendent and especially exponential equations, the further refinement of the method of series of functions and numerical quantities, the development of a general method for Diophantic arithmetic, and the projecting of a 'geometry of the situation' or a *Characteristica geometrica*.[10] Leibniz also sought to develop further his *Hypothesis Physica Nova* in order to refine his method

---

8. Leibniz to Conring, 24 August 1677, A II, 1:563. Still, not all was lost, for Cartesian assumptions could initiate the development of demonstrative arguments: 'Nevertheless I would always advise a lover of the truth to deepen it [the Cartesian system], because we see in the latter an admirable skill of the mind, and his physics, uncertain as it is, can serve as a model for the true one, which must at the very least be as clear and as concerted as his; for a novel [*Roman*] can be beautiful enough to be imitated by a historiographer. To sum it up: Galileo excels at reducing mechanics to science; Descartes is admirable at explaining by fine conjectures the reasons for nature's actions' (draft letter to unidentified recipient, n.d., in Leibniz (1875–90), I, 335–36).

9. Leibniz to Fabri, early 1677, A II, 1:464: according to Descartes, everything was reduced to the blind arbitrariness of a world founder ('caecum arbitrium conditoris'); Leibniz to Philipp, end January 1680, A II, 1:786.

10. See introduction to A III, 2.

of scientific analysis, responding to Descartes's laws of motion and refraction, debating 'geometric reason' with Simon de la Loubère, or sending a copy of his quadrature to La Roque.[11] Such discussions of mathematics with a wide range of contacts, including Berthet, Gallois, Eckhard, Fabri, Hooke, Tschirnhaus, and Huygens, would, however, take a backseat after Oldenburg's death.

Crucially, Leibniz often inquired after information that he could draw on in his official capacity at court. There was often, in fact, an overlap between his official duties and his own intellectual pursuits and interests. He was keen to gather information for the purposes of introducing reforms at the Hanoverian court, pumping both Hansen and Justel for a wide range of political, economic, and technical material. From Hansen, he requested information relating especially to policing, government, the judiciary, manufacturing, commerce, and ceremonies.[12] From Justel, he requested information about political changes since the death of Cardinal Mazarin, and a list of ordinances and regulations concerning commerce and navigation. In addition, he sought French technical dictionaries and books on mechanical arts, manufacturing, and commerce.[13] Justel's response, however, was not as enthusiastic as Leibniz might have hoped: there was no 'catalogue of the inventions of the French', and moreover the French, Justel continued, had 'neither invented much nor possessed the gift of invention, being better suited to perfecting what others ha[d] invented', notably through recourse to English publications on commerce. In short, the disenchanted Huguenot told Leibniz, there was no one in Paris he did not already know who could contribute significantly to his knowledge.[14]

Sharing Scientific and Technical Curiosities

Much of the correspondence, however, was of a broader scientific nature, and much of the information exchanged concerned the circulation of scientific or technical 'curiosities'. These ranged widely: particular drugs 'with marvellous effects' (which caused part of the mind to lose its senses), Leibniz's design for a frictionless new scale, the printing of geographical maps on silks, the publication of Mariotte's *Physical Treatise* and *Treatise*

---

11. Leibniz to La Rocque, 25 April 1678, A III, 2:386–89.
12. Leibniz to Hansen, 7/17 July 1679, A I, 2:497.
13. Justel to Leibniz, April 1678, A I, 2:337.
14. Justel to Leibniz, 22 July 1678, A I, 2:355.

on Plants, a recipe for invisible ink (*encre de sympathie*).[15] There was also discussion of various optical instruments, such as magnifying glasses and microscopes, and new works on dioptrics, as well as Huygens's new treatise in which he attempted to account for double refraction.[16] Huygens, Leibniz was informed, had recently returned from Holland with microscopes with which 'one [could see] in the water an infinite number of small moving animals', and from which it was possible to conclude that all nature was animated.[17] Microscopes were 'one of the most useful discoveries for the advancement of physics and medicine' and would, through play, allow for 'an infinity of considerable observations'.[18] Good ones, however, were difficult to obtain, prompting Leibniz to devise a process for making microscope lenses in under an hour:[19]

> By this means you will dispose of a small theater of nature in miniature or a mine rich in delightful experiences. This is how we will be able to advance [science] marvellously in the space of a few years, now that not only philosophers and the curious, but also women and workers will be able to make microscopes immediately and for very little, and to closely examine the materials in which they are immersed since everyone will be able to have access to what was previously reserved to a few curious people.[20]

With Edme Mariotte, in particular, Leibniz discussed questions of physics and geometry, especially Jacques Ozanam's work on the quadrature of the circle and prime numbers.[21] Both also shared news of experiments on air resistance and traded meteorological observations obtained from barometers, thermometers, and hygrometers.[22]

---

15. Christian Albrecht Walter to Leibniz, 20/30 January 1678, A III, 2:334; Leibniz to Jean Paul de la Rocque, December 1677, A III, 2:298; Christoph Pratisius to Leibniz, 29 June 1677, A III, 2:465; Edme Mariotte to Leibniz, 23 July 1679, A III, 2:793; Leibniz to Mariotte, 30 June 1679, A III, 2:773; Leibniz to Schuller, 2 October 1677, A III, 2:240.

16. Christian Albrecht Walter to Leibniz, 20/30 January 1678, A III, 2:334; Hansen to Leibniz, 22 November 1677 A III, 2:274; Edme Mariotte to Leibniz, 23 July 1679, A III, 2:794.

17. Edme Mariotte to Leibniz, 6 March 1678, A III, 2:354.

18. Leibniz to Chevreuse, 19 December 1678, A III, 2:295.

19. Leibniz to Jean Paul de la Rocque, end of 1677, A III, 2:259.

20. Leibniz to Jean Paul de la Rocque, 2:261. Leibniz would in this regard demonstrate a continuing interest in Dutch pioneer of microscopy Antoni van Leeuwenhoek and his work.

21. Edme Mariotte to Leibniz, 6 March 1678, A III, 2:351.

22. Mariotte to Leibniz, 5 July 1680, A III, 3:221.

Of all the types of knowledge that were to be promoted, one stood out: the 'practicalities of life' (*commoditez de la vie*), as Henri Justel labelled them, namely, practical matters and inventions that benefited the public good. These included such contraptions as a kind of pressure cooker invented by Denis Papin that could boil fish and caused quite a stir in London at the Royal Society in May 1679, a machine 'able to grind enough wheat for 1500 men in twelve hours', and Comiers's invention of a machine that could knead 'as much dough as one wishe[d]'.[23] Others were of broader practical application, including new water-powered and water-lifting machines in Paris and Copenhagen, or of military application—Leibniz was always eager for information relating to 'artillery and firearms', such as a canon that could advance '300 paces farther than the enemy . . . without heating up or moving back'.[24] A curious invention mentioned several times in Leibniz's correspondence with Justel is his infamous project for a 'coach-boat' (*carosse-bateau*), a coach that could be transformed into a tent and a boat and whose practicality and usefulness Leibniz sought to defend, not altogether successfully, against Justel's scepticism.[25]

Leibniz assumed that progression towards truth from falsehood would be possible only by filling the gaps in existing knowledge, and he regularly mobilized his network to that effect. Ideally, Henri Justel wrote to Leibniz, an 'able man', in the likeness of the Roman historian and author Pliny, would undertake to record for posterity the 'useful and curious digressions on an infinity of rare things that had come to light' in their time, in such a manner as to constitute a 'genuine history of the world.'[26] Leibniz's cor-

---

23. Frederick Slare to Leibniz, 8 July 1680, A III, 3:231; Hansen to Leibniz, 20 December 1677, A I, 2:296.

24. Leibniz to Noel Douceur, Summer 1680, A III, 3:223; Hansen to Leibniz, 20 December 1677, A I, 2:296.

25. Justel to Leibniz, 7 March 1678, A I, 2:324–25: 'I really appreciate your kindness for sending me a drawing of the coach-boat which is a curious object. We find that it would probably take a long time assembling and dismantling it, and that it would be difficult preventing water from entering it. Moreover it is difficult to imagine where the wheels of the coach would fit when the latter is turned into a boat. . . . For these reasons, which are only conjectures made on the basis of your drawing, we do not believe that this invention can be useful. I only care about machines when they are inexpensive, comfortable and convenient. I hope that in your response you can address these small difficulties and doubts which may be unfounded, since I have never seen the Machine.'

26. Justel to Leibniz, A I, 2:317: 'I care very much about this volume on commodities of life, which is truly worthy of you and of this century. But I beg you not to spare us from useful and curious digressions on an infinite number of rare things which have come to your knowledge. I have hoped for a long time that a clever man undertake to do for our century what Pliny did for his, because in Pliny we find an infinite number of observations on the origin of the arts and the conveniences of life which men used in his time, and even

respondence with the various members of his network offers a catalogue of the 'rare and the marvellous' replete with eight-foot men, an 'extraordinary' deer, diamonds that sparkle in the dark, oversized birds, an armless woman, and apples that ripen at Christmas.[27]

Remaining cautious about the various 'boasts' (*fanfaronnades*) and 'illusory promises' (*propheties*) in circulation, Leibniz often expressed scepticism and asked for the confirmation (or refutation) of various rumours, including, for instance, those surrounding the manufacture of water-resistant cloth.[28] In their discussion on the subject of Comiers's 'talking head' and broader research interests, Leibniz and Hansen articulated their epistemological methodology most clearly: it consisted in trying to distinguish the true from the false, the possible from the impossible, in a context of considerable uncertainty. Yet 'groping' was insufficient, for the truth should be self-evident even to a blind person. Establishing the 'solidity' of an assertion was key to 'delivering oneself from popular errors'.[29]

---

on the corruptions that arose as a result of too many delights, whose knowledge is also useful. There are many things that would be lost without Pliny. This is why I wish that a capable person leave a true portrait of our time to posterity; with regard to its manners, customs, discoveries, coins, commerce, arts and manufactures; luxury . . . vices, corruption, diseases . . . and their remedies. He would neglect what can be learned from history, and he would only focus on what tends to be forgotten and yet deserves to be remembered, more so perhaps than what is ordinarily noticed. If so many other great designs prevent you from thinking of them, you will nevertheless be able to include a number of conveniences useful for that purpose in your volume on commodities. If we had such a Pliny in our century, posterity would follow his example and the continuation of this work would provide a true history of the world.'

27. Justel to Leibniz, 17 February 1677, A I, 2:248; Jean Paul de la Roque to Leibniz, 10 July 1677, A III, 2:189; March 1674, A VIII, 2:696.

28. Leibniz to Philipp, 25 February/7 March 1679, A I, 2:432; Justel to Leibniz, 27 September 1677, A I, 2:294; Leibniz to Philipp, 25 February/7 March 1679, A I, 2:432: 'I beg you to recommend me to Mr. Gudius when the opportunity arises. I earnestly wish to enjoy his conversation, as well as yours. I have conceived such a high esteem for M. Gudius . . . and I hold him above Mr. Vossius despite this man's great bluster. I recently leafed through his last book. What a marvelous speculator [*O le beau raisonneur*]. If he is as intrepid in history and fine literature, as he is in physics and mathematics, I would hardly trust what he says without having examined it well myself'; Justel to Leibniz, 27 September 1677, A I, 2:294: 'There is a German who claims to have admirable prophecies and the secret to making taffeta water-resistant. As he has a prophetic spirit, I do not place a lot of faith in what he offers.'

29. Justel to Leibniz, 17 February 1677, A I, 2:248: 'I only study to free myself from popular errors and to believe only what I see concerning nature and art, and to make a compilation of the commodities of life. If you know any, please share them with me.'

Setting Himself up as a Master Informer

Leibniz thus sought to carve out for himself a particular position not only at court, but also within the Republic of Letters, which, although mentioned explicitly few times, clearly forms the backdrop to many of his letters.[30] He went beyond simple adhesion to its codes to strive to develop, from his outpost in Hanover, an information network that would serve both his professional and personal scientific pursuits. Leibniz proved to be particularly skilful at leveraging the precious commodity of information, positioning himself as a master informer, omniscient about the activities of the Republic of Letters and uniquely able to report on them. Seeking to be perceived as irreplaceable, and mindful to calibrate his interactions with each member of a wide network of correspondents, he quickly became valued for his skill in publicizing verified discoveries and inventions.[31] As a key purveyor of information, Leibniz was regularly consulted by his various correspondents for his opinion or advice on a wide range of issues, or to verify pieces of information or rumours, even of a political nature.[32] On two occasions, for instance, Henri Justel asked him to confirm or deny whether 'trees [had been] planted in Germany by the head rather than by the root', and whether 'an artificial man with a lung and arteries' had been produced in Germany.[33]

Bemoaning the dearth of publications in Germany, Leibniz exchanged with Justel updates about Huet's collection, new books, and *commodités*.[34] He seems to have found in Justel, with whom he exchanged a dozen of letters

---

30. Louis Ferrand to Leibniz, 23 March 1677, A I, 2:260: 'The current news among us in the Republic of Letters is that Mr Baluze has given us his chapter in two folio volumes which cost 10 ecus. It is undoubtedly one of the most beautiful collections that we can have. The first volume on the monuments of the Greek Church by Mr Cotelier will be published soon. As for Mr Cotelier, he kisses your hands, and told me that there are many manuscripts relating to mathematics in the King's library... [even though] he was uncertain whether there were any relating to the analysis of the ancients.' With Oldenburg's death, 'the whole Republic of Letters has lost a good, enlightened, and well-intentioned subject'; Abbé de la Rocque to Leibniz, A I, 2:290.

31. Abbé de la Rocque to Leibniz, A I, 2:290.

32. See, for example, Christian Albrecht Walter to Leibniz, A III, 2:338: 'I would like to be even two hours with you to hear your judgement on all this, and your opinion on the course we should take. Because I will never lose the confidence I place in your considerable knowledge.'

33. Justel to Leibniz, 27 September 1677, A I, 2:293; 31 January 1678, A I, 2:316. Compare the 'strange reports' in the *Philosophical Transactions*, as discussed by Hoppen, 'The Nature of the Royal Society, Parts I and II' (1976).

34. Leibniz to Justel, A I, 2:481: 'What is printed in Germany is generally very little'; Justel to Leibniz, A I, 2:337; 28 June 1677, A I, 2:277.

over the period 1676 to 1679, an intellectual peer and a philosophical sparring partner. Their philosophical discussions included biblical criticism and Spinoza's posthumous works, Leibniz cautioning Justel against the dangerous 'paradoxes' in the Dutch philosopher's *Ethica*.[35]

Leibniz also requested news about the Republic of Letters from Louis Ferrand, and he regularly updated the Abbé de la Rocque with new discoveries.[36] From Christian Philipp he sought help in constructing a special compass that could discern the most subtle changes in position, and in obtaining a copy of Richard Simon's *Critical History of the Bible*. Through his network Leibniz hoped to receive 'the composition of artificial stones' (which Hansen hoped to extract from Comiers), news from Tschirnhaus's trip to Italy and of the progress made by the Abbé Mariotte with his *Logique* (which Leibniz would eventually praise as 'extraordinary') and his *Treatise on Colours*.[37] Other pieces of information he received concerned the return of Vansleb to Paris, an extraordinary sighting of a lunar eclipse, the English royal physician's successful cure of Bossuet and the duke of Orleans's eldest daughter, and Vicquefort's escape from Zelle.[38] Of course, this information network could also serve the purposes of control and surveillance at a distance, enabling Leibniz to ascertain the location and activities of various key individuals,[39] as in his use of Crafft to report on Becher and of Schuller to report on Crafft. And just as he himself was occasionally sought out by others, Leibniz remained alert for opportunities to establish new contacts, at times mobilizing his correspondents to extend his circle of correspondents, not least in Germany: 'As soon as I will be in Berlin I will try to introduce you to those you refer to in your letters.'[40] He requested Justel put him in contact with the French astronomer Adrien

---

35. Leibniz to Justel, 4/14 February 1678, A II, 1:154.

36. Louis Ferrand to Leibniz, 23 March 1677, A I, 2:260.

37. Hansen to Leibniz, 28 February 1678, A I, 2:321; Ehrenfried Walther von Tschirnhaus, 17 April 1677, A III, 2:60; Mariotte to Leibniz, 29 April 1677, A III, 2:75; Justel to Leibniz, 4 October 1677, A I, 2:297; Mariotte to Leibniz, 28 April 1678, A III, 2:407.

38. La Rocque to Leibniz, April 1678, A III, 2:385: 'Father Wansleb has forsaken his beard and his Darmenian habit to take back the Jacobin one. He is in a convent near Paris where he goes sometimes'; Hansen to Leibniz, 31 October 1678, A I, 2:375: 'The lunar eclipse that we witnessed on Saturday was very extraordinary, and we have not had one like it in four hundred years from what our scientists tell us'; Mariotte to Leibniz, 13 July 1678, A III, 2:468.

39. See, for example, Schuller to Leibniz, 19/29 March 1678, A III, 2:539: 'Mr Mohr was recently said to be still living in France. Becher has gone to Osnabrück for matters not unknown to me.'

40. Heino Heinrich von Fleming to Leibniz, 22 February/4 March 1678, A I, 2:323.

Auzout, and Philipp with the famous antiquarian and epigraphic collector Marquard Gudius, whose reputation Leibniz considered equal to that of the Scaligers and the Saumaises.[41]

Although he rebuked Père Cherubim, a natural philosopher and author of *La nature et presage des Cometes* (1665), for falsely seeking to claim credit for the invention of a telescope (actually invented by the German Capuchin Anton von Rheita), and he insisted on crediting the discovery of phosphorus to its true author, Brand, Leibniz himself was not immune to occasionally plagiarizing ideas for his own purposes, especially from practitioners he considered socially inferior to himself.[42] Nor was he averse to exaggerated self-praise, referring as early as 1675, when he had yet to make most of his discoveries, to 'the renown that [he] ha[d] acquired in the courts of Princes and among a good part of the learned and illustrious of Europe.'[43]

Leibniz had a robust sense of his worth, and even after his departure from Paris he was keen to uphold a certain image of himself as enjoying special favours from and proximity to the duke in Hanover. This self-presentation may have contributed to the high esteem in which many of his correspondents held him. Justel, for instance, along with other '*curieux*' never failed to 'render justice to [his] merits' and wrote in the autumn of 1677 that what Leibniz had conceived in the field of natural law, by providing definitions on which everyone could agree, 'exceeded anything that was available on the topic'.[44]

Thévenot, for his part, inquired after Leibniz's work in the hope of incorporating some of his findings into his own: 'I would like to know which of your good and great designs you are working on now, what you have advanced in the knowledge of nature since you departed from here [Paris], if you have discovered a few things that I could add to my own volume. This is the answer, the memory, and the compliment that I ask

---

41. Leibniz to Justel, April 1678, A I, 2:337: 'I wish to be known to Mons. Auzout by your recommendation'; Leibniz to Philipp, 25 February/7 March 1679, A I, 2:432.

42. On the telescope, see Hansen to Leibniz, 22 November 1677, A III, 2:273; on phosphorus, Leibniz to La Roque, mid-March 1678, A III, 2:355: 'I tell you this only because it is just to take care that no harm be done to those who are fortunate enough to be successful in something considerable after having worked [at it] all their lives.' Leibniz also made clear to Oldenburg as early as 1677 that Kunckel was not the discoverer of phosphorus: Leibniz to Oldenburg, 3/13 May 1677, A III, 2:117. On taking others' ideas, see Wakefield, 'Leibniz and the Wind Machines' (2010).

43. Leibniz to Johann Friedrich, February 1677, A I, 2:19.

44. Hansen to Leibniz, 28 May 1677, A I, 2:272; Justel to Leibniz, 4 October 1677, A I, 2:297.

you.'⁴⁵ The same Thévenot had also lavished praise on Leibniz's 'genius' as well as 'the surprising effects of [his] inventions' on another occasion.⁴⁶ Leibniz's erudition was widely celebrated, with Christian Walter finding in Leibniz alone 'what Italy and France produce[d] day by day that [was] rarer'.⁴⁷ In his various correspondences, Leibniz was described as a 'generous' soul, disseminating 'tenderness' in his letters, a man 'of profound knowledge, and rare merits' who was remembered fondly.⁴⁸

While some of these statements should obviously be taken with a grain of salt and merely abide by the conventions of intellectual discourse, collectively they support the idea that Leibniz's correspondents' appreciation of his (real or perceived) position in Hanover seemed to coexist with genuine admiration for him as a valuable member of the Republic of Letters.

Some of Leibniz's correspondents were keen to point out that their letter writing went beyond the requirements of polite epistolarity, remarking that

> if our letters were based only on a reciprocal exchange of fine curiosities, I confess to you in earnest that I would have abandoned it. For you have told me in your March 25 letter of various things, which I could not undertake to match without the greatest co-operation in the world. I undertake it however, with disappointment and confusion, with pleasure, to show you that this business proceeds from a sincere friendship, and on my side, from a deep reverence for your merit.⁴⁹

As Leibniz entered the service of the duke of Hanover, he acted as a counsellor advising the latter on various purveyors of knowledge and ventures. Leibniz henceforth positioned himself as a gatekeeper who would seek to recruit inventors in the service of his employer and induce them to 'discover' themselves—ironically, a position he had once occupied. From his place in the Republic of Letters and at court, Leibniz was solicited to recommend and put people in contact. But by positioning himself as a gatekeeper at the court of Hanover, however unofficially, on scientific and technological matters, he also attracted demands and requests from a wide range of people, including Georg Schuller, to help them secure a

---

45. Thévenot to Leibniz, 12 January 1678, A I, 2:309.
46. Hansen to Leibniz, 24 January 1678, A I, 2:312.
47. Christian Walter to Leibniz, 28 June 1677, A III, 2:184.
48. Christian Walter to Leibniz, 20/30 January 1678, A III, 2:334; Hansen to Leibniz, 10 July 1679, A I, 2:494; 28 June 1677, A I, 2:276.
49. Christian Walter to Leibniz, 27 April 1678, A III, 2:411.

position at court.[50] When Eckhard was forced to leave the University of Rinteln in Lower Saxony in 1678, Leibniz was able to intervene with the duke to obtain for Eckhard the post of pastor and superintendent in Jeinsen. The German medical physician Kornmann, for his part, had presented the duke with a list of inventions and remedies.[51] He described himself to Leibniz as a '*curieux*' who had 'invested much time and money' in his labours but 'expected nothing'; he would nonetheless happily serve the duke as his personal physician and provide the recipe for moxa, a Chinese technique, in exchange for a title and protection as well as the freedom to 'travel to [continue] collect[ing] curiosities'.[52] Crafft, some of whose research Leibniz had presented to the duke during the summer of 1677, hoped that the ruler would promote him on account of the growing esteem in which Leibniz himself was held.[53]

Leibniz regaled the editors of scholarly publications such as the *Journal des sçavants* with his discoveries, which were considered of great value, at least in some quarters. He determined the value of each piece of information he received and then repackaged and disseminated it back to the network as he deemed appropriate, thereby simultaneously advancing the public good and his private interests.[54] According to the Abbé de la Roque, the journal's editor, news of Leibniz's anomalous deer with an excrescence on its head seemed to have caused quite a stir at court: 'I took it myself to Monseigneur the Dauphin who regularly reads my newspapers. . . . Never has a gift been more appropriate for making a splash since the prince was about to give an audience . . . or received with more marks of esteem. M. de Montausier and M. de Condom, one of whom is governor, as you know, and the other preceptor of this young prince have been charmed, and we still only speak of this deer at court.'[55]

Leibniz's 'learned thoughts', which La Rocque requested he communicate to him, seem to have appeared to the latter as the perfect vehicle to gain access—and ingratiate himself to higher circles.[56] Leibniz was all too

50. Georg Schuller to Leibniz, 6 July 1678, A I, 2:15.
51. Leibniz to Kornmann, 8 September 1678, A III, 2:491.
52. Kornmann to Leibniz, 4 October 1678, A III, 2:511.
53. Leibniz to Crafft, end of July 1677, A III, 2:201; Crafft to Leibniz, 7 December 1677, A III, 2:289.
54. Gädeke, 'Gottfried Wilhelm Leibniz' (2005), 293, 297.
55. La Rocque to Leibniz, 10 July 1677, A III, 2:188. See Ariew, 'Leibniz on the Unicorn' (1998).
56. Hansen to Leibniz, 12 July 1677, A I, 2:279: 'Mr l'abbé de la Roque . . . paid court to Monseigneur le dauphin by presenting him with the story the deer.'

happy to take advantage of La Roque's eagerness to cultivate a relationship with both himself and the duke:[57]

> You would however make the duke happy and oblige me greatly, if you would share from time to time, and more often than you deem appropriate, any curious piece of information from Paris or elsewhere. For you frequent people of the mind, among whom you hold such a considerable rank... and who offer you an infinite number of observations which do not find their way in your journal; I hope you will do us this special grace... and you will have reason to receive praise from us.[58]

In addition to promoting Leibniz in his journal and sending the duke all his journals, La Roque kept Leibniz and the duke abreast of news (including that which did not feature in his journal) as well as reports of various French inventors and craftsmen, such as of a skilled Parisian gentleman who had devised a 'mechanical dove that could fly thirty feet', a musket that operated 'without powder or air compression', and 'a machine to help the handicapped work', and had harnessed the principles of hydraulics for manufacturing purposes that he was due to present at the Académie.[59] In exchange, La Roque hoped to locate a bookseller in Frankfurt who could publish his journal through the mediation of Hansen and Leibniz.[60] Such an 'advantageous commerce' was obviously not disinterested, and La Roque's aims and attempts to ingratiate himself with the duke through Leibniz struck the latter as transparent, leading him to deride them on at least one occasion.[61] Leibniz clearly had limited use for La Roque, even confiding that the latter was 'not as mentally powerful as Mons. L'Abbé Gallois who had since joined Colbert's service'.[62] Still, in an economy of exchange and mutual use, Leibniz was happy to extract

---

57. La Roque to Leibniz, 10 July 1677, A III, 2:188; La Rocque to Johann Friedrich, end of October 1678, A III, 2:527.

58. La Roque to Leibniz, 27 September 1677, A III, 2:224.

59. La Roque to Leibniz, 2:224; Leibniz to Johann Friedrich, 29 March/8 April 1679, A I, 2:157; La Roque to Leibniz, 9/19 December 1678, A III, 2:561; 3 June 1678, A III, 2:456.

60. Hansen to Leibniz, 16 August 1677, A I, 2:290.

61. La Roque to Leibniz, 3 June 1678, A III, 2:456; Leibniz to Johann Friedrich, November 1678, A I, 2:99: 'Here is the letter from the abbé de la Roque, author of the *Journal des sçavans*, for His Majesty. I add the one he sent me, so that you can better judge his intention. I find it quite amusing that people defend themselves of desiring what they so obviously do, and every time this happens I am reminded of George Dandin's maid who held out her apron while pretending to refuse the money that her mistress's *galant* offered her.'

62. Leibniz to Christian Philipp, 6/16 December 1678, A I, 2:393.

whatever he could from La Roque. Whether any of this would actually result in tangible collaborations or advancement was an altogether different matter, as Leibniz would himself soon find out.

Within this vision of learning, Leibniz ascribed a key role to rulers in the promotion of the sciences, particularly in financing and guiding inventors: 'The will of a monarch alone would be more effective than all our Methods and all our knowledge' and could 'enable us to obtain in a few years what would otherwise have taken several centuries.'[63] Scientific improvement could be promoted through the beneficial alliance of inventors and dedicated princes under peacetime conditions.[64] In this context Louis XIV appeared as a model for the patronage of learning: only he was in 'the state and mood' to advance the sciences,[65] and this made Leibniz all the more determined to seek some kind of affiliation with the Académie des Sciences.

63. Wahl, '"Im tunckeln ist ein blinder"' (2015), 238; Leibniz, 'Discours touchant la méthode de la certitude et l'art d'inventer', August 1688 to October 1690, A VI, 4:955.
64. Wahl, '"Im tunckeln ist ein blinder"', 240.
65. Leibniz to Jean Gallois, mid-October, A III, 3:726.

CHAPTER THIRTEEN

Facing Dead-ends in France

Instrumentalizing Phosphorus

By late 1677 Leibniz had settled in his new role in Hanover and could not have been more content—or so he claimed. To Gallois he painted a rosy picture of his position and life at court, where, as a freshly appointed court counsellor, he was valued and benefited from several advantages, including a higher stipend than the one initially offered by the duke as well as accommodation and dining rights.[1] Crucially, too, Leibniz boasted of the company of and close proximity to a prince 'whose extraordinary talents and great virtues caused a stir in the world', and of whose 'goodness and generosity' he was a beneficiary. This prince, contrary to rumours, had not 'deviated from the straight path' but demonstrated 'discernment that went beyond what one could expect' and was guided not by interest but 'the public good', which had led him to extend to Leibniz an unusual degree of freedom to attend to his own scientific pursuits.[2]

It is difficult, however, to escape the impression that Leibniz had mixed feelings about his new life in Hanover. In his letters to Gallois, he looked back at his time in Paris nostalgically, conceding that he regretted having had to leave, his stay having been cut short 'before it had been possible to pluck its fruits', but that he had had little other choice.[3] Had he not

---

1. Leibniz to Jean Gallois, first half of December 1677, A III, 2:294.
2. Leibniz to Jean Gallois, September 1677, A III, 2:226; first half of December 1677, A III, 2:294; A III, 2:566.
3. Leibniz to Jean Gallois, first half of December 1677, A III, 2:294; September 1677, A III, 2:226: 'In fact, Sir, I blush when I think of the pain I caused you and Mons. the Duc de Chevreuse: and yet you were kind enough not only to favour me, but even to invite me to seek your assistance again. The whole mistake I made was that I did not do sooner what I had to do in the end, because I would not have bothered you so often, and I would

[225]

grasped this opportunity in Hanover, he 'ran the risk of giving up the certain for that which still remained uncertain'.[4] In Paris, Leibniz had had the 'honour of making the acquaintance of people of great merit, including [Gallois]', and this had brought him 'consolation' for a time that may otherwise have been considered lost, especially since his attempt to join the Académie on the strength of his mathematical merits and calculating machine had failed, despite Huygens's and Gallois's support.[5]

Leibniz was determined to reestablish conversation in particular with Jean-Baptiste Colbert through his intermediaries, and to secure some kind of associate membership from the Académie. He had written to Duke Johann Friedrich that he had 'always carefully avoided any attachment other than that of the duke since [he] owe[d] everything to his goodness alone'.[6] This statement, however, seems to be contradicted by Leibniz's zealous effort to maintain personal and working connections to the Parisian scientific and courtly circles, not only by pitching some of his own proposals but—in a series of letters to Gallois, Chevreuse, Huygens, and even Louis XIV's almighty minister Colbert—by positioning himself as a conveyor of promising and potentially lucrative ventures.

The discovery of phosphorus, whose news he also shared with the duc de Chevreuse and on which he had published an article in the *Journal des Sçavans*, presented Leibniz with a perfect opportunity. This matter, Leibniz explained, sparkled at night and was bound to yield an 'infinity of beautiful applications' given the correct encouragement and the attribution of a 'reward commensurate with a discovery of this magnitude'.[7] This 'new species of art', as Leibniz wrote to Gallois, was 'well worth the perpetual light sought for so many centuries', could 'compete with the most beautiful productions of nature' and would help reveal 'a new country for

---

not have wasted so much time.... Indeed I do not regret having stayed so long in Paris since I got to know a few people whose extraordinary merit I will always honour, among whom you are one of foremost, which one can advance without flattering you. Perhaps the time will come that your kindness will not be entirely without effect, that the goodwill I enjoyed will be recognized, and the damage that I suffered through my fault will be repaired.'

4. Leibniz to Jean Gallois, first half of December 1677, A III, 2:293.

5. Leibniz to Jean Gallois, 2:293; Leibniz to Huygens, end of November/beginning of December 1679, A III, 2:903: 'My arithmetic *quadrature* has been clarified and demonstrated: I have preserved it for the *académie*, in the eventuality that its author be somewhat placed in relation to it, or that this treaty be deemed worthy enough of being placed with more important ones.'

6. Leibniz to Johann Friedrich, Autumn 1678, A, I, 2:85.

7. Leibniz to the duc de Chevreuse, 19 December 1678, A III, 2:578.

us', especially if placed in the hands of a king whom Leibniz perceived as directing the advancement of the sciences.[8]

When vaunting the merits and potentialities of phosphorus for the French state in several letters to Gallois, Leibniz once again set out to establish himself as a potential asset to the court of France, where his research on motion and the quadrature as well as his project for a universal characteristic had been widely praised.[9] Now he presented himself as a privileged purveyor of information, faithfully relaying back to France, and ultimately for Colbert's consideration, news of various 'curiosities' from across Europe.[10] This included news about a plant-based elixir (*eau vulnéraire*) that healed wounds in the promptest fashion, and the manufacturing technique (mentioned earlier) that allowed ten silk stockings to be produced for the price of one.[11] In dioptrics, too—a topic of particular interest to Colbert[12]—Leibniz had been able to identify a man able to build microscopes 'in one day', thereby potentially rendering discoveries 'common and inexpensive', especially in the hands of a scientific academy.

Emboldened, as he stated, only because Gallois had led him to believe that he might in time be rewarded and recognized for his talents, Leibniz suggested that it was appropriate for the Académie exceptionally to offer him membership, since he deserved it and would have been offered it had he remained longer in Paris.[13] He felt all the more entitled to such a special

---

8. Leibniz to Jean Gallois, September 1678, A III, 2:506. 'In fact, it is the king's greatness to properly recognize beautiful discoveries, and to encourage the inventors of whatever country they may be, since his glory is for all countries and he has been entrusted with the welfare of mankind.' A III, 2:580. Charlotte Wahl speaks of phosphorus's 'symbolic capital' in '"Im tunckeln ist ein blinder"' (2015), 234: 'Because of their sensational effects and the role played by light symbolism in the mythology and self-representation of the princes, phosphorus seemed ideally suited to inspire princes to science.'

9. Leibniz to Jean Gallois, September 1677, A III, 2:230: 'I work sometimes in the field of motion, and I cannot find an author who has not issued faulty rules, as I can demonstrate and even verify by experience'; Leibniz to Gallois, first half of December 1677, A III, 2:295: 'I left my Manuscript on the quadrature in Paris. Mons. Soudry, no doubt a clever man in geometry and mechanics, promised to arrange its printing. I beg you to facilitate this'; Leibniz to Gallois, 21 July 1679, A II, 1:733.

10. Leibniz to Gallois, first half of December 1677, A III, 2:295: 'I do not know if Monsieur Colbert would consider this at the present moment'; Leibniz to Gallois, 21 July 1679, A III, 2:789.

11. Leibniz to Gallois, September 1677, A III, 2:230: 'A healing water (*eau vulnéraire*) produced in these countries, cures and soothes the pain with marvelous speed'; Leibniz to Gallois, first half of December 1677, A III, 2:294.

12. Leibniz to Gallois, September 1677, A III, 2:230.

13. Leibniz to Gallois, September 1678, 21 July 1679, A III, 2:789: 'You gave me to understand, when I was leaving, that some day an occasion might present itself that would

treatment because the king had in the past 'honoured some individuals of repute in Germany with marks of his kindness', even though some of them contributed only minimally to the work of the Académie and were no better known in Paris than Leibniz himself—not, Leibniz quickly interjected in a weak attempt to maintain a semblance of modesty, that he numbered himself 'among the ranks of these gentlemen'.[14]

Having received no response to his letters, Leibniz renewed his claim a few months later, laying it on even thicker than before: he wrote in the hope of 'obtaining the abolition of the crime of silence and ingratitude to which [Gallois] may already have condemned [him]', and he thanked Gallois for 'suffering his importunities' and 'going to so much trouble for [his] sake'.[15] Leibniz described himself as a 'stranger, a man who was of no use to [him]', redeemed solely by Gallois's opinion that he could 'contribute something to the advancement of science', an accomplishment that he had unfortunately had to leave unfinished in Paris. With Gallois's support, Leibniz would 'immortalize' phosphorus, a sample of which he possessed, and demonstrate its potential 'as one of the royal academy's more important discoveries'.[16] In the summer of 1679, having still received no reply in six months, Leibniz stated yet more explicitly his aspiration to gain some kind of associate membership, from which position he would, like his peers, be able to celebrate the French king and his minister, especially now that peace had returned to the hemisphere.[17] Not receiving a royal pension would risk 'damaging [his] reputation'. Leibniz dared to trouble Gallois, who was in 'a position to promote beautiful things', only out of a 'sense of gratitude' and because, he claimed, he felt 'compelled to do so'.[18] Despite his plea, Leibniz was careful to defer to Gallois's discretion and judgement and presented himself as a mere mouthpiece without any personal stake who simply relayed information, and whose sole interest consisted in serving the 'public good'.[19] In fact, as with all the information he relayed back to Gallois and Colbert, Leibniz presented himself as an unambitious and disinterested messenger with little competence or relevant opinion of his

---

be of greater honour and advantage to me than what had been resolved at the time'. Leibniz to Gallois, September 1678, A III, 2:505.

14. Leibniz to Gallois, 21 July 1679, A III, 2:789; 19 December 1678, A III, 2:570; September 1678, A III, 2:505.

15. Leibniz to Gallois, 19 December 1678, A III, 2:570.

16. Leibniz to Gallois, 2:570.

17. Leibniz to Gallois, 21 July 1679, A III, 2:789.

18. Leibniz to Gallois, 19 December 1678, A III, 2:570.

19. Leibniz to Jean Gallois, first half of December 1677, A III, 2:295: 'I'm sending you all this, Sir, because you know what to make of these things.'

own—so useless in fact that, were it not for these curiosities, the letter would have been 'left blank'.[20] He reckoned that he 'would be remembered' in Paris in this manner, and perhaps even invited to showcase 'the little knowledge' he possessed.[21] Alternating between gratitude, deference, and self-abnegation, Leibniz sought to maintain his connection with Gallois and through him with Colbert—the real addressee of Leibniz's letters—in whose 'good graces' Leibniz begged Gallois to 'keep him'.[22]

From 1679 onward, Leibniz's determination to find a position in, or at least some sort of connection to, France seems to have acquired greater urgency. Now that he had left Paris, he was desperate to maintain a connection to the Académie—despite his working for a foreign power—and phosphorus appeared to him as providing the ideal means of doing so. After failing to hear from Gallois or Chevreuse, and drafting (but not sending) a letter in early September 1679 to the Académie in which he promoted phosphorus's remarkable qualities and promised its composition,[23] Leibniz turned to his former teacher Christiaan Huygens in a last-ditch attempt at gaining a foothold in the prestigious French institution.

Failing recognition by the members of the Académie of his mathematical innovations and especially of his quadrature whose manuscript had been mislaid, Leibniz set out once again to instrumentalize the discovery of phosphorus, promptly sending Huygens a sample wrapped up in wax. Even a small piece, Leibniz explained to his trusted mentor, would suffice for many experiments because it would glow for several hours. Not only that, but it displayed unexpected and remarkable potentialities such as serving as an invisible ink that would reappear upon being rubbed after a few hours.[24] Leibniz, who claimed to have begun production of phosphorus, begged Huygens to demonstrate it to Colbert and at the Académie in the hope that the communication of its composition may serve to

20. Leibniz to Gallois, 2:295, 294: '[Phosphorus] . . . is revealing a new world for us, and . . . is susceptible of an infinite number of wonderful applications. Only a great King can benefit the public, for the inventor needs to be encouraged. . . . I leave it to you to judge, Sir, whether we can still reflect on things of this nature. The author has asked me to write to you, and I am only serving as an interpreter.'

21. Leibniz to Gallois, September 1678, A III, 2:506; Leibniz to the duc de Chevreuse, 19 December 1678, A III, 2:580.

22. On several occasions, Leibniz reiterated his indebtedness for the 'goodness that [Gallois] lavished on him', which he hoped Gallois would not come to regret even though, because of Leibniz's departure, 'time had prevented them from seeing its fruits'. Leibniz to Gallois, first half of December 1677, A III, 2:295.

23. Leibniz to the Académie Royale des Sciences, first half of September 1679, A III, 2:835.

24. Beeley, 'A Philosophical Apprenticeship' (2004), 68.

'facilitate what [he had] reason to hope for one day'.[25] In a later letter, Leibniz reiterated his plea for Huygens's intercession on his behalf in reestablishing some sort of connection with the Académie, even remotely from Germany, either on the basis of his mathematical merits or as a kind of scientific informant who would relay useful information in a timely fashion. Such a 'regulated correspondence' (*correspondence réglée*) would help initiate a new kind of rapport with the Académie whereby Leibniz could inform it occasionally of matters worthy of mention—including, he still hoped, of his new relational theory of a space, the *analysis situs*, that he had conjured up and would 'one day produce extraordinary consequences', as well as various other 'considerable experiments'.[26] In this manner, Leibniz would be able to maintain some association with the Académie without having to be physically present in Paris.

Huygens answered Leibniz a few days later and seemed better disposed to assist him than Gallois had. After apologizing for his silence, he explained that he had been writing to Colbert, who was then out of town, and actively engaged trying to promote Leibniz at the Académie, where he had not only broken the news of phosphorus just a few days after receiving the sample but had even demonstrated it publicly.[27]

Like Leibniz, Huygens also recognized phosphorus's possible military applications, which were bound to attract Colbert's interest.[28] All this, however, was strictly contingent on the accuracy of Leibniz's set of instructions and the successful demonstration of phosphorus's ability to ignite gunpowder before members of the Académie and Colbert.[29]

---

25. Leibniz to Huygens, 8 September 1679, A III, 2:848: 'I beg you to show the effect to Mons. Colbert, the Duc de Chevreuse and at the academy. If you find that it is well received, I am happy to communicate its composition to the academy, even if obtaining it was very difficult for me.'

26. Leibniz to Huygens, 8 September 1679, A III, 2:848, 850. Leibniz was explicit in his ambitions: 'It may well be that what I have accomplished in other disciplines might still seem appropriate to count, one day, among the things which belong to the academy and particularly my arithmetic quadrature of which I even left the manuscript in Paris.' On the relational space theory, see De Risi, 'Analysis Situs' (2014).

27. Huygens to Leibniz, 22 November 1679, A III, 2:887: 'I have placed the shiny piece in a small flask with a little water as you told me, and have shown the effect to many inquisitive people, as well as last Wednesday to the gentlemen of our academy, who assembled for the first time since the holidays.'

28. Leibniz to Huygens, end of November/early December, A III, 2:893: 'Gunpowder is easily lit in the sun and by movement with a little of this phosphorus; to ignite gunpowder, all you have to do is take a piece the size of a pinhead or less . . . mix this little piece with crushed powder and grind the whole using a knife for example . . . and the powder will soon ignite.'

29. Huygens to Leibniz, 22 November 1679, A III, 2:887: 'I have a great desire to

After several failed attempts and dashed promises, members of the Académie greeted with considerable enthusiasm Leibniz's claim to possess the formula for phosphorus.[30] As the prospect of demonstration before Colbert and the Académie drew nearer, however, Leibniz made a curious suggestion as to how Huygens should proceed, namely, by performing the experiment without revealing Leibniz's name, in the interest of not making the latter's renewed request for membership too obtrusive. Another, more plausible explanation is that Leibniz did not wish this new venture to risk being tainted by his past series of failed promises, hoping instead that it would benefit from Huygens's authority and in the process eventually help rehabilitate him:

> If you find it convenient to demonstrate this new fire to Mr Colbert, and find it appropriate to take this opportunity to raise my interests, I would like that, without mentioning my name, you indicate that this curiosity comes from a German well versed in such matters, as well as geometry and mechanics, who offers to communicate their composition; and that he offers to maintain a correspondence with the Academy, and to inform its members of the various beautiful curiosities that are discovered from time to time in Germany ... so long as they extend him the honour of considering him in some way part of the Royal Academy with some appointment.[31]

It was imperative that Leibniz's name be mentioned only once the experiment had succeeded, he explained, so that Colbert, who had already heard about Leibniz (as well as his various, unevenly successful schemes) and frequently received requests from him, not be prejudiced against it. Crucially, it was necessary not to let this novelty's potential be dashed by 'some disgust' at hearing the same proposal again:

> For [Monsieur Colbert] having often heard my name in times not suited to it and grown tired of it, will be repelled if he is reminded of it. Because great men having initially had difficulties with one particular matter do not give up their opinion easily, and in those cases one succeeds only by proposing it as brand new. I hope that if the Duc

---

conduct an experiment with gunpowder, but I hold this matter to be so precious that before having another sample, I cannot resolve myself to consume what I have. Besides wanting to perform this experiment before Mr Colbert, I would like to be sure that it will not fail. And to this effect, I beg you to send me the instructions.'

30. Huygens to Leibniz, 2:888.

31. Leibniz to Huygens, end of November/beginning of December, A III, 2:894.

de Chevreuse and the Abbé Gallois take up this matter, they also be warned to refrain from presenting Colbert with an old request.[32]

In his final letter to Huygens on this matter, sensing Huygens's uneasiness, Leibniz laid out the reasoning behind his previous comment and, recognizing perhaps that he had overstepped the mark, qualified it, ultimately leaving the matter up to Huygens's discretion—and hoping, too, that the latter would overlook his somewhat embarrassing previous misstep:

> With regard to my last note you remain at liberty to act as you see fit. I thought that a new solicitation would be more welcome than an old one, and that it would be easier to sound out their intentions. . . . If it is possible to refrain from mentioning my name and to speak in general terms, it may be good to do so: but if this turns out to be impossible, it is better to be honest if the name is indeed requested. Please dear Sir, have the kindness to keep this opinion to yourself: my confidence in your benevolence has allowed me to raise this point.[33]

Huygens did in fact go on to perform the experiment before the Académie, but without being able to reproduce the effects described by Leibniz,[34] whose attempts to become affiliated with the Académie remained unsuccessful, at least until 1699. After Leibniz's letter of 26 January 1680, an eight-year hiatus in his correspondence with Huygens followed, mainly on account of Huygens's illness.[35] This gap, combined with the lack of references to Leibniz elsewhere in his correspondence suggests that perhaps Leibniz was not as important to Huygens as he thought himself to be. His eagerness and insistence—and one dares say his presumption and the pressure he may have placed on Huygens to assist him—compounded with the often unintelligible and unrealizable projects he submitted to Huygens, not to mention the increasingly dubious pretexts invoked by Leibniz, is likely finally to have exhausted both parties.[36] Moreover, the

---

32. Leibniz to Huygens, 2:900–901.

33. Leibniz to Huygens, 1/11 December 1679, A III, 2:908. Huygens would in fact severely criticize Leibniz's strategy on the grounds that it was likely to cause him serious trouble with Colbert. 11 January 1689, A III, 3:48.

34. Leibniz to Huygens, 8 September 1679, A III, 2:847–50; end November/beginning December, A III, 2:899–902; 11 January 1689, A III, 3:48.

35. It is interesting to note, however, that Huygens barely mentions Leibniz in his correspondence with others during those years.

36. Salomon-Bayet, 'Les académies scientifiques' (1978), 169, speaks of a 'laborious dialogue' and of Leibniz being 'skillfully clumsy' (*malhabile dans son habilité*).

unsuccessful demonstration of phosphorus at the Académie may have led Leibniz to retreat from the correspondence in embarrassment.[37]

Their correspondence would resume only in 1688—when they were on more equal terms—after Huygens's solution to the problem of isochromic curves was published in the *Nouvelles de la République des Lettres*. From this point on, they exchanged letters on a number of scientific themes, including the integral calculus, the problem of the structure of the universe, and the theory of light.[38]

That his successive failures with Gallois and Huygens did not deter Leibniz from going to the very top reveals much about his personality. Far from being discouraged, Leibniz, possibly in a mix of desperation and extreme tenacity, addressed in December 1679 directly to Colbert a letter that may have been intended to coincide with Huygens's presentation of phosphorus at the Académie and was certainly intended to remind the minister of Leibniz's merits.[39] His discoveries, including his solution to the century-old problem of the 'quadrature of the circle', which had 'fructified' since being 'born in France', had been received and acknowledged 'as legitimate by some outstanding savants'.[40] Thus he had, he told Colbert pointedly, 'earned the right not to be abandoned'. Geometry being the 'spirit of the Sciences', it was 'of vital interest to the Republic that [it] not be neglected, for it sharpen[ed] the minds and [taught] to argue strictly'. France had done much for the sciences: 'our age is indebted to you for the rebirth of Geometry, since your Viète and Descartes have resurrected the art of discovery, hidden by the ancients, like a holy fire', and it took great care to 'preserve this glory' by 'request[ing]', 'prais[ing]', and 'support[ing] so many excellent specimen of ingenious minds'.[41]

Colbert's lack of response did not discourage Leibniz permanently but merely prompted him to draft another letter three years later. This time Leibniz opted for a different approach: he now undertook, with the minister's permission, to expand the bounds of knowledge with a study of soils and stones, plants and animals—objects of which humankind had thus far only a superficial grasp, even though the comprehension of

37. Heinekamp, 'Christiaan Huygens' (1981), 100.
38. Heinekamp, 100.
39. Leibniz to Colbert, first half of October 1682, A III, 3:718–21. I was unable to find any reference to Leibniz in Colbert's private papers.
40. On Leibniz's *De quadratura arithemetica* (1676; text in A VII, 6:520–626), see Crippa, *The Impossibility of Squaring the Circle* (2019), 93–156.
41. Leibniz to Colbert, December 1679, A III, 2:919.

them 'naturally preceded' all other forms of knowledge. Such an up-close examination of the material, something he was then ideally positioned to undertake, would help grasp the principles and mechanical processes from which everything else derived.[42] From his favourable position in Germany, Leibniz offered himself as an instrument towards the accomplishment of Louis XIV's metaphysical destiny, in the advancement of the sciences and the general welfare of mankind, to which all 'civilized peoples on earth and rational centuries to come' were indebted and on which knowledge otherwise 'impenetrable to men'—including 'the secret of the heavens, the greatness of the earth, and the measurement of time'— had been bestowed.

Leibniz would not gain admittance to the Académie—in fact, it is uncertain that he received replies to any of his letters to either Gallois or Colbert. A mere two years later, however, Ehrenfried Walther von Tschirnhaus would be admitted to the Académie on the strength of Colbert's recommendation and his recipe for the composition of phosphorus, which Leibniz had happily communicated to him in support of his candidacy.[43] The Académie's decision not to grant Leibniz membership has traditionally been ascribed to his Protestant faith or German nationality.[44] While these explanations are both perfectly plausible, I would like to suggest that, at this stage in his life and career, Leibniz, with his various unsuccessful schemes, lack of an outstanding track record in the sciences, suspicious loyalties, and unorthodox ways of proceeding, was perhaps not the most desirable candidate in the eyes of the prickly and demanding Colbert, but rather someone to be used but ultimately kept at a distance.

From Hanover, Leibniz would pursue his research nonetheless. The next few years witnessed the elaboration of a logic capable of "weighing" probabilities, the devising of health care policies including to combat the spread of plague, forays in medical statistics and life insurance, the innovation of new methods in the areas of tangent determination and quadratures, Leibniz's explicit rehabilitation of substantial forms,[45] and his proposal for a *scientia generalis* (general science). Through the latter mankind would be able to establish a record of all truths and human knowledge. He also composed his *Meditationes de Cognitione, Veritate, et Ideis*, in which he distinguished different kinds of knowledge, and his *Brevis demonstratio erroris*

---

42. Leibniz to Colbert, first half of 1682, A III, 3:721.
43. Prinzler, 'Aus der Geschichte' (1993), 12.
44. For example, Antognazza, *Leibniz* (2009), 174. He was eventually elected in 1700.
45. Antognazza, 217.

memorabilis Cartesii et aliorum circa legem naturae, in which he set forth an alternative law of the conservation of force to Descartes.

But overall, the period of remarkable innovation that had characterised Leibniz's life since his return from Paris came to a close with the death of Johann Friedrich in December 1679 and the accession of his brother Ernst August (1629–98). The new duke, irked by the debacle of the windmill project in the Harz mountains, which he summarily drew to a close in 1685, tasked Leibniz instead with composing a history of the Guelf house and of the dynasty of Braunschweig-Lüneburg in particular,[46] in the hope of establishing a connection to the Italian noble house of Este—a task Leibniz would eventually creatively turn into his posthumously published history of the Earth, *Protogaea sive de prima facie telluris* (1749). In January 1680 Leibniz drafted a memorandum proposing a new range of enterprises, including a princely printing company, laboratories, and an archive that, ideally, he himself would be called on to supervise. But all these projects, as well as his scheme for an imperial society devoted to the study of natural sciences—within a decentralized Germany—were to remain unrealized.[47] He also failed to obtain the position of librarian and privy counsellor at the imperial court in Vienna in 1680—a position he had been coveting since 1673—and from which he could have implemented some of his economic projects. Although he sought to adapt his projects to the requirements and conventions of the court and to spearhead reform from within it, the transnational communication and collaboration he advocated remained largely anathema to court culture.[48]

46. Antognazza, 230.
47. Klopp, *Die Werke von Leibniz* (1864–84), 3:329; cf. 308–9.
48. See Ramati, 'Harmony at a Distance' (1996), 451: 'Thus Leibniz's simultaneous professional socialization in the officialdom of princely courts and in the Republic of Letters prevented him from becoming a more successful participant in either community.'

Conclusion

LEIBNIZ ARRIVED IN Paris in early March 1672 on a diplomatic mission on behalf of his patron, Johann Christian von Boineburg, aiming with his infamous Egyptian plan to divert Louis XIV's plans for military expansion in Europe. Although ultimately unsuccessful, this mission allowed Leibniz to remain in Paris and take full advantage of the city's blossoming intellectual and scientific life. His sojourn in Paris, at the time the most sophisticated and advanced centre of European culture, was a revelatory experience for him. There, he invented the calculus, took an active interest in the debates of his time, whether they concerned Orientalism, Cartesianism, or even the ongoing *Querelle des anciens et des modernes*. Leibniz revealed himself to be a veritable social animal, butterflying from one savant and project to the next and encountering a unique environment, his status as an outsider affording him a privileged vantage point on the Parisian intellectual scene and courtly circles, and especially on Jean-Baptiste Colbert's fledgling state Republic of Letters. Both Leibniz and Colbert were administrators who thought in systems—the first as a savant, the second as a merchant—and Leibniz was a prime witness to Colbert's newly fashioned absolutist and centralized vision of *académies* as institutions of learning. Although Leibniz was not central to Colbert, he understood Colbert's project better than most at the time, hoping in fact to secure for himself a position in this network, ideally at the Académie des Sciences, from where, he was convinced, he would be able to implement his projects.

Leibniz's hope was disappointed, but his Paris period nonetheless acted as a catalyst, increasing his renown in the scholarly world. Furthermore, he and his ideas were, so to speak, 'transmuted', helping spawn a period of remarkable innovation upon his return to Hanover. In this sense, Leibniz's return to Germany spelled his passage from theory to practice, experimentation, and implementation, building on the knowledge and know-how he had acquired during his Parisian stay. In Paris and later in Hanover, Leibniz was socially mobile, moving in and out of multiple intellectual and scientific circles, assessing the validity of 'curiosities', advocating his scientific agenda, promoting technical and commercial ventures, and conjuring up new procedures or methods—always steered by a vision of reform and seeking practical applications from theoretical knowledge, even though this was not always possible.

He believed in the power of 'new philosophers' to exploit the productive potential of science and sought to vindicate his own claims to superior expertise through his various schemes. In Hanover in particular, he emerged as a technological and commercial projector who competed in a marketplace of new schemes and inventions for the attention of powerful patrons who in their turn competed for fame, reputation, and power. This often required couching his proposals and inventions in terms of 'reason of state' and utility as well as careful strategizing and no small artifice and intrigue.[1] The staggering range of Leibniz's activities included publishing articles, promoting scientific and philosophical societies as well as a universal language, inventing a calculating machine, and advocating engineering projects as well as seeking to acquire the secret of phosphorus.[2] All these, he hoped, would accrue great benefits to Johann Friedrich and grant him access to the highest realms of power, from which he would then be able to carry out even greater schemes for the benefit of mankind. As Leibniz understood, his idealism could be sustained only by a certain dose of realism: once inside powerful political structures, he hoped to deploy them towards the implementation of his own vision.

In Paris and Hanover, Leibniz met a group of people who influenced him and who in turn were influenced by him, including Huygens, Huet, Crafft, and Conring. He acted as an intermediary between different worlds, that of artisans and that of scholars, that of inventors and that of patrons, often linking the holders of productive knowledge and the court. Crucially, he hoped to create a place for himself within the Republic of Letters and to establish his own brand of knowledge and sphere of influence. In person and especially through correspondence, he established the relationships and networks that would sustain him as a savant for many years to come, and he amassed a significant symbolic capital by gathering

---

1. See Keller, *Knowledge and the Public Interest* (2015), 271, 275.

2. Ohnsorge, 'Leibniz als Staatsbediensteter' (1966), notes that Leibniz's mind could not content itself with only one activity, and it was always seeking to disperse itself in various directions.

information that he sought to leverage into a position at court. But if his efforts were generally successful within the Republic of Letters, they were rather less well received at court.

Leibniz aspired to live an 'amphibious' existence in which he would be able to combine scientific and philosophical with political, legal, and diplomatic roles. His activities as a savant, a philosopher, a projector, and a theologian often doubled with his legal, political, and diplomatic activities as he passed seamlessly from one function and register to another. In the face of Louis XIV's aggressive and expansionistic foreign politics, the German savant aspired to a key political role which would see him implement reforms, reconcile Catholicism and Protestantism within the empire, mediate conflicts, and broker peace in the hope of helping create a new defensive alliance of German states under the leadership of the Holy Roman Empire.

Leibniz's case goes beyond mere self-fashioning: he was simultaneously one and multiple figures, whose various identities did not exclude each other but often overlapped and fundamentally cannot be disentangled from one another: as a client concerned with securing a stable position and pleading for recognition, a young scientist eager to develop his mathematical skills, a trouble-shooter for his employers, a self-effacing adviser, a skilled agent and negotiator always on the lookout for new information, and finally, a young optimist moved by dreams of religious reconciliation and projects of a universal characteristic and the promotion of the public good. Personal ambition and the genuine desire to improve the condition of mankind and advance the cause of science went hand in hand.

More generally, the framework in which Leibniz operated was made up of interesting paradoxes. His correspondence brings to light a courtier and scholar keen to evolve within established models of power and science while attempting to effect change from within. In this regard, a paradox of his correspondence is particularly striking: the coexistence of various performances of transparency and an adherence to the canons of epistolarity with an obsession with secrecy. In 1710 the German scholar Zacharias von Uffenbach visited Leibniz in Hanover and later reported that, although the librarian was eager to show books and manuscripts from the ducal library, he was very reluctant do so from his own, behaving 'secretively' (*geheimnisvoll*) about books Uffenbach asked to see.[3] From early in his career, Leibniz developed an acute sense of observation, which

---

3. Uffenbach, *Herrn Zacharias Conrad von Uffenbach* (1753–54), 1:409–11; see also Bertram, 'G. W. Leibnizens Beziehungen' (1906–7), 196–98.

he refined through his constant collection of information. He was keen to acquire new information, which he would then keep secret. Through his hyperactivity and his omnipresence, Leibniz hoped to always remain in control and be, so to speak, one step ahead. Calculating, always weighing his options, probing different avenues of maximizing his chances of realizing his plans, Leibniz sought to impose his will and vision on events in the face of conditions that he understood to be precarious and highly mutable.[4] He was a man on a mission, often at odds with the rest of the court, within a competitive and mutually exploitative world.

And yet, while Leibniz was in certain respects luckier than most, benefitting early on from powerful patrons, a cursory survey of his life between 1672 and 1679 reveals surprisingly mixed results. Ultimately, his prospects were limited and he inevitably encountered obstacles, especially so early on in life when he had yet to make a significant intellectual impact. Despite his undeniable talent and energy, Leibniz suffered from the considerable disadvantage of not being of noble extraction, a fact that would have limited his entrée into courtly circles. Much as he was admired, he also regularly left perplexed or dismayed some of his interlocutors. He was an audacious young man, eager to proclaim his great ideas and implement his schemes, but his overconfident language and lack of disclosure of information sometimes left even his best-inclined correspondents, such as Carcavy, frustrated and unimpressed. He had a habit for promising much but delivering little, and many of his projects failed. And despite his best efforts, Leibniz remained at the mercy of events and circumstances, which at times he misread.

Leibniz often enjoyed testing the norms of his time, and his particular skill here lay in simultaneously embodying and transgressing them. He was perceptive, quick to discern networks, circles, power dynamics, and to seek to place them in his service. This exemplary exceptionality served to assert Leibniz's particular place within a system of norms while also signalling his position as a favourite. However, his behaviour could come across as arrogant and capricious, and, if it often succeeded, it sometimes backfired spectacularly. Back in Hanover, Leibniz wanted success at court, but only on his terms. Ultimately, one concern drove him: how to make the establishment and structures in place work for him and his projects, while always maintaining a foot firmly outside of them. Yet, as he discovered, despite his attempts to exercise an influence well above his official

---

4. Pocock, *The Machiavellian Moment* (1975), in this regard offers a useful lens through which to read Leibniz's trial and tribulations. See also Buchenau, 'Leibniz' (2010), 199, who argues that every political problem became a 'game that could be solved like a mathematical equation'.

station, Leibniz could bend the rules only so far, especially in such tightly controlled settings as Colbert's state Republic of Letters. In his zeal, he sometimes failed to grasp that his vision and way of doing things did not necessarily align with those of his employers or potential patrons.

Crucially, as he set out to navigate their codes in the hope of gleaning information and making useful contacts, he was not always the accomplished operator he thought himself to be, and he was himself aware that he often failed to produce a favourable first impression. This proved to be a significant problem in a context in which social realities always remained hierarchical, and in a culture in which the rhetorical dance of patronage and clientelism constituted the very oxygen that sustained learned and courtly life. Leibniz's failure to master certain social norms sometimes made him appear gauche and insincere, while his doggedness and even brashness, especially in the face of difficulties, were bound to make him appear ambiguous at the very least. Personal missteps, too, may even have alienated potential patrons, as in his unfortunate correspondence with the Duke of Mecklenburg. Leibniz's impatience and tendency to seek to bypass hierarchies and conventions of courtly behaviour, on account of his belief in his own exceptionality, would have made him an undesirable recruit. Ironically, his very dedication to reform and the felicity of mankind could leave him looking like a dubious character.

Seeking to maintain his independence and pursue his own goals and interests led him to offer his services nonexclusively to various political masters. In France, Leibniz's mobility—especially as a foreign agent eager to collect information and offer his services in exchange of powerful patronage—his tendency to associate with everyone and anyone, would have cast doubt on the strength of his loyalty, for which reason the French establishment preferred to confine him to rendering particular and limited services.[5]

For many at the time—and indeed later—Leibniz was difficult to pin down. Similarly, in Germany he seems not to have been perceived as a fully reliable political agent, and consequently to have remained restricted to the antechambers of power, acting as an informant and adviser only in an unofficial capacity, at times exaggerating his involvement in political matters.

The years 1672–79 were perhaps the most important years of Leibniz's life, laying the epistolary and courtly foundations for his future intellectual

---

5. Salomon-Bayet, 'Les académies scientifiques' (1978), 170, characterizes Leibniz as first and foremost 'a man of movement'.

work. But for all their catalyzing effect, these were hard years, full of disappointments and unrealized ambitions, and his trajectory was anything but uniformly upward. Ironically, Leibniz's very drive and fluidity also proved to be his Achilles' heel, hampering his progress. Though undeniably seen as promising and talented by some in these years, Leibniz would have to wait a few more years—until the publication of his differential calculus under the title *Nova Methodus pro Maximis et Minimis* in 1684 and that of his integral calculus two years later—truly to consolidate his reputation as a savant and reap the benefits of the seeds he had sown in his early career.

ACKNOWLEDGEMENTS

THE RESEARCH FOR this book was conducted principally in Oxford, but also at the Leibniz-Archiv in Hanover, the Max Planck Institute for the History of Science (MPIWG), the Leibniz-Zentrum für Literatur- und Kulturforschung (ZfL) in Berlin, the Interdisziplinäres Zentrum für die Erforschung der Europäischen Aufklärung (IZEA) at the Martin-Luther-Universität Halle, the Lichtenberg-Kolleg (Göttingen Institute for Advanced Study) at the Georg-August-Universität Göttingen, the École des Hautes Etudes en Sciences Sociales (EHESS) in Paris, the Early Modern Cosmology Group at the Università Ca' Foscari in Venice, and the Centre for Mathematical Philosophy (MCMP) at the Ludwig-Maximilians-Universität in Munich.

Over the course of my research I was fortunate to receive support from the Royal Society, the British Society for the History of Philosophy, the British Society for the History of Science, the German Academic Exchange Service (DAAD), the German History Society, the Wyndham Deedes Memorial Trust, the Society for the Study of French History, the Stapley Trust, the Royal Historical Society, and the Association for German Studies.

I thank Laurence Brockliss and Richard Arthur for their comments, and Christian Badura for assistance with Latin translations. I am particularly grateful to Daniel Garber and Jacob Soll for their advice as well as their kindness and encouragement.

I am indebted to various friends and colleagues who have accompanied me and my work over the years in one capacity or another, including Jeffrey Barash, Claudine Cohen, Lorraine Daston, Morderchai Feingold, Peter Fenves, Daniel Fulda, Markus Gabriel, Nora Gädeke, Eva Geulen, Bernard Gowers, Kathrin Gowers, Stephan Hartmann, Joel Horowitz, Sophie Horowitz, Guy Jackson, Vera Keller, Michael Kempe, Marie-Louise Lillywhite, Dana Jalobeanu, Julika Mimkes, Alexandra Paddock, Pietro Omodeo, Siegmund Probst, Celeste Schenck, Andrea Thiele, Matteo Valleriani, Charlotte Wahl, Daniel Weidner, Charles Wolfe, and Richard Wolin. At Princeton University Press I am grateful to Josh Drake, Anita O'Brien, and especially Ben Tate. My warmest thanks also go to Stephen Harrison and Louise Locock for their generosity in providing me with a home for nearly two years during my time in Oxford. Tigger and especially Finn proved to be the best and most affectionate research assistants I could possibly wish for.

Four people in particular accompanied me and this project over the years and deserve special mention: Davide Crippa for his indefectible friendship; John Davis at the Queen's College, Oxford, who went above and beyond his role as college adviser in supporting and encouraging me; Nicholas Halmi for putting up with my various idiosyncrasies; and Bernard, my rock and faithful companion of many years.

APPENDIX: TRANSLATIONS OF
KEY TEXTS DISCUSSED

EXTRACTS FROM THE LETTERS QUOTED in the book and this appendix have been translated primarily from French and Latin. I am responsible for all the translations.

1 *Society and Economy* (Societät und Wirtschaft), *1671*
SOURCE: LEIBNIZ-ARCHIV, LH XXXIV, FOLS. 228-29 [A IV, 1:559-61]

By ensuring that manufactured goods are produced locally rather than imported, our society will avoid monopolies since it will always be inclined to supply products at their fair price, or in many cases even cheaper. It will also strive to prevent the formation of any monopoly by merchants or corporations of artisans, and to avoid the excessive accumulation of wealth by merchants or the impoverishment of artisans—as is the case in Holland, for instance, where the merchants live well while the craftsmen are subject to continuous poverty and hard work. This is harmful to the Republic since even Aristotle acknowledges that craftsmanship should be one of the better-compensated activities. 'Nam Mercatura transfer tantum, Manufactura gignit' [since markets can only offer what factories produce].

And why indeed should so many people find themselves in poverty and misery for the benefit of such a small handful? The farmer does not live in need since his bread is guaranteed, and the merchant has more than enough. The others either have no employment [*nichts nuz*] or are servants of the state. Society can satisfy all the needs of the farmer too, provided that it always buys his produce at a sufficiently fair price, whether low or high. In this manner society can preserve itself from food shortages of natural origin for all eternity since it can create a general reserve of cereals.

The establishment of such a society seeks the elimination of a deep-seated source of regression in many of our republics, namely, people's ability to earn a living as they see fit, to get rich at the expense of a hundred others, or to go bankrupt, dragging down with them the hundreds of other people who placed themselves under their responsibility. Nothing prevents an individual from ruining his own family, just as nothing prevents him from squandering his own funds or those of others.

Should the money be invested in other countries? This is out of the question. Each country should equip itself with the capacities enabling it to produce these necessary goods and manufactured products which previously had to be imported so that it does not have to rely on others for what it can produce itself; each country will be instructed on how to exploit its own resources appropriately. In a country that possesses sufficient wool, factories will have to be built for the preparation of the fabric; another country with an abundance of flax will occupy its population with the production of clothing, and so on. Thus, no country among those which offer the necessary degree of freedom to society will prevail over another; rather, each will see itself prospering in the areas where God and nature intended it to excel.

Manufactures will therefore always be based at the point of origin of raw materials, while commerce ... will occur along rivers and oceans—an arrangement that is abandoned (with manufactures near centres of trade, far from raw materials) only when the necessary society and cohesion are lacking in many places, especially in the absence of republics.

A great flaw in republics and countries is that many regions have more scholars (not to mention the unemployed) than craftsmen. But no one is idle in the society we envision; it requires scholars for their continual discussions and joyful discoveries. This society can leave to other professions the task of caring for the poor. Likewise that of detaining criminals, which is also of great benefit to republics.

It could be objected that artisans today work only out of necessity; if tomorrow all their needs were satisfied, they would no longer have to work at all. I maintain, however, the opposite, namely, that they would be happy to produce beyond what they do now out of mere necessity. For if a man is not assured of his own subsistence, he has neither the heart nor the mind for anything, but produces only what he hopes to sell (which is not a great quantity, owing to the small number of customers), is preoccupied with trivialities, and does not have the heart to undertake anything big and new. He therefore earns little, often indulges in drink to dispel his own desperation and drown his sorrow, and is tormented by the malice of his workers. But then everything will be different: everyone will work with pleasure because each will know what he has to do. He will never find himself without work involuntarily, as is presently the case, since no one will work for himself, but rather in cooperation; and if one has too much and the other not enough, then the first will give to the second. On the other hand, no artisan will suddenly be compelled—as is now sometimes the case—to torment himself or his men with excessive toil, since the amount of work will always be kept more or less constant.

Workers will work together, completing their work to the best of their ability in the common joy of a job well done, and the masters will themselves undertake work that requires more expertise. No master will be distraught that an intelligent worker might desire to become a master himself, for how can that harm him? Workers' provisions will be provided to them without charge. No master will have to worry about providing for his children or marrying them off well. Parents will be relieved of the task of educating their own children, since that will be undertaken by society: all children, when young, will automatically be raised by women in public institutions. Scrupulous care will be taken that they are not too crowded, that they are kept clean, and disease-free. Who could live a happier life than that? Craftsmen will work merrily together ... singing and conversing, except for those whose work requires more concentration.

The bulk of the work will be carried out in the mornings.

Every effort will be made to provide pleasures other than drinking—for example, discussions about the trade and the telling of all sorts of funny stories, by means of which they should be able to find something to quench their thirst.... There is no greater pleasure for a reasonable man, or for any man accustomed to it, than to find himself in an assembly where agreeable and useful things are discussed, and therefore every group, including artisans, should have someone taking note of the useful remarks that might be made. But the highest rule of the society will be to encourage true love and trust among its members, and to express nothing irritating, contemptuous, or insulting to others. Even heads of state should avoid making hurtful insults, except in cases where nothing else is effective, since such behaviour prevents trust from being established. No man will be ridiculed for making a mistake, however serious; on the contrary, he should be reprimanded in a brotherly way and, at the same time, punished justly at the next opportunity. Punishment will consist of greater and heavier work, the master being assigned the work of a worker, and the workman that of an apprentice.

Moral virtues will be promoted to the highest standard and, as far as possible, according to the principle 'Octavii Pisani per gradus' [according to Octavius Pisa, by successive degrees]. If it is found that two people cannot resolve their dispute, they will be separated. Lies will also be punished.

2 Leibniz to Carcavy, November 1671

SOURCE: LEIBNIZ-ARCHIV, LBR 143, FOL. 11 [A II, 1:287-90]

I intend the following section of this letter to be treated separately from what preceded it and confidentially for, with your permission, I will address certain things that concern my own private business. A short while ago I recommended to you vendors of books and machines, and now with increasing boldness [I recommend] also myself.

I see that you are not only diligent, but also aspire to great and outstanding things and embrace with your mind the progress of the arts and sciences, since the great Colbert, the most apt minister for the true glory of the Great King, encourages such an excellent plan. And truly, what can concern the State more than to raise its own power and human happiness? ... A new light arises, but it is sparse, uncertain, and dusky, similar to the first day of creation before it had gathered in the sun, as recounted in Genesis. The knowledge accumulated is colossal, akin to a forest fit for felling, as it were, but scholars lack coordinated action.

What the English Royal Society has accomplished so far, too, is great in itself but scant in comparison to what would lie in our power if we wanted it enough. And I believe that this [goal] is pursued by you whenever I imagine the excellent mind of Your King, who—if, as he has begun, plans to improve human affairs—will make his memory blessed to mankind. You will not only win over the English, but all men: if you would make a further noble effort worthy of your grandeur and of a King who seeks his glory in the happiness of mankind.

But where does all this lead to, you will ask? I will tell you. Some kind of trade is necessary to those who drive forward such excellent and far-reaching affairs, and where both inventions and experiments, and ... outstanding minds are born. Among the merchant arts it is fundamentally important to transfer the raw material of others to oneself, and to cultivate it by one's own artifices. The same has to be achieved in the sciences.

There is a class of man in Germany—I do not know whether there are many outside Germany—harmful to himself, useless to all, and currently neglected, but of greater use for the republic in the future than one would assume, if only it were known. These men alternate between insanity and wisdom. And because there is no great mind without this mixture, at least the *appearance* of stupidity has prevailed among them, or, to state it philosophically, has taken residence. Mostly, they do not breathe or speak anything but inventions, experiments, new ideas, and not always in vain:

often there is a lot of gold hidden in this manure, if one has the patience to look. Arguably we owe this class of people the most important discoveries in art and nature, argumentation, metallurgy, gunpowder, as well as advances in medicine.

A section of this class of men searches for the Philosopher's stone, another section perpetual motion, yet another malleable glass or incombustible oil, or even—if it please the gods—an elixir of immortality and other fooleries. And without doubt, for the most part, such Phaethon-like ventures go astray, but some stumble upon heroic medicines . . . or machines useful for life. . . . Some of these experiments are delivered in a way that they are useless in themselves; but combined with others that are hidden in another part of this dung heap, they can in the end yield a huge gain.

These men, however, cannot help themselves: most of them wither away with ruined faculties and lost reputations, neglected, despised, poor, angered by their misfortunes and, as they believe, an ungrateful world. They take to their grave, along with their kind, what it was in the state's interest not to allow to perish, for while nobody cared for them, of course, nobody listened or bought from them what they did not want to give out for free after it had been created with great labour. As for me, I did not abandon my work in jurisprudence and my chosen way of life, by which I know I can adequately support all other neglected and passing things, and took great care not to expose myself excessively to these men's contagion. Still, I have had a most pleasant and, what is generally rare with suspicious people, an intimate conversation with great numbers of such men, which will perhaps not prove altogether fruitless.

I know those who have used up ten—who would believe it?—even twenty full years wandering throughout Europe in order to compile hidden and curious discoveries; and yet these good men, once they returned home, had hardly enough to sustain their lives; not on account of a flaw within their venture, of which they have many and excellent ones, but a flaw of character . . . of someone who is rolling that Sisyphian rock in the vain hope of finding something, neglecting everything else. The arts of Englishmen are known to me, and I know in what manner so many famous experiments have reached them; [I know] for how long these experiments were despised as long as they were conducted by lowly German artisans, and how, now that they have reached the grandees of the Royal Society, they are deemed the highest inventions or secrets.

Since this is so, do not receive with ill humour what I now propose: that I can render myself—whatever I am to you for advertising many outstanding things—an instrument perhaps not useless, and can compile excellent things not only from chemists and noted mechanics, but also from doctors, mathematicians, and learned men of all kinds.

I have cultivated ... a correspondence with most learned men throughout Germany ... [and] people have a favourable opinion of me throughout courts everywhere, also among German and foreign ministers; most famous men, not only Germans, but also Italians, Englishmen, Belgians are accustomed to responding to my letters very kindly, and send back not only idle talk but also substance; learned men of all kinds, theologians, jurisconsults, doctors, mathematicians, as well as historians and linguists, are not displeased about their exchange of letters with me. Appointments have been frequently offered to me by great leaders, or certainly by the greatest men in the state, often in person, which I have declined only so that I might have time to conduct my scientific investigations more freely, and perhaps with greater benefit to the public.

And I have never asked anything more ardently than that it be allowed for me to spend my time quietly perfecting the sciences and this so-called trade of the arts. If your recommendation, Illustrious Lord, can achieve something, you will perhaps feel that you were not mistaken in your benefaction: you too will recognize—since your judgements are of such wisdom and substance—the importance of what I propose, which bears on the acquisition of state secrets, namely, arts through which human power is increased and which are to be sought everywhere.

If the pontiffs had passed a bill that all inventions by monks should be disclosed only to them under the obedience of faith, they would without doubt be the rulers of the world, for even Barthold Schwarz's invention of gunpowder would belong to them alone.

There is no mortal who knows this better than the great Colbert, to whom you will easily recommend me with your wisdom and most persuading loyalty, and you will also ensure that there will be a place for me among the foreigners whom the Greatest King adorns with his generosity. I, in return, will send you, with every book fair [i.e., in spring and autumn], not only books and indices of books, but weekly and monthly news about ingenious minds, thinkers, inventions, experiments, and attempts as well as results, upcoming publications (including news of the hidden, neglected, and dead), and news of libraries and museums—news, more generally, of all matters.

You will easily find someone more learned than I, but no one better prepared for this plan through experience, education, and finally through the joy itself which I have almost from childhood on received from the variety of manifold acquaintances. ... Nobody regretted having recommended me, I have shown that I could achieve something in that realm, not only perhaps by vain words until now, but also sometimes by deeds. And you will not meet anyone who knows me who ... won't agree. But a substantial and honest mitigation is necessary for the expenses which have to be borne, which must be spent on ... secretaries, but also sometimes for journeys to interesting people.

It must also be considered that the time which is spent on these matters ... (especially since it is mainly through exchanges that something can be obtained from curious people) must be deducted from other occupations. Furthermore, I wish that my proposition at the very least be kept secret, whatever you may think of it: if you discard it, then because I do not wish to be held accountable for things not said; if you lend your ears to it, then because in that case it may happen, once news of the matter has been disseminated, that the men I can make use of will resist me, motivated by envy.

I hope for a swift reply, and for the rest entrust the whole matter to your loyalty and wisdom, for I have not written this but for your virtue, reticence, candour, and trustworthiness. If a negotiation comes to happen, and if I can put the beginning of this winter to use as soon as possible ... and my plan be fully approved by you, even greater things than expected could take place. I do not dare undertake a journey for something uncertain, people being as they are here, a journey could bring about suspicion (for nobody will believe that I only travel for the sake of a scholarly matter) or even harm.

3 Leibniz to Pierre-Daniel Huet, 15 April 1673

SOURCE: LEIBNIZ-ARCHIV, LBR 428, FOLS. 4-7 [A II, 1:363-65]

... You will recall that we spoke recently of the great work which was undertaken by order of the King, at the instigation of the very illustrious Duke of Montausier, under your direction, and to general approval: to restore a disappearing world of letters [*revovare literas fugientes*], reviving the light of an almost dying antiquity, and giving new life to the best authors, while the age of contempt has returned after barely a century in the shape of a new kind of barbarism, and these authors have begun to close their eyes again as though they were tired of living. Indeed, I see some [of our contemporaries] taking advantage of the opinions and complaints of great men, Bacon, Galileo, and Descartes, to put down ancient wisdom and conceal their own ignorance, to the point of seeming rightly to despise a knowledge unworthy of being known. They deprive themselves, and to a certain extent the world, by dispensing with all the benefits and trials of so many centuries ... as if a kind of great restoration, if we are to believe Bacon, or wiping the slate clean, if Descartes is to be believed, were needed in order to think correctly; but one should perhaps forgive the indolence of those who made discoveries more freely in their own times, and the idle idolatry of those who accept dogmas, for they produce from their own resources important findings comparable to the discoveries of the ancients. One should not recuse those who produce their own experiments, even if in general they judge other disciplines unfairly.... Certain disciples, who are just as sectarian as before since they have only changed masters, violently protest against all the dogmas of antiquity.... That they consider a man sufficiently philosophical, quite erudite, preferable to Aristotle ... who explains the phenomena of nature only by means of subtle matter, and ... vortices—all this clearly ends in the ruin of holy doctrine.

Youth in fact allows itself to be carried away by an ignorance so sweet, so seductive, and willingly pretend to ignore, with presumption, so many things.... [This ignorance entails] a greater loss: I think, in fact, that if we disregard all the other disciplines and only engage with the experiences and productions of the language of our time, religion, whose truth is confirmed either by visible miracles ... or by the oracles of antiquity, finds itself in danger; this warning was reiterated recently by Meric Casaubon in a letter to a friend published in English. To this disease of the century which creeps in so perniciously, you oppose a remedy which will be effective for the second or the third generation; for if you imbue a prince with such great hope, destined for such great things, with mysteries that touch the heart of doctrine, you will have worked towards posterity for many centuries. The world has grasped from the example of François I, whose effects endure today, how much power the encouragement of a single prince wields; all study to please the prince, and the fate of letters is linked to his will, which directs the flow of valuable spirits. But noblemen too, and those whose birth destine them to play a role in [public] affairs, will no longer be deterred by notorious criticism ... it will be given to each, by the power of reading alone and without the tedious step of going to beg for help elsewhere, to penetrate into the sanctuaries to which only the greatest minds could hope to gain access thanks to painful work in the past.

That eminent men have been chosen to contribute to your monument, no one who knows you doubts. I am all the more amazed at the benevolence of your judgement, you who thought recently that a man like me could be of any use. I first attributed this idea to your natural benevolence ... but when I saw you stand firm—you whose judgement is so penetrating that I could not doubt it—upon reflection I found a way to reconcile my scruples with your wishes. I admit, in fact, although I do not claim either the mind or the knowledge, that I have nonetheless sought, through my zeal, to win the praise of impartial censors: and what else could be expected from a native of Germany, a nation whose sole gift of the spirit is perseverance?

It remained to choose the writer on whom I would test my strength, and such was your kindness that you allowed me to know what was still available: among those in whom one could find the example of best philosophy, you also gave me your recommendations; of this kind, as you said, remained Pliny, Mela, the agronomists, Apuleius, Capella, Boethius.

But since you have done me the grace of letting me choose, I will say, with your permission, what I think suits best my own work. Pliny easily turned me away from him: he needs a man of stature, and what I believe first and foremost, a doctor, who is familiar with the whole gamut of medical matters. I did not dare immerse myself in the agronomists, because unless one is initiated in the mysteries of economics, it is not worth taking on the challenge, especially in order to compare our techniques with those of the ancients (which is the only way to benefit from the light that we will have shed on these writers). Thus I chose, so as not to detain you any longer, Martianus Capella: a widely used author, of pleasing variety, who not only skims the sciences but penetrates them; he is the only writer among those who survive to present a sort of encyclopedia of the liberal arts. The most difficult and disfigured? Perhaps, but there is nothing I dare not undertake under your guidance. I hasten to add that I would still have chosen him even if other authors had been available: I have always burned to unite the wisdom of the ancients with the discoveries of our time: this is why I impute to good fortune the fact that he remains available until now.

4 Leibniz to Christian Habbeus, 5 May 1673

SOURCE: LEIBNIZ-ARCHIV, LBR 347, FOLS. 15–18 [A I, 1:416–17]

... During my time in France I have observed that the factories there are for the most part in the most flourishing state that can be desired, on account of that nation's talent as well as the particular care of the King, who has brought in the best workers from everywhere, and spares nothing to draw from them their secrets and inventions, which often are not worth much in the hands of an individual but, when they are placed under the authority of a great prince, are capable of enriching many people. ...

As Paris is the Metropole of Gallantry [*Métropolitaine de la Galanterie*], it would be important to fish for the fine and delicate aspects of workers' secrets, which can sometimes be done with skill mixed with a little liberality. They are marvelous in the art of Chinese varnishing; and there is nothing more beautiful than their gilding which is carried out inexpensively and even without gold. A certain kind of very beautiful earthenware is produced that resists fire without losing its lustre. The art of casting or pouring all kinds of materials into sand moulds is refined to its ultimate perfection, as is also the casting of medals. The machine that weaves silk stockings and fabrics like a [human] worker is in vogue here. People here take particular care in perfecting dyes. They work well with iron, and there are people who possess the secret of moulding iron into shapes and then filing it away, which is not commonly done. There are others here who crush marble and stones with machines and then produce the shape that pleases them with the paste produced which then becomes as hard as the original marble. Another has a mill with which one man can grind in a day what could otherwise only be ground in many. A certain person from Montpellier came to present to the King a wonderful essence that can stop blood instantaneously even if an artery is cut. Public experiments have been carried out by order of the King. There is, finally, an infinity of curiosities—such as in goldsmithing, enamelling, glassware, watchmaking, leather manufacture, pewter pottery—to which most foreigners do not pay attention and yet only one of which would repay their expense if reported.

For my part I have had the opportunity not only to come into contact with a large number of good workmen, but also to get something out of them as well, and had I been willing to spend a little, I would have learned much more. For wanting to have my arithmetic machine built, with which the four species of calculation are carried out without the slightest work of the mind—a model made a long time ago and shown in France and England to general praise and which is on the verge of being completed—I had occasion to get to know these people. And I will be able to multiply these contacts if my plan is approved, and the knowledge gained thereby can only benefit the country.

5 Leibniz to the Dauphin, end of 1675

SOURCE: LEIBNIZ-ARCHIV, LH IV, 8, FOLS. 66–67 [A II, 1:394–98]

Since the time is coming, my Lord, when mankind, whose concern You now are, will be placed in Your care, it will be in the general interest that

You be instructed in its abilities which it will be Your duty to increase, and reliably pass on undiminished to posterity. From everywhere You see people flock to You to provide accounts of how the World has been governed thus far, You see that the Greatest King, who, in Yourself, will last many years, does this in order that the inventory of the world be impressed upon You; whence You may understand not only how things presently stand, but also through which steps it has come to this point. . . .

There are people who disclose to You the wonderful secrets of destiny, who will remind You of that red land animated by the breath of life and of the apple ill-fated for posterity, of the land purged by the disastrous flood, of the world just fresh out of the [*sc.* Noah's] ark, of the fathers of peoples, the creators of tribes, and the holder of nations, Babylon; and they will remind You further of the World's old days. . . . Thence [*sc.* They will remind You of] Abraham's people chosen by God, not on account of its own merit but so that from this the heavenly restorer of human matters, so often foretold by prophecies, may at length come forth, and a people scattered all over the world may stand as sure witness of the continued tradition of such miraculous historical events.

You will marvel at a number of fishermen, animated by some higher power, declaring war on mankind, and at the mind of philosophers, and at those triumphing over the power of Emperors, in the end putting up a victory sign in Rome itself under Constantine. Soon [*sc.* You will marvel at] how the Barbarians are roused by God from the outermost north, so that they may be taught. You will become aware of Clovis's vow, and of Martel's famous victory, to which the preservation of religious belief and power is owed, and of Charles the Great's campaigns for the faith. And when, in a fatal revolution, the East washes away the West's fortune under the Saracens, You will see Your ancestors, marked by the cross, rehearse for those great matters . . . whose beginnings our time will possibly see, when the fates turn.

There are others who instruct You on how to conduct matters, who convey the rise and fall of empires, and above all that which makes the Republic durable, what makes it prosperous, how subjects' love prevents terror. . . . [Also] since an invention in physics will establish clearly that the whole character of the war remains unchanged, there will always be men who describe to You the laws of motion, and the mysterious mechanisms of nature discovered or enhanced by the various inventions of men. Those same men will teach that man's greatest instrument is the human mind itself, which exercises a reign as great as some small deity in these matters.

Since this is so, it will be worth the trouble to expound the history of that war which man conducts with hidden and rebellious nature ... so that it may be clear by which skills [*artes*] those ancient guides on this battlefield of the mind, whom we call philosophers, have confronted it [*sc.* nature], how far the boundaries of our empire over the physical world have been extended, and what is that new intellectual world revealed to us in our age by Galileo, Bacon, Descartes, and other argonauts.

I took upon myself the duty of writing the history of philosophy, which is neither less delightful or nor less useful than politics. You will see that the Oriental peoples, as the originators of religion, were also the safekeepers of truth, but that much was kept secret among the Egyptians, Babylonians, and Indians, until Pythagoras brought philosophy to the West. I hold him to have been a great man, for I see he did this to make men better and rulers wiser, which is why, for a long time, the administration of the state in the cities of Greater Greece was in the hand of the Pythagoreans. Pythagoras seems to have practised two disciplines especially, mathematics and ethics. Democritus spent his life with experiments, Socrates spoke amiably about human matters; Parmenides, Plato, Aristotle and Chrysippus developed from metaphysics and dialectics a certain technique of disputation, that was subtle and ingenious and not to be despised. The reign of the Stoics was long, foremost in free Greece and in Rome. For that sect with its fierce liberty was a nuisance to those who reigned. The ancient Christians did not shy away from Plato, because they judged he sometimes said things that did not deviate from the arcane doctrines of religion. But St Augustine joined the peripatetic acuteness of disputing. Hence when the Barbarians had plundered East and West, I do not know by which fate it occurred that Aristotle was the only one of the ancients whose body of philosophy remained unscathed. Thus when Charles the Great considered the restoration of the Schools, the renewed philosophy was necessarily Aristotelian. For the discoverer of new doctrines was not to be expected from the barbarians of those times. ... For Aristotle was the first to bind reasoning to certain strict laws after the model of mathematics. His Organon will always, so long as there is fair judgement, be held among the most beautiful inventions of mankind. Nor does his physics really deviate as much from the experiments and inventions of our time as many think. Harvey, who informed the world about the circulation of blood, found pleasure especially in Aristotle's writings on procreation [*De generatione animalium*], the parts [of animals, *De partibus animalium*], and the history of animals [*Historia animalium*]. And Aristotle left us, if I am to judge anything, with his logic, rhetoric, and political science, admirable monuments. Thus, since in the Monasteries, which were then the only refuges of learning, almost nothing but Aristotle and Augustine was read, it is no surprise that people of profuse intellect and spare time, but remote from experience and the world, have held certain abstract ideas, acquired from these authors, longer than necessary. Whence arose the scholastic doctrine, which itself has its uses; for there are, among the older scholastics, Thomas Aquinas, Scotus, Guillaume Durand, Ockham, and others, some outstanding reflections that those who grasp the merit of the first philosophy will appreciate. In this situation Aristotle held dominion, until, among Greece's shipwrecked remains stranded in Italy, some Platonists set themselves up. From this time on, wars befell the Republic of philosophy, and Aristotle competed with varied success, wavering in single combats but not defeated in war. He held back the Platonists easily—for what is easier than to repel words with words?—but once he saw that machines were advancing and that trenches were dug, and that physics marched into the fight, with mechanics and chemistry on the rise, then he finally began to doubt his reign's perpetuity. Therefore, he abandoned part of his jurisdiction undefended to his enemies, receded from the sky altogether; and confined within the sublunar sphere, he not without success offered the four elements in response to the enemy, who hurled nothing but atoms. But when Descartes's vortices appeared—a new and strong kind of war machine—the Elements betook themselves to flee on foot. Fire, already a friend to chemistry, deserted to the enemy; water slipped away under his hands like eels; he was astonished to discover that the air had become dense; earth feared lest it should experience the fate of a Comet, carried away and tossed from whirlwind to whirlwind. The matter now fell back to the *trarii* [*sc.* the eldest and most experienced soldiers in Roman legions], and the reign of philosophy was at stake, had it not been for one outstanding youth famed in weapons—they call him ether—who traces his origins back to the sky, and corresponds to the element of the stars. Like a second Achilles coming from the ships, he brought his men succour; around him, the vortices buzzed in vain. Ether, assuming now the name of world soul, now of universal spirit, set out as an explorer [or 'a spy'] to the enemy's camp. He brought back captive fire to the [*sc.* Aristotelian] standards, aroused dissent among the enemies with many skills, caused the atoms to whirl. With Archaeus [*sc. archaeus*, the lowest, densest aspect of the astral plane, as named by Paracelsus], his brother, serving with great authority and fame among the chemists, he secretly returned to favour and with his help, as the companions of Paracelsus and Helmontius, produced salt, sulphur, and mercury, the threefold liquid, gas and blas [*sc*. Van Helmont's terms for occult principles in bodies], acid and alkali—scuffled with subtle matter, bits of dust, debris, furrowed and knotted particles.

Thus the Peripatetics breathed again. Now peace is at stake, and the negotiators Hippocrates and Archimedes hope that this will be obtained from those exhausted by war . . . they turn their weapons against the common enemy, once the pact has been negotiated. . . . so that nature's barriers will be shattered by the various devices set in motion by Drebbel, Bacon, Torricelli, Guericke, Boyle. And those hidden enemies who prepare an ambush on our life every day, unless they are destroyed by total annihilation, shall certainly be reduced to a smaller number.

6 Account of the Present State of the Republic of Letters (Relation de l'état présent de la République des Lettres)

SOURCE: LEIBNIZ-ARCHIV, LH XXXIX, FOLS. 33–34 [A IV, 1:568–71]

DRAFT A *ACCOUNT OF THE PRESENT STATE OF THE REPUBLIC OF LETTERS*

Here is an account of the trips I have undertaken to a country in which all great men want to live. I did not visit all the provinces, nor did Father Martinus explore all of China, or Olearius, who speaks so well of Muscovites, reside among the Samoyeds. I note that each country has a manner of practising erudition that will not be valued elsewhere.

It is necessary not only to provide an account of the present state of the Republic of Letters, but also the conjectures of what is to come. In matters of religion, we now work towards reconciling reason with revelation. There are people who have too poor an opinion of Holy Scripture. One day we will have rigorous geometric demonstrations of *Deo* and *Mente*. The ideas of Plato [will be] resurrected. Mons. des Cartes should have lived longer. I would gladly have given him the age of Mons. Roberval. . . . There will come a time when the mechanical arts will be exhausted, and we will only rely on the secret properties of natural bodies. Geometric speculations will serve the soul, rather than the body. . . . The soul, if dreams lasted a long time, would at last learn to reason perfectly while musing.

Medicine [will be] greatly altered by these great new diseases, such as that venereal pestilence scurvy. In ten years we may have amassed more new medicines than in all previous ages. . . . In geometry, I have formulated an analysis for the geometry of squares, as well as another for the inverse method of tangents, and in algebra, an arithmetical operation. There remains to be had a good demonstration for arithmetical analysis. . . .

Mons. the Duke of Montausier, Mons. the President de la Moignon support the *belles lettres.* Large monuments [*bastiments*] are built on ideas. They will one day be overthrown, but there will remain something solid and admirable.

A beautiful musical tune that will always be sung is like a beautiful geometrical theorem.

It will be appropriate here to touch on the question whether the inscriptions and medals should be made in such modern languages as French. Medals are for all places and times; inscriptions [are] for all times. The King must have more regard for His glory than for His language. In order to make inscriptions in French, it would be necessary to find the secret of stabilizing the French language so that it does not evolve. This characteristic belongs only to the dead or dying languages, whose peoples no longer exist or whose empires are in decline. Too many good volumes will cause them to all perish, and since not so many good volumes are produced in Latin, those will survive, because it will be easy to distinguish the good ones from the bad. Time is like the King, who takes it as a maxim to reward only those who are quite unique in each species.

Wanting to dedicate it to the King, we could begin as follows: Sire, I present to Your Majesty the account of a country, where His immortality will be guaranteed. These are the Elysian fields of the Heroes, and it is necessary to pass through there to have any hope of posterity.

We must conclude with a pathetic speech to excite devotion. . . . The comparison of Galileo and of Descartes can be adroitly included.

I urged Mons. Huygens to work in physics, and Mons. de Mariotte in Medicine. . . .

That the Arithmetic machine calculates, irrespective of Pythagoras's belief that counting is a property unique to man. That beasts are soulless, or their souls immortal. . . .

That men are too *savant*; that the Chinese excel by hiding things. . . . Hobbes makes as much noise in morality as does Descartes in physics.

People are foolish everywhere, and I never would have believed that in a city as refined as Paris children today would be kidnapped to bathe the sick in their blood. . . .

I know clever people who never bother to read any demonstration. . . . I am however thoroughly pleased that the task of demonstrating rigorously has been taken up.

Mons. Huet must render a great service unto Christians, by proving that the works of Holy Scripture belong to the authors to whom they are attributed.... Had Queen Isabella not been moved in particular by a few persuasive conversations, Columbus would not have discovered the new world.

I would not want the rigour of demonstrations to slacken. Someone I know purports to demonstrate the divisibility of magnitude to infinity, very poorly. The Chinese lacked geometry because they did not appreciate demonstrations of what one must and can establish a priori.

DRAFT B *ACCOUNT OF THE PRESENT STATE OF THE REPUBLIC OF LETTERS, AND CONJECTURES OF WHAT WILL COME*

The Republic of Letters is an otherworldly colony led by a certain Greek adventurer, named Pythagoras. He went to seek new countries after the example of the Argonauts, having learned in Egypt and among the Brahmans, where he had stayed for some time, that there were an infinite number of worlds to discover. For on his return to Greece, he preached so much to those of his nation of the riches of these unknown countries that he finally found himself in the position of being able to equip a small fleet. He received only those who submitted themselves to a rather rough novitiate of five whole years, and who resolved not to speak until after Pythagoras had opened their mouths. The time to embark having come, there was such intense competition from all over Greece, from people who had come.

DRAFT C

The Republic of Letters is an otherworldly colony led by a certain Greek adventurer, named Pythagoras. It was he who began to clear part of the country, and to plant a certain drug there, which Mercury had shown him and which we call glory. It is similar to tobacco, in that it feeds on smoke; but it has the sweetness of sugar, when the latter is turned into powder. This is why this colony soon prevailed over all the plantations of America; and Europe having given in, a great trade was established between it and our country, which was later disrupted by the savages originating from the mainland of ignorance and misery, who surprised the inhabitants, and destroyed their houses.

This war prevents the colony from subsisting on its own, and every year we are obliged to send refreshments called pensions from Europe, which is not always done with the greatest regularity since the goods sent in return are often wasted, and sent in excessive quantities through the bad management of the inhabitants, who lavish great praise indifferently on everyone, as attested by these dedications which result in one beginning to feel contempt for all the glory which comes through this channel. The source of this evil lies in the need of the inhabitants, which makes them give away this merchandise too cheaply and results in European shops being full of it, diminishing its first reputation....

However, there is nothing so noble as the essence that is extracted from this drug, as long as its preparation is known. It is the true nectar of the gods and the liqueur of immortality, which Apollo made Augustus drink through the ministry of Virgil, and Pallas Alexander through that of Aristotle. A great prince of our time having been warned by one of these ministers, who is in charge of the inspection of commerce, of this disorder and of the damage which followed the desecration of this celestial gift of immortality, took the resolution to remedy it: He has judged well that true glory is due only to the Heroes, that only they should live forever, for they provide happiness to others and they enliven their century. This is the only reward the world can grant them for their work. But they must not be deceived, and it is in their interest that their portraits be in bronze rather than in wax.

7 'Funny Thought' (Drôle de Pensée), September 1675

SOURCE: LEIBNIZ-ARCHIV, LH XXXVIII, FOLS. 232–33 [A IV, 1:562–68]

The Representation which took place in Paris on the river Seine, of a machine used to walk on water, elicited in me the following meditation which, however odd, cannot fail to be of consequence if carried out.

Suppose that some persons of means with an interest in curiosities, and especially in machines, agree to organize public exhibits of such things. To this end, it would be necessary for them to raise funds in order to meet the necessary expenses.... It would be all the better if private individuals were able to defray these costs and we could dispense with great noblemen and even powerful people at court, for a powerful nobleman would take over the venture if he found it successful. Success would ensure protectors at court.

People capable of constantly inventing new things would also be required. But as too many would give rise to disorder, it would be better to have no more than two or three associates, masters of the privilege, who would employ all others and determine the conditions for certain exhibits.... Among the people employed would be painters, sculptors, carpenters, watchmakers, and other such folk, as well as mathematicians, engineers, architects, boat-builders, entertainers, musicians, poets, bookbinders, typographers, engravers, and others that would be added gradually.

The exhibits would include magic lanterns ... kites, artificial meteors, all sorts of optical wonders; a representation of the sky and the stars; comets; a globe like that of Gottorp or Jena; fireworks, water fountains; strangely shaped vessels; mandrakes and other rare plants. Extraordinary and rare animals.... A royal machine displaying races of artificial horses.... Representations of battle scenes and the display of fortifications made of wood on elevated stages that would be explained by an expert.... Infantry drills.... Cavalry exercises. Naval battles on a miniature canal. Extraordinary concerts. Rare musical instruments. Speaking trumpets. Counterfeit gems and jewellery.

The performance could always be combined with some story or comedy. Theatre of nature and of art. Swimming. Extraordinary tightrope dancers. Perilous jumps. Demonstrations of a child raising a heavy weight with a thread. Anatomical theatre ... laboratory.... In addition to the public representations there would also be private ones, displaying small adding machines [*machines de Nombres*], paintings, medals, or libraries. New experiments on water, air, vacuum. Mons. Guericke's machine of twenty-four horses could be included in the larger exhibits; for the little ones, his globe.... One could even distribute certain rarities ... and perform transfusions and infusions. One would also predict the weather for the next day. Kircher's cabinet.

We would bring the fire-eater from England if he is still alive and, with a telescope, show the moon as well as other heavenly bodies. We could test machines that can throw things with precision as well as others representing the human body with its muscles, nerves, and bones. Swammerdam's insects ... Thévenot's arts. Pleasant disputations. Dark rooms. Paintings offering different depictions depending on the angle one views them from. Public distractions.... Grotesque images painted on oiled paper with lamps inside.... Magic lanterns ... would depict extraordinary and grotesque movements impossible for men to make. Horse ballets, ring races.... Chess games staging men.... Other sorts of elaborate games could be taught and performed.... It would be possible to play tennis ... and even invent a new useful game. It would be possible to establish some training academies and colleges for youth which could then be incorporated in the College of the Four Nations [in Paris]. Comedies of different styles from all over the world, including Hindu, Turkish, and Persian ones. Comedies that would represent each trade, with its skills, tricks, jokes, masterpieces, rules, and particular ridiculous styles.... French buffoons would be sought out.... Fire-flying dragons ... ships navigating against the wind.... Self-playing instruments. Chimes, etc.... Breaking glasses by screaming.... Monsieur Weigel's inventions. Sleights of hands. Card tricks....

This venture's usefulness to the public, as well as to the individual, would be greater than might be imagined. As to the public, it would ... open people's eyes, stimulate inventions, provide beautiful sights, instruct people in an endless number of useful or ingenious novelties. All those with new inventions or ingenious designs could come there to publicize them and earn a living. It would be a general clearing-house for all inventions staging all things imaginable.... All *curieux* would turn to it.... Academies, colleges, tennis courts and other games, concerts, galleries of paintings would be added to it. Conferences and lectures. Profits, it seems, would be substantial. Optical marvels would cost little but constitute a large section of these inventions.

All respectable people [*tous les honnêtes gens*] would want to have seen these curiosities in order to be able to talk about them, and even ladies of quality would want to be taken there, and more than once. There would always be an inducement to improve things further, and it would be good if those who undertook them were assured of their secrecy in other large cities such as Rome, Venice, Vienna, Amsterdam, Hamburg, or the main courts....

This would serve to lay everywhere the foundations for the assembly of an Academy of Sciences that would maintain itself and not fail to produce beautiful things, and towards which perhaps curious princes and illustrious persons would contribute some of their wealth for the public satisfaction and the growth of the sciences. Everyone would be aware of it ... and this venture could yield such beautiful and significant results as are imaginable which, some day perhaps, would be admired by posterity.

There could be several houses in different parts of the city which would represent various things. Or rather different rooms, akin to shops ... in the same building, whose rarities would be displayed in rooms that would be rented out.... Privileges could compel all those who wish to present their novelties to do so in the Academy of Representations. The General Address Office could be revived and put to much better use ... if pushed forward properly.

This Academy would not incur any expense, simply by providing others with the possibility of exhibiting against a certain amount of money. In this manner, the Academy would always be profitable.

This Academy would host lotteries . . . and the sale of quantities of small curiosities.

We could also establish an Academy of Games (*Academie des jeux*) there or more generally an Academy of Pleasures (*Academie des plaisirs*) . . . where one could play cards and dice including chess and checkers. . . . There would be such houses or academies spread throughout the city. These houses or rooms would be built in such a way that the master of the house could hear and see everything that is said and done, without being noticed, by means of mirrors and pipes. This would be of great importance to the state, serving as a kind of political confessional. . . .

People at the Academy should be forbidden from swearing and blasphemy, for under that pretext, Academies of the sort we are describing have been abolished. One would seek to make it fashionable to admire fine players or performers, namely, those behaving without outbursts. And those who broke the rules should give something back . . . to the game, for in this manner it would be to the players' advantage to observe the law. Troupes of players who behaved in an unrestrained manner would be denied access. . . . One should evoke fashion and trendiness—rather than piety, which the vulgar despise—to justify this state of affairs.

Cheating would most often be permitted. . . .

This house would become a palace over time, and that house would even contain . . . shops of all sorts of things imaginable.

Gambling would serve as the finest pretext in the world to engage in something as useful to the public as this. One could take advantage of people's weakness, and deceive them to better cure them. . . . Is there anything so just as to make extravagance serve the dissemination of wisdom . . . and to turn the poison into an elixir?

8 Leibniz to Duke Johann Friedrich, Autumn 1679

SOURCE: LEIBNIZ-ARCHIV, LBR F12, FOL. 160 [A II, 1:760–61]

While in Paris I met a person of religion, whose merit was generally recognized. This man had meditated a very long time on controversies, he was well versed in the knowledge of antiquity, and the reading of the Holy Fathers was one of his greatest pleasures: he venerated them, but not excessively. He was the best man in the world to explicate a passage and show its true meaning, and he did so with force and singular clarity. He mastered perfectly what we call the humanities [*humanités*], and when he composed verses, which happened to him only very rarely . . . it was difficult to believe he had done anything else in his life. His style was simple and natural in both Latin and the vernacular, but strong and incisive in a few places. It was concise but clear, pleasant without artifice. He did not like borrowed mannerisms [*couleurs empruntez*], and he believed that the beauty of a speech should consist in the strength of its argumentation. Everyone agreed that he was a master in the art of reasoning.... When we had become friends, he revealed to me a few times something about the course of his studies. He had cultivated the humanities and historical studies until the age of thirteen to fourteen. From fifteen to seventeen he had so refined the subtleties of the scholastics that he embarrassed his masters. Many people find this study unnecessary, but he often shared with me how grateful he was for it, and that he recognized in it how far the refinement of the human mind can go. He told me that he found in the scholastics many things so solid and beautiful that they would be admired in the wider world if they were stated clearly.

But he did not stop there, as his friends feared, and from the ages of eighteen to twenty-one he studied jurisprudence with such great success that he was publicly praised, and a great prince well versed in these matters believed him capable of working towards the reform of this science. He was appointed at a princely court, and he demonstrated that he was equally predisposed to practice and theory if necessary. This lasted until the age of twenty-five, and during that time he had had the opportunity to study controversies. He was present when the murmur of new discoveries in mathematics and physics aroused his curiosity. Could he have passed for something and yet not contributed to the progress of science? He grew disappointed with all his past studies, for he clearly saw that an invention of importance in mathematics is the most assured mark of a solid mind.... He left both his studies and office to spend some time in Paris, which is the centre of fine curiosities. It was then that he showed what he could, for in two years' time he stood out from all the rest. He was recognized by the greatest men in Paris as one of the first geometers, capable of making discoveries of consequence. He demonstrated machines of his invention which were considered surprising. And we can say that never was a foreigner of his kind (for such he was) received more favourably by people of merit.

It was at this time that I got to know him; his approach did not promise anything extraordinary, his ordinary conversations were rather weak, he did not have, or did not affect the art of showing off. And I was surprised not to recognize in him the marks of what I had been told [about] him. But I was much mistaken afterwards. I surprised him one day reading controversial books; I told him of my astonishment, because he had passed for a mathematician by profession, having done hardly anything else in Paris. It was then that he told me that we were mistaken, for he had many other preoccupations—that his principal meditations concerned theology; that he only applied himself to mathematics as to scholastics to perfect his mind and to learn the arts of invention and demonstration, which he believed he had pushed as far as possible.

9 Leibniz to Colbert, December 1679

SOURCE: LEIBNIZ-ARCHIV, LH 37, V, FOL. 194 [A III, 2:917-20]

My most illustrious Lord,

I believe that this sample of my work will be of interest to You ... for it was born in France three years ago now, and since then it has been received and acknowledged as legitimate by some outstanding men who attend to the increase of the sciences—often with their own auspicious inventions. Under Your guidance, it has earned its right not to be abandoned [and] its content ... has given me the courage to speak to You. ... Which geometer has not addressed the problem of the quadrature of the circle? ... I have seemed luckier than others, for I have grasped this method arithmetically and do not know whether the nature of things offers a simpler one, since it expresses itself most simply in a series of numbers ... this is the duty of the geometrical quadrature.

That this subject, researched for so many centuries, has been enriched with an altogether fruitful addition will seem a matter not entirely to be spurned—especially by You, my Lord, who have accustomed Yourself to determine the worth of things with great judgement. Geometrical discoveries bring immortality, and often survive even the more useful experiments of nature (if you look at their applications to daily life). Pythagoras's duplication of the square endures to this day, as does Plato's duplication of the cube; but the result of Democritus's diligence has not been passed down to us.

Indeed, for geometrical knowledge our mind is indebted only to itself, but physical knowledge is owed partly to luck; but each and everyone loves his own children more ardently. In the end, we look at the inner workings of geometry, but we regard only the outer shell of nature. To our body, physics is more useful; to the perfection of the intellect, geometry is more efficacious. Thus it is no wonder how much more everyone with a stronger intellect delights in geometry, even when it comes to physics itself. After Archimedes had found so many other things of greater use for life, he only had a sphere and cylinder engraved in his tombstone, the relationship between which he had discovered. It is of vital interest to the Republic that geometry not be neglected, for it sharpens the mind and teaches it to argue strictly. To sum things up in one sentence, there is as big a difference between a simple physician and one instructed in geometry as there is between a Chinese philosopher and a European one. For it is certain that the Chinese are superior to us in experiments of nature but are defeated in the art of geometry. Thus they have no strong grasp of the description of the earth, nor of the course of the sky, nor the wonders of our machines. But they have preserved in perpetual tradition the properties of different bodies which have been brought forth by the fortuitous sequence of so many centuries. Their spirits have the same softness as their bodies, and just as their practical life is removed from the military, so their contemplative [life] is removed from geometry, since for both a certain strict rigour is needed. In the Occident, only Your France can be compared to the oriental empire of China; it is self-sufficient, fully sovereign, and not exposed to invasion, being protected by mountains, rivers, and the sea, and is ruled by one monarch, as well as admirable in its political institutions—indeed, since You, my Lord, under the greatest prince, have had the management of it. If you [sc. the French] take second place to the Chinese when it comes to the size of your lands, you easily surpass them in excellence of the sciences and in military strength. Therefore it is clear that you earn praise especially in geometry, which is the essence of the sciences, and the military, which is the Republic's muscle.

Others may speak of the arts of war; to me it suffices to say that our age is indebted to You for the rebirth of geometry, since your Viète and Descartes have resurrected the art of discovery, hidden by the ancients, like a holy fire. But You now preserve outstandingly this glory as well; for the nation neither falls short itself, nor do You fall short of it, being ever watchful to promote its dignity; and this way many excellent specimens of ingenious minds arise daily in Your country, by Your request, praise, and support. In addition, it is free of envy, the greatest sign of Your felicity; for as everyone has great trust in his own virtue, so he judges that of others with equanimity.

This I see happen among Your people who are favourably disposed to praise others just as much as Your own countrymen. These things might seem unrelated to my purpose—for I am not addressing Your France, my Lord, but You yourself—were it not clear that all the present praises of Your people refer to You, by whose brilliant counsels everything is given life, and geometry itself has attained new vigour.

But the benefit of this has reached even me, for I gladly confess that my petty geometry was certainly born in France and almost reached the age of four years [*sic*, see beginning of letter].

Therefore accept its first fruits, my Lord, which are owed to You, and consider the possibility that Your approval can also motivate a mediocre man to greater and more useful things in the future.

10 Leibniz to Colbert, Mid-1682

SOURCE: LEIBNIZ-ARCHIV, LBR 167, FOL. 1 [A III, 3:719–21]:

The advancement of the hard sciences constitutes a true increase in mankind's legacy, and the heroic intentions of the greatest of Kings are aided by the incomparable zeal which You demonstrate for His glory and for the general good; we can say that You have found a way to considerably oblige all that there will be of rational centuries to come and all civilized peoples on earth, peoples whose praise will lend You special recognition.

The genuine discoveries of which You are the promoter are for all places and all times. A Persian king will be able to marvel at what the telescope reveals, and a Chinese mandarin will be delightfully astonished when he understands the infallibility of the demonstrations of a missionary geometer. What will these peoples say when they see this marvelous machine which You had made, and which produces an accurate representation of the state of the sky at any given moment?

I believe they will recognize that the spirit of man takes after the divine; that this divinity is communicated more particularly to Christians, and that France is today happier than all nations since she possesses a King whom God has taken pleasure in showering with so many graces that it seems that what has hitherto remained impenetrable to men must be reserved for His reign. The secret of the heavens, the greatness of the earth, and the measurement of time are of this nature.

... I had the good fortune to communicate to Your Royal Academy two things not insignificant. The first is the composition of a light that can be called perpetual and that chance presented to a chemist of my acquaintance, and to which I have myself contributed something since; and the other is true proportion such as it exists in assignable numbers, between the circle and its quadrangle, which we have sought since Archimedes and which I have discovered and demonstrated.

In fact, Monseigneur, everything that is found everywhere must be Yours and all those who cultivate the sciences are indebted to You for their support, since the benign influences of this Sun, whose rays emanate from You, lend life and vigour to the fine arts. This makes me hope that You will approve, Monseigneur, of a task that I now put to You and of which I am providing samples.

I have always believed that of all the sciences, physics is both the least advanced and the most necessary, especially when it comes to terrestrial bodies, which are commonly divided into three kingdoms, the mineral, the vegetable, and the animal. It must be conceded, however, that a number of excellent men have worked on the nature of plants and animals. But we have only scratched the surface so far when it comes to our knowledge of minerals; and the reason for this, I believe, is that we would not be able to advance much without considering those sites from which we draw minerals; for the cabinets of the curious only contain dead pieces detached from their matrix, and can only be of use to those who have examined what is happening in these places and are able to dispel the prejudices of miners, which the best authors follow blindly. As for the chemists, having simply mingled these bodies together, destroying or disfiguring them in a thousand ways, for their part they have rather obscured the knowledge of the origin of minerals—even if they have acquired considerable experience in the process. However, the situation in the country where I find myself has provided me with the opportunity to visit mines and to observe many things there; but ... what suddenly alerted me to a possibility of considerable consequences was a natural wonder that fell into my hands. It is a stone that I own and that I intend for You, Monseigneur, a stone on which nature has perfectly drawn two animals with features in a metallic material.

As it is not currently convenient for me to send the stone itself, I shall send in advance its description, as well as the mechanical explanation of its generation, so that You may judge by this sample if I can help further the knowledge of these materials, which one does not have the opportunity to cultivate in France, but that I have had the occasion to examine up close. This study is of no small consequence, because—without speaking of the economy and agriculture, the plants and animals born and nurtured from the soil—I believe that the knowledge of soils and stones naturally precedes that of the other natural productions, even though I must admit that necessity ... dictates that we start backwards. However, we shall never go far enough if we neglect the principles. But it is from Your judgement that I must learn whether this research deserves further investigation.

BIBLIOGRAPHY

Leibniz's Writings

Dutens, Louis, ed. *Gotholfredi Guillelmi Leibnitii . . . opera omnia*. 6 vols. Geneva: Fratres de Tournes, 1768.
Foucher de Careil, Alexandre, ed. *Oeuvres de Leibniz*. 7 vols. Paris: Didot, 1859–75.
Klopp, Onno, ed. *Die Werke von Leibniz*. 11 vols. Hanover: Klindworth, 1864–84.
Leibniz, Gottfried Wilhelm. *The Art of Controversies*, ed. and trans. Marcelo Dascal, Dordrecht: Springer, 2006.
——. *Die philosophischen Schriften*, ed. Carl Immanuel Gerhardt. 7 vols. Berlin: Weidmann, 1875–90.
——. 'Historia Inventionis Phosphori'. In *Gotholfredi Guillelmi Leibnitii . . . opera omnia*, ed. Louis Dutens, vol. 2. Geneva: Fratres de Tournes, 1768. First published in *Miscellanea Berolinensia* 1 (1710): 91–98.
——. Letter to Friedrich Hoffmann, 27 September 1699. In *Opera Omnia physico-medica*, Friedrich Hoffmann, supp. vol. 1, 51. Geneva: Fratres de Tournes, 1749.
——. 'New System of Nature', trans. A. E. Kroeger. *Journal of Speculative Philosophy* 5 (1871 [1695]): 209–19.
——. *Sämtliche Schriften und Briefe*, ed. Preussische [later Deutsche] Akademie der Wissenschaften. 68 vols. in 8 series to date. Darmstadt [later Berlin]: Reichl [later Akademie Verlag and De Gruyter], 1923–. (Series 2, vol. 1, is cited from the revised 2nd ed., Berlin: Akademie, 2006. Forthcoming volumes are cited where possible from PDF *Vorausgaben*, available on the website *Leibniz-Edition: Die Akademie-Ausgabe*, https://leibnizedition.de.)

Other Works

Abdel-Halim, Mohamed. *Antoine Galland: Sa vie et son oeuvre*. Paris: Nizet, 1964.
Abou-Nemeh, Catherine. 'Daring to Conjecture in Seventeenth- and Eighteenth-Century Sciences'. *Isis* 113, no. 4 (2022): 728–46.
Alsted, Johann Heinrich. *Encyclopaedia septem tomis distincta, Tomus sextus: Artes mechanicae*. Herborn, 1630.
Altwicker, Tilmann, and Francis Cheneval, eds. *Rechts- and Staatsphilosophie bei G. W. Leibniz*. Zürich: Mohr Siebeck, 2020.
Antognazza, Maria Rosa. *Leibniz: An Intellectual Biography*. Cambridge: Cambridge University Press, 2009.
——, ed. *The Oxford Handbook of Leibniz*. Oxford: Oxford University Press, 2014.
Ariew, Roger. 'Leibniz on the Unicorn and Various Other Curiosities'. *Early Science and Medicine* 3 (1998): 267–88.
Armgardt, Matthias. 'Leibniz as Legal Scholar'. *Fundamina*, special issue 1 (2014), *Meditationes de iure et historia. Essays in honour of Laurens Winkel*, 27–38.
Arnauld, Antoine. *Oeuvres*. 43 vols. Paris: Sigismond d'Arnay, 1775–83.

Arthur, Richard. *Leibniz*. Malden, MA: Polity, 2014.
Artosi, Alberto, and Giovanni Sartor. 'Leibniz as Jurist'. In *The Oxford Handbook of Leibniz*, ed. Maria Rosa Antognazza, 641–62. Oxford: Oxford University Press, 2014.
Ash, Eric. 'Expertise and the Early Modern State'. *Osiris* 25, no. 1 (2010): 1–24.
Azouvi, F. 'Entre Descartes et Leibniz: l'animisme dans les "Essais de Physique" de Claude Perrault'. *Recherches sur le XVIIe siècle* 5 (1982): 9–19.
Badalo-Dulong, Claude. *Trente ans de diplomatie française en Allemagne*. Paris: Plon, 1956.
Barber, W. H. *Leibniz in France, from Arnauld to Voltaire: A Study in French Reactions to Leibnizianism*. Oxford: Clarendon Press, 1955.
Becher, Johann Joachim. *Närrische Weißheit und weise Narrheit*. Frankfurt: Zubrod, 1682.
———. *Psychosophia, das ist Seelen-Weiszheit*. Güstrow: Chritsian Scheippel, 1678.
———. *Trifolium Becherianum Hollandicum oder . . . Drey Neue Erfindungen, bestehende in einer Seiden-Wasser-Muble und Schmeltz- Wercke*. Frankfurt: Johann David Zunner, 1679.
Beeley, Philip. 'Learned Discourse Between Ambition and Resignation: Leibniz's Scientific Networks in England, 1670–1696'. Unpublished essay, 2018.
———. 'Mathematics and Nature in Leibniz's Early Philosophy'. *International Archives of the History of Ideas* 166 (1999): 123–45.
———. 'A Philosophical Apprenticeship: Leibniz's Correspondence with the Secretary of the Royal Society, Henry Oldenburg'. In *Leibniz and His Correspondents*, ed. Paul Lodge, 47–73. Cambridge: Cambridge University Press, 2004.
Beiderbeck, Friedrich. 'Leibniz's Political Vision for Europe'. In *The Oxford Handbook of Leibniz*, ed. Maria Rosa Antognazza, 664–82. Oxford: Oxford University Press, 2014.
Beiderbeck, Friedrich, Irene Dingel, and Wenchao Li, eds. *Umwelt und Weltgestaltung: Leibniz's Politisches Denken in seiner Zeit*. Göttingen: Vandenhoeck & Ruprecht, 2015.
Berkowitz, Roger. *The Gift of Science: Leibniz and the Modern Legal Tradition*. Cambridge, MA: Harvard University Press, 2005.
Bertram, Friedrich. 'G. W. Leibnizens Beziehungen zu Z. K. von Uffenbach'. *Zeitschrift für Bücherfreunde* 10 (1906/7): 195–99.
Bertrand, Simon. 'Leibnitz et ses réseaux: des voies de la connaissance au commerce des lumières'. *Quaderni* 39 (1999): 77–85.
Bertrand, Alexis. *Mes vieux médecins*. Lyon: Storck, 1904.
Bertucci, Paola. *Artisanal Enlightenment*. New Haven, CT: Yale University Press, 2017.
Bevilacqua, Alexander. *The Republic of Arabic Letters: Islam and the European Enlightenment*. Cambridge, MA: Harvard University Press, 2018.
Biagioli, Mario. 'Etiquette, Interdependence, and Sociability in Seventeenth-Century Science'. *Critical Inquiry* 22 (1996): 193–238.
———. 'Galilée bricoleur'. *Actes de la recherche en sciences sociales* 94 (1992): 85–105.
———. *Galileo, Courtier: The Practice of Science in the Culture of Absolutism*. Chicago: University of Chicago Press, 1993.
Blanning, Tim. *The Pursuit of Glory: Europe 1648–1815*. London: Allen Lane, 2007.

Blay, Michel. *Reasoning with the Infinite: From the Closed World to the Mathematical Universe*. Chicago: University of Chicago Press, 1999.

Blay, Michel, and Robert Halleux, eds. *La science classique (XVIe–XVIIIe siècle): Dictionnaire critique*. Paris: Flammarion, 1998.

Bodemann, Eduard. *Die Leibniz-Handschriften der Königlichen öffentlichen Bibliothek zu Hannover*. Hanover: Hahn, 1895.

Bodéüs, Richard, ed. *Leibniz-Thomasius Correspondance, 1663–1672*. Paris: Vrin, 1993.

Böger, Ines. *'Ein Seculum . . . da man zu Societäten Lust hat'. Darstellung und Analyse der Leibnizschen Sozietätspläne vor dem Hintergrund der europäischen Akademiebewegung im 17. und 18. Jahrhundert*. 2 vols. Munich: Utz, 1997.

Boislisle, A.A.G.M. de, ed. *Correspondance des contrôleurs généraux des finances*. 3 vols. Paris: Imprimerie nationale, 1874–97.

Borowski, Audrey. 'Perrault, Claude'. In *Encyclopedia of Early Modern Philosophy and the Sciences*, ed. D. Jalobeanu and C. T. Wolfe. Cham, Switz.: Springer, 2021. https://doi.org/10.1007/978-3-319-20791-9_625-1.

———. 'Projectors'. In *Encyclopedia of Early Modern Philosophy and the Sciences*, ed. D. Jalobeanu and C. T. Wolfe. Cham, Switz.: Springer, 2021. https://doi.org/10.1007/978-3-319-20791-9_624/1.

———. 'Republic of Letters'. In *Encyclopedia of Early Modern Philosophy and the Sciences*, ed. D. Jalobeanu and C. T. Wolfe. Cham, Switz.: Springer, 2021. https://doi.org/10.1007/978-3-319-20791-9_627-1.

Bos, H. J. M. 'The Influence of Huygens on the Formation of Leibniz' Ideas'. In *Studia Leibnitiana: Supplementa 17 [and] 18—Leibniz à Paris (1672–1676)*, ed. Kurt Müller, Heinrich Schepers, and Wilhelm Totok, vol. 1, 59–68. Wiesbaden: Steiner, 1978.

———, ed. *Studies on Christiaan Huygens*. Lisse, Neth.: Swets & Zeitlinge, 1980.

Bredekamp, Horst. *Die Fenster der Monade. Gottfried Wilhelm Leibniz' Theater der Natur und Kunst*. Berlin: Akademie, 2004.

Breger, Herbert. 'Becher, Leibniz und die Rationalität'. In *Kontinuum, Analysis, Informales—Beiträge zur Mathematik und Philosophie von Leibniz*, ed. Herbert Breger and Wenchao Li, 29–41. Berlin: Springer, 2016.

Breger, Herbert, and Friedrich Niewöhner, eds. *Leibniz und Niedersachsen*. Stuttgart: Steiner, 1999.

Briggs, Robin. 'The Académie Royale and the Pursuit of Unity'. *Past and Present* 131 (1991): 38–88.

Brockliss, Laurence. *French Higher Education in the Seventeenth and Eighteenth Centuries: A Cultural History*. Oxford: Clarendon Press, 1987.

Brown, Harcourt. 'Un cosmopolite du grand siècle: Henri Justel'. *Bulletin de la Société de l'Histoire du Protestantisme français* 82 (1933): 187–201.

Brown, Stuart. 'Foucher's Critique and Leibniz's Defense of the "New System"'. In *Leibniz: Reason and Experience*, ed. Stuart Brown, 96–104. Milton Keynes, UK: Open University Press, 1983.

———. 'The Leibniz-Foucher Alliance and Its Philosophical Bases'. *Leibniz and His Correspondents*, ed. Paul Lodge, 74–96. Cambridge: Cambridge University Press, 2004.

———, ed. *The Young Leibniz and His Philosophy (1646–76)*. Dordrecht: Kluwer, 1999.

Buchenau, Stefanie. 'Leibniz. Philosophe-diplomate, Le traité sur la sécurité publique de 1670'. *Discussions* 4 (2010). https://perspectivia.net/publikationen/discussions/4-2010/buchenau_leibniz.

Bury, Emmanuel. 'L'humanisme de Huet: paideia et érudition à la veille des Lumières'. In *Pierre-Daniel Huet (1630–1721): Actes du Colloque de Caen (12–13 novembre 1994)*, ed. Soûad Guellouz, 197–209. Paris: PSCL, 1994.

Chapelain, Jean. *Lettres*, ed. Philippe Tamizey de Larroque. 2 vols. Paris: Imprimerie Nationale, 1880–83.

Charles, Sébastien. 'Entre réhabilitation du scepticisme et critique du cartésianisme: Foucher lecteur du scepticisme académique'. *Astérion* (2013). https://doi.org/10.4000/asterion.2382.

Colbert, Jean-Baptiste. *Lettres, instructions et mémoires*, ed. Pierre Clément. 7 vols. Paris: Imprimerie Impériale, 1861–67.

Cole, C. W. *Colbert and a Century of French Mercantilism*. New York: Columbia University Press, 1939.

Collas, Georges. *Jean Chapelain (1595–1674): étude historique et littéraire d'après des documents inédits*. Paris: Perrin, 1912.

———. *Un poète protecteur des lettres au XVIIe siècle, Jean Chapelain, 1595–1674*. Paris: Perrin, 1912.

Condren, Conal. *Argument and Authority*. Cambridge: Cambridge University Press, 2006.

Condren, Conal, Stephen Gaukroger, and Ian Hunter, eds. *The Philosopher in Early Modern Europe: The Nature of a Contested Identity*. Cambridge: Cambridge University Press, 2006.

Cook, Daniel J. 'Leibniz and "Orientalism"'. *Studia Leibnitiana* 40, no. 2 (2008): 168–90.

———. 'Leibniz on "Advancing toward Greater Culture"'. *Studia Leibnitiana* 50, no. 2 (2018): 163–79.

———. 'The Young Leibniz and the Problem of Historical Truth'. In *The Young Leibniz and His Philosophy (1646–76)*, ed. Stuart Brown, 103–22. Dordrecht: Kluwer, 1999.

Courtès, Huguette. 'Arnauld et Leibniz: un dialogue impossible'. *Chroniques de Port-Royal* 47 (1998): 303–21.

Couturat, Louis. *La Logique de Leibniz: d'après des documents inédits*. Paris: Alcan, 1901.

Crippa, Davide. *The Impossibility of Squaring the Circle in the 17th Century: A Debate among Gregory, Huygens and Leibniz*. Basel: Birkhäuser, 2019.

Dainville, François de. *L'Éducation des Jésuites: XVIe–XVIIIe siècles*, ed. M.-M. Compère. Paris: Minuit, 1978.

Dascal, Marcelo. *Leibniz: What Kind of Rationalist? Logic, Epistemology, and the Unity of Science*. Dordrecht: Springer, 2008.

Davillé, Louis. 'Le séjour de Leibniz à Paris'. *Archiv für Geschichte der Philosophie* 32 (1920): 142–49.

———. 'Le séjour de Leibniz à Paris'. *Archiv für Geschichte der Philosophie* 33 (1921): 67–78, 165–73.

———. 'Le séjour de Leibniz à Paris'. *Archiv für Geschichte der Philosophie* 34 (1922): 14–40, 136–41.

———. 'Le séjour de Leibniz à Paris'. *Archiv für Geschichte der Philosophie* 35 (1923): 50–61.

———. *Leibniz historien. Essai sur l'activité et la méthode historiques de Leibniz*. Paris: Alcan, 1909.

De Risi, Vincenzo. 'Analysis Situs, the Foundations of Mathematics, and a Geometry of Space'. In *The Oxford Handbook of Leibniz*, ed. Maria Rosa Antognazza, 247–58. Oxford: Oxford University Press, 2014.

Desmaizeaux, Pierre. *Recueil de diverses pièces, sur la philosophie, la religion naturelle, l'histoire, les mathematiques*. 2nd ed. 2 vols. Amsterdam: Changuion, 1740.

Dessert, Daniel. *Colbert ou le serpent venimeux*. Paris: Complexe, 2000.

———. *Le royaume de monsieur Colbert 1661–1683*. Paris: Perrin, 2007.

Dew, Nicholas. *Orientalism in Louis XIV's France*. Oxford: Oxford University Press, 2009.

———. 'Reading Travels in the Culture of Curiosity: Thévenot's Collection of Voyages'. *Journal of Early Modern History* 10 (2006): 39–59.

Dilthey, Wilhelm. 'Leibniz und sein Zeitalter'. In *Studien zur Geschichte des deutschen Geistes*, ed. Paul Ritter, 3–82, Leipzig: Teubner, 1926.

Dingel, Irene. 'Leibniz und seine Überlegungen zu einer Kirchlichen Reunion'. In *Leibniz in Mainz: europäische Dimensionen der Mainzer Wirkungsperiode*, ed. Irene Dingel, Michael Kempe, and Wenchao Li, 93–104. Göttingen: Vandenhoeck & Rupprecht, 2019.

Dingel, Irene, Michael Kempe, and Wenchao Li, eds. *Leibniz in Mainz: europäischer Dimensionen der Mainzer Wirkungsperiode*. Göttingen: Vandenhoeck & Rupprecht, 2019.

Döring, Detlef. 'Korrespondenten von G. W. Leibniz. 10 Christian Philipp'. *Studia Leibnitiana* 21, no. 1 (1989): 101–23.

Dreitzel, Horst. 'Hermann Conring und die politische Wissenschaft'. In *Hermann Conring (1606–1681): Beiträge zu Leben und Werk*, ed. Michael Stolleis, 135–72. Berlin: Duncker & Humbolt, 1983.

Droste, Heiko. *Im Dienst der Krone: schwedische Diplomaten im 17. Jahrhundert*. Münster: Lit Verlag, 2006.

Duchesneau, François. 'Leibniz et les hypothèses de physique'. *Philosophiques* 9 (1982): 223–38.

Dumas Primbault, Simon. 'Leibniz en Créateur'. In *Actes du colloque Matières à raisonner*, ed. Françoise Briegel. Paris, 2022.

Emsley, John. *The Shocking History of Phosphorus: A Biography of the Devil's Element*. London: Macmillan, 2000.

Fasolt, Constantin. *The Limits of History*. Chicago: University of Chicago Press, 2004.

———. *Past Sense—Studies in Medieval and Early Modern European History*. Leiden: Brill, 2014.

Feingold, Mordechai. *Jesuit Science and the Republic of Letters*. Cambridge, MA: MIT Press, 2003.

Finster, Richard, et al., eds. *Leibniz Lexicon: A Dual Concordance to Leibniz's 'Philosophische Schriften'*. Hildesheim, Ger.: Olms, 1988.

Fischer, Kuno. *Geschichte der neuern Philosophie*, vol. 2: *Leibniz und seine Schule*. Mannheim, Ger.: Bassermann, 1855.

Fontenelle, Bernard de. *Leibniz, vie et oeuvre suivi de l'éloge de Fontenelle*, ed. Jean-Michel Robert. Paris: Agora Pocket, 2002.

Forberger, Rudolf. 'Johann Daniel Crafft. Notizen zu einer Biographie (1624 bis 1697)'. *Jahrbuch für Wirtschaftsgeschichte* 5 (1964): 63–79.

Foucher de Careil, Alexandre, ed. *Nouvelles lettres et opuscules inédits de Leibniz*. Paris: Druand, 1857.

Freudenberger, Herman. 'Introduction'. In *State and Society in Early Modern Austria*, ed. Charles Ingrao, 141–53. West Lafayette, IN: Purdue University Press, 1994.

Fumaroli, Marc. *La République des Lettres*. Paris: Gallimard, 2015.

Furetière, Antoine. *Dictionnaire universel*. The Hague: A. and R. Leers, 1690.

Gädeke, Nora. 'Gelehrtenkorrespondenz der frühen Aufklärung: Gottfried Wilhelm Leibniz'. In *Handbuch Brief*, vol. 2: *Historische Perspektiven—Netzwerke—Zeitgenossenschaften*, ed. Marie Isabel Matthews-Schlinzig, Jörg Schuster, Gesa Steinbrink, and Jochen Strobel, 799–811. Berlin: de Gruyter, 2020.

———. 'Gottfried Wilhelm Leibniz'. In *Les grands intermédiaires culturels de la République des Lettres*, ed. Christiane Berkvens-Stevelinck, Hans Bots, and Jens Hässeler, 257–306. Paris: Champion, 2005.

———. 'Leibniz lässt sich informieren—Asymmetrien in seinen Korrespondenzbeziehungen'. In *Kommunikation in der Frühen Neuzeit*, ed. Klaus-Dieter Herbst and Stefan Kratochwil, 24–46. Frankfurt: Lang, 2009.

Gantet, Claire. 'Leibniz' Sicht von Krieg und Gewalt in der Staaten- und Völkergemeinschaft'. In *Umwelt und Weltgestaltung: Leibniz' politisches Denken in seiner Zeit*, ed. Friedrich Beiderbeck, Irene Dingel, and Wenchao Li, 231–54. Göttingen: Vandenhoeck & Ruprecht, 2015.

Garber, Daniel. *Leibniz: Body, Substance, Monad*. Oxford: Oxford University Press, 2009.

———. 'Leibniz: Physics and Philosophy'. In *The Cambridge Companion to Leibniz*, ed. Nicholas Jolley, 270–352. Cambridge: Cambridge University Press, 1995.

———. 'Leibniz's Reputation, the Fontenelle Tradition'. In *Insiders and Outsiders in Seventeenth-Century Philosophy*, ed. G. A. J. Rogers, Tom Sorell, and Jill Kraye, 281–93. Abingdon: Routledge, 2009.

———. 'Motion and Metaphysics in the Young Leibniz'. In *Leibniz: Critical and Interpretive Essays*, ed. Michael Hooker, 160–84. Minneapolis: University of Minnesota Press, 1982.

———. 'Novatores: Negotiating Novelty in Early Modern Philosophy'. Lecture, All Souls College, Oxford, 5 February 2018.

Gaulmier, J. 'À la découverte du Proche-Orient: Barthélemy d'Herbelot et sa *Bibliothèque orientale*'. *Bulletin de la Faculté des Lettres de Strasbourg* 48 (1969): 1–6.

Gerland, Ernst, ed. *Leibnizens und Huygens' Briefwechsel mit Papin, nebst der Biographie Papin's*. Berlin: Königliche Akademie der Wissenschaften, 1888.

Goethe, Norma, Philip Beeley, and David Rabouin, eds. *G. W. Leibniz: Interrelations between Mathematics and Philosophy*. New York: Springer, 2015.

Goldenbaum, Ursula. 'Transubstantiation, Physics and Philosophy at the Time of the Catholic Demonstrations'. In *The Young Leibniz and His Philosophy (1646–76)*, ed. Stuart Brown, 79–102. Dordrecht: Kluwer, 1999.

Goldgar, Anne. *Impolite Learning, Conduct and Community in the Republic of Letters, 1680–1750.* New Haven, CT: Yale University Press, 1995.

Golinski, J. V. 'A Noble Spectacle: Phosphorus and the Public Cultures of Science in the Early Royal Society'. *Isis* 80 (1989): 11–39.

Graber, Frédéric. 'Du faiseur de projet au projet régulier dans les Travaux Publics (XVIIIe-XIXe siècles): pour une histoire des projets'. *Revue d'histoire moderne et contemporaine* 58, no. 3 (2011): 7–33.

Grell, Chantal. *Histoire intellectuelle et culturelle de la France du grand siècle.* Paris: Nathan Université, 2000.

Griard, Jérémie. 'Le meilleur régime selon Leibniz'. *Philosophiques* 31, no. 2 (2004): 349–72.

Gruber, Johann Daniel. *Commercii Epistolici Leibnitiani ad omne genus eruditioni comparati, per partes publicandi. Tomi prodromi pars altera.* Hanover: Schmid, 1745.

Guellouz, Soûad, ed. *Pierre-Daniel Huet (1630–1721): Actes du Colloque de Caen (12–12 novembre 1993).* Paris: PSCL, 1994.

Guerin, M. F. E. *Dictionnaire pittoresque d'histoire naturelle et des phénomènes de la nature*, vol. 7. Paris: Imprimerie de Cosson, 1838.

Guiffrey, Jules, ed. *Comptes des bâtiments du roi sous le règne de Louis XIV.* 5 vols. Paris: Imprimerie nationale, 1881–1901.

Haase, Carl. 'Leibniz als Politiker und Diplomat'. In *Leibniz. Sein Leben, sein Wirken, seine Welt,* ed. Carl Haase and Wilhelm Totok, 195–226. Hanover: Verlag für Literatur, 1966.

Haase, Carl, and Wilhelm Totok, eds. *Leibniz. Sein Leben, sein Wirken, seine Welt.*

Habermas, Jürgen. *The Structural Transformation of the Public Sphere: An Inquiry into a Category of Bourgeois Society*, trans. T. Burger and Fr. Lawrence. Cambridge, MA: MIT Press, 1989.

Hahn, Roger. *The Anatomy of a Scientific Institution: The Paris Academy of Sciences, 1666–1803.* Berkeley: University of California Press, 1971.

———. 'Huygens and France'. In *Studies on Christiaan Huygens*, ed. H. J. M. Bos, 53–65. Lisse, Neth.: Swets & Zeitlinge, 1980.

Hall, A. Rupert, and Marie Boas Hall, eds. *Correspondence of Henry Oldenburg*, vol. 8. Madison: University of Wisconsin Press, 1971.

Hall, Marie Boas. 'Leibniz and the Royal Society 1670–1676'. In *Studia Leibnitiana: Supplementa 17 [and] 18—Leibniz à Paris (1672–1676)*, ed. Kurt Müller, Heinrich Schepers, and Wilhelm Totok, vol. 1, 171–82. Wiesbaden: Steiner, 1978.

Hamilton, Alistair. *Johann Michael Wansleben's Travels in the Levant, 1671–1674: An Annotated Edition of His Italian Report.* Leiden: Brill, 2018.

Hammerstein, Notker. *Jus und Historie: Ein Beitrag zur Geschichte des historischen Denkens an deutschen Universitäten im späten 17. und 18. Jahrhundert.* Göttingen: Vandenhoeck & Rupprecht, 1972.

Hassinger, Herbert (1953). 'Becher, Johann Joachim'. In *Neue Deutsche Biographie*, vol. 1, 689–90. Berlin: Duncker & Humboldt, 1953.

———. *Johann Joachim Becher, 1635–1682. Ein Beitrag zur Geschichte des Merkantilismus.* Vienna: Verlag Adolf Holzhausens, 1951.

Hatch, Robert. 'Between Erudition and Science: The Archive and Correspondence of Network of Ismaël Boullau'. In *Archives of the Scientific Revolution: The Formation and Exchange of Ideas in Seventeenth-Century Europe*, ed. Michael Hunter, 49–72. Woodbridge, UK: Boydell, 1998.

Hecht, Hartmut, and Jürgen Gottschalk. 'The Technology of Mining and Other Technical Innovations'. In *The Oxford Handbook of Leibniz*, ed. Maria Rosa Antognazza, 526–40. Oxford: Oxford University Press, 2014.

Heinekamp, Albert. 'Christiaan Huygens vu par Leibniz'. In *Huygens et la France*, ed. René Taton, 99–113. Paris: Vrin, 1981.

Henkemans, Francisa Snoeck. 'La prétérition comme outil de stratégie rhétorique'. *Argumentation et analyse du discours* 2 (2009). https://doi.org/10.4000/aad.217.

Hermann, Wolfgang. *The Theory of Claude Perrault*. London: Zwemmer, 1973.

Hirsch, Eike Christian. *Der berühmte Herr Leibniz: Eine Biographie*. Munich: Beck, 2000.

Hochstrasser, Tim. 'G. W. von Leibniz and "Court Philosophy"'. *Court Historian, Newsletter of the Society for Court Studies* 3 (1998): 2–8.

Hoffmann, Friedrich. *Opera Omnia physico-medica*. 6 vols. Geneva: Fratres de Tournes, 1749.

Hofmann, Josef Ehrenfried. *Aus der Frühzeit der Infinitesimalmethoden: Auseinandersetzung um die algebraische Quadratur algebraischer Kurven in der zweiten Hälfte des 17. Jahrhunderts*. Berlin: Springer, 1965.

———. *Leibniz in Paris 1672-1676: His Growth to Mathematical Maturity*. Cambridge: Cambridge University Press, 1974.

Hoppen, Theodore. 'The Nature of the Royal Society, Part I'. *British Journal for the History of Science* 9 (1976): 1–24.

———. 'The Nature of the Royal Society, Part II'. *British Journal for the History of Science* 9 (1976): 243–73.

Hotson, Howard. 'Leibniz's Network'. In *The Oxford Handbook of Leibniz*, ed. Maria Rosa Antognazza, 563–90. Oxford: Oxford University Press, 2014.

Huet, Pierre-Daniel. *Demonstratio evangelica*. Paris: Daniel Hortemels, 1679.

———. 'An Extract of a Letter from a Learned French Gentleman, Concerning a Way of Making Sea-water Sweet'. *Philosophical Transactions of the Royal Society* 5 (1670): 2048.

———. *Mémoires*, ed. Philippe-Joseph Salazar. Toulouse: Société de Littératures Classiques, 1993.

Huygens, Christaan. *Oeuvres complètes*, ed. Société Hollandaise des Sciences. 22 vols. The Hague: Nijhoff, 1888–1955

Jalobeanu, D., and Wolfe, C. T., eds. *Encyclopedia of Early Modern Philosophy and the Sciences*. Cham, Switz.: Springer, 2021.

Johns, Christopher. *The Impact of Leibniz's Geometric Method for the Law*. Abingdon, UK: Routledge, 2019.

Jones, Matthew L. *The Good Life in the Scientific Revolution, Descartes, Pascal, Leibniz, and the Cultivation of Virtue*. Chicago: University of Chicago Press, 2006.

———. *Reckoning with Matter: Calculating Machines, Innovation, and Thinking about Thinking from Pascal to Babbage*. Chicago: Chicago University Press, 2016.

Jurgens, Madeleine, and Johannes Orzschig. 'Korrespondenten von G. W. Leibniz: 7. Christophe Brosseau'. *Studia Leibnitiana* 16, no. 1 (1984): 102–12.

Kangro, Hans. 'Joachim Jungius und Gottfried Wilhelm Leibniz: Ein Beitrag zum geistigen Verhältnis beider Gelehrten.' *Studia Leibnitiana* 1, no. 3 (1969): 175–207.

Keller, Vera. 'Happiness and Projects between London and Vienna: Wilhelm von Schröder on the London Weavers' Riot of 1675, Workhouses, and Technological Unemployment'. *History of Political Economy* 53, no. 3 (2021): 407–23.

———. *Knowledge and the Public Interest, 1575–1725*. Cambridge: Cambridge University Press, 2015.

———. '"A Political *Fiat Lux*." Wilhem von Schroeder (1640–1688) and the Co-production of Chymical and Political Economy.' In *'Eigennutz' und 'gute Ordnung': Ökonomisierungen der Welt im 17. Jahrhundert*, ed. Sandra Richter and Guillaume Garner, 353–78. Wiesbaden: Harrassowitz, 2016.

Keller, Vera, and Ted McCormick. 'Towards a History of Projects.' *Early Science and Medicine* 21 (2016): 423–44.

Kempe, Michael, ed. *Der Philosoph im U-Boot: praktische Wissenschaft und Technik im Kontext von Gottfried Wilhelm Leibniz*. Hanover: Gottfried-Wilhelm-Leibniz-Bibliothek, 2015.

———. 'Dr. Leibniz, oder wie ich lernte, die Bombe zu lieben. Zum Verhältnis von Wissenschaft und Militärtechnik in Europa um 1700'. In *Der Philosoph im U-Boot: praktische Wissenschaft und Technik im Kontext von Gottfried Wilhelm Leibniz*, 113–45. Hanover: Gottfried-Wilhelm-Leibniz-Bibliothek, 2015.

———. 'In 80 Texten um die Welt. Globale Geopolitik bei G. W. Leibniz'. In *Umwelt und Weltgestaltung: Leibniz' politisches Denken in seiner Zeit*, ed. Friedrich Beiderbeck, Irene Dingel, and Wenchao Li, 255–73. Göttingen: Vandenhoeck & Ruprecht, 2015.

———. *Sept Jours dans la Vie de Leibniz*. Paris: Flammarion, 2023.

Kettering, Sharon. 'Gift-giving and Patronage in Early Modern France'. *French History* 2 (1988): 131–51.

———. *Patrons, Brokers and Clients in Seventeenth Century France*. Oxford: Oxford University Press, 1986.

Knecht, Herbert. *La logique chez Leibniz: essai sur le rationalisme baroque*. Lausanne: L'Age d'homme, 1981.

Krafft, Fritz. 'Phosphorus: From Elemental Light to Chemical Element'. *Angewandte Chemie—International Edition in English* 8 (1969): 660–71.

Krajewski, Markus, ed. *Projektemacher. Zur Produktion von Wissen in der Vorform des Scheiterns*. Berlin: Kulturverlag Kadmos, 2004.

Kuhn, Thomas. *The Structure of Scientific Revolutions*. Chicago: University of Chicago Press, 1962.

Laerke, Mogens. 'À la recherche d'un homme égal à Spinoza: G. W. Leibniz et la Demonstratio evangelica de Pierre-Daniel Huet'. *Dix-septième siècle* 232 (2006): 387–410.

———. '*Ignorantia inflat*. Leibniz, Huet and the Critique of the Cartesian Spirit'. *Leibniz Review* 23 (2013): 13–42.

———. *Les Lumières de Leibniz. Controverses avec Huet, Bayle, Regis et More*. Paris: Classiques Garnier, 2015.

Laurens, Henry. *Aux sources de l'orientalisme: la Bibliothèque Orientale de Barthélemi d'Herbelot*. Paris: G. P. Maisonneuve et Larose, 1978.

Lazardig, Jan. 'The Machine as Spectacle: Function and Admiration in Seventeenth-Century Perspectives on Machines'. In *Theatrum Scientiarum—English Edition*, vol. 2: *Instruments in Art and Science: On the Architectonics of Cultural Boundaries in the 17th Century*, ed. Jan Lazardig, Ludger Schwarte, and Helmar Schramm, 152–75, Berlin: de Gruyter, 2008.

———. '"Masque der Possibilität". Experiment und Spektakel barocker Projektmacherei'. In *Spektakulare Experimente: Praktiken der Evidenzproduktion im 17. Jahrhundert*, ed. H. Schramm, L. Schwarte, and J. Lazardig, 176–212. Berlin: de Gruyter, 2006.

———. 'Universality and Territoriality: On the Architectonic of Academic Social Life Exemplified by the *Brandenburg Universität de Völker, Wissenschaften und Künstel* (1666 / 67)'. In *Collection—Laboratory—Theater: Scenes of Knowledge in the 17th Century*, ed. Jan Lazardig, Ludger Schwarte, and Helmar Schramm, 176–98, Berlin: de Gruyter, 2008.

Lebrun, Charles-François. *Opinions . . . sur le projet de remboursement de la dette exigible en assignats forcés*. Paris: Imprimerie nationale, 1790.

Leibniz-Connection, Die: Personen- und Korrespondenz-Datenbank der Leibniz-Edition. Göttingen: Universität Göttingen, 2020. https://leibniz.uni-goettingen.de.

Lennon, Thomas. *The Plain Truth: Descartes, Huet, and Skepticism*. Leiden: Brill, 2008.

Lenzen, Wolfgang. 'Leibniz and the Calculus Ratiocinator'. In *Technology and Mathematics*, ed. Sven Ove, 47–78. Cham, Switz: Springer, 2018.

Lindenfeld, David. *The Practical Imagination: The German Sciences of State in the Nineteenth Century*. Chicago: University of Chicago Press, 1997.

Lodge, Paul, ed. *Leibniz and His Correspondents*. Cambridge: Cambridge University Press, 2004.

Look, Brandon, ed. *The Continuum Companion to Leibniz*. London: Continuum, 2011.

Lopez, Denis. 'Huet pédagogue. L'humanisme de Huet: paideia et érudition à la veille des Lumières'. In *Pierre-Daniel Huet (1630–1721): Actes du Colloque de Caen (12–13 novembre 1994)*, ed. Soûad Guellouz, 197–209. Paris: PSCL, 1994.

Lorber, Michael. *Theatrum Naturae & Artis—Johann Joachim Bechers Reformpädagogik als alchemisches Unterfangen*. Berlin: De Gruyter, 2017.

Lux, David. *Patronage and Royal Science in Seventeenth-Century France: The Académie De Physique in Caen*. Ithaca, NY: Cornell University Press, 1989.

Maber, Richard. 'Colbert and the Scholars: Menage, Huet, and the Royal Pensions of 1663'. *Seventeenth-Century French Studies* 7, no. 1 (1985): 106–14.

———. 'Knowledge as Commodity in the Republic of Letters, 1675–1700'. *Seventeenth-Century French Studies* 27 (2005): 197–208.

Mackensen, Ludolf von. 'Die Vorgeschichte und die Entstehung der 4 Spezies Rechenmaschine von Gottfried Wilhelm Leibniz nach bisher unerschlossenen Manuskripten und Zeichnungen mit einem Quellenanhang der Hauptdokumente'. PhD dissertation, University of Munich, 1968.

Malcolm, Noel, ed. *The Correspondence of Thomas Hobbes: Volume II: 1660–1679*. Oxford: Oxford University Press, 1998.

———. 'Private and Public Knowledge: Kircher, Esotericism, and the Republic of Letters'. In *Athanasius Kircher: The Last Man Who Knew Everything*, ed. Paul Findlen, 297–308. New York: Routledge, 2004.

Marras, Cristina. 'Leibniz Citizen of the Republic of Letters: Some Remarks on the Interconnection Between Language and Politics'. *Studia Leibnitiana* 43, no. 1 (2011): 54–69.

Marten, Maria, and Carola Piepenbring-Thomas. *Fogels Ordnungen: Aus dem Werkstatt des Hamburger Mediziners Martin Fogel (1634–1675)*. Frankfurt: Klostermann, 2015.

Martin, Henri-Jean. *Livre, pouvoirs et société à Paris au XVIIe siècle (1598–1701)*. Geneva: Droz, 1999.

McDonough, Jeffrey. 'Leibniz's Philosophy of Physics'. In *The Stanford Encyclopedia of Philosophy*, ed. Edward N. Zalta. 2021. https://plato.stanford.edu/archives/fall2021/entries/leibniz-physics/.

Meder, Stephan. *Der unbekannte Leibniz: die Entdeckung von Recht und Politik durch Philosophie*. Cologne: Böhlau, 2018.

Mendonça, Marta. 'Leibniz *vs.* Foucher: Is There Anything Wrong with the *Système nouveau?*' In *The Practice of Reason. Leibniz and His Controversies*, ed. Marcelo Dascal, 187–221. Amsterdam: Benjamins, 2010.

Mercer, Christia. 'The Young Leibniz and His Teachers'. In *The Young Leibniz and His Philosophy (1646–76)*, ed. Stuart Brown, 19–40. Dordrecht: Kluwer, 1999.

Mesnard, Jean. 'Les premières relations parisiennes de Christiaan Huygens'. In *Huygens et la France: Table ronde du Centre national de la recherche scientifique, Paris, 27–29 mars 1979*, ed. René Taton, 33–40. Paris: Vrin, 1981.

Meyer, Morgan. 'Les courtiers du savoir, nouveaux intermédiaires de la science'. *Hermès, la Revue* 57 (2010): 165–71.

Meyer, Paul. *Samuel Pufendorf. Ein Beitrag zur Geschichte seines Lebens*. Grimma, Ger.: Edelmann, 1894.

Moll, Konrad. '*Deus sive harmonia universalis est ultima ratio rerum*: The Conception of God in Leibniz's Early Philosophy'. In *The Young Leibniz and His Philosophy (1646–76)*, ed. Stuart Brown, 65–78. Dordrecht: Kluwer, 1999.

———. 'Von Erhard Weigel zu Christiaan Huygens, Feststellungen zu Leibnizens Bildungsweg zwischen Nürnberg, Mainz und Paris'. *Studia Leibnitiana* 14, no. 1 (1982): 56–72.

Montcher, Fabien. *Mercenaries of Knowledge. Vicente Nogueira, the Republic of Letters, and the Making of Late Renaissance Politics*. Cambridge: Cambridge University Press, 2023.

Mugnai, Massimo. '*Ars Characteristica*, Logical Calculus, and Natural Languages'. In *The Oxford Handbook of Leibniz*, ed. Maria Rosa Antognazza, 177–208. Oxford: Oxford University Press, 2014.

Müller, Kurt, and Gisela Krönert. *Leben und Werk von Gottfried Wilhelm Leibniz, Eine Chronik*. Frankfurt: Klostermann, 1969.

Müller, Kurt, Heinrich Schepers, and Wilhelm Totok, eds. *Leibniz à Paris (1672–1676)* (*Studia Leibnitiana*, Supplementa 17/18). 2 vols. Wiesbaden: Steiner, 1978.

Mulvaney, Robert. 'The Early Development of Leibniz's Concept of Justice'. *Journal of the History of Ideas* 29 (1968): 53–72.

Nadler, Steven. *The Best of All Possible Worlds: A Story of Philosophers, God, and Evil in the Age of Reason*. Princeton, NJ: Princeton University Press, 2010.

Niderst, Alain. 'Comparatisme et syncrétisme religieux de Huet'. In *Pierre-Daniel Huet (1630–1721): Actes du Colloque de Caen (12–12 novembre 1994)*, ed. Soûad Guellouz, 75–82. Paris: PSCL, 1994.

Nipperdey, Justus. '"Intelligenz" und "Staatsbrille": Das Ideal der vollkommenen Information in ökonomischen Traktaten des 17. und frühen 18. Jahrhunderts.' In *Information in der Frühen Neuzeit. Status, Bestände, Strategien*, ed. Arndt Brendecke, Markus Friedrich, and Susanne Friedrich, 277–99. Münster: LIT Verlag, 2008.

O'Hara, James. 'Leibniz, Leeuwenhoek und die Entwicklung der experimentellen Naturwissenschaft'. In *1716—Leibniz' Letztes Lebensjahr. Unbekanntes zu einem bekannten Universalgelehrten*, ed. Michael Kempe, 145–75. Hanover: Gottfried Wilhelm Leibniz Bibliothek, 2016.

———. *Leibniz's Correspondence in Science, Technology and Medicine. (1676–1701). Core Themes and Core Texts*. Leiden: Brill, 2024.

Ohnsorge, Wilhelm. 'Leibniz als Staatsbediensteter'. In *Leibniz. Sein Leben, sein Wirken, seine Welt*, ed. Carl Haase and Wilhelm Totok, 173–94. Hanover: Verlag für Literatur, 1966.

Olazo, E. de. 'Leibniz and Scepticism'. In *Scepticism from the Renaissance to the Enlightenment*, ed. R. H. Popkin and Charles B. Schmitt, 133–68. Wolfenbüttel, Ger.: Herzog August Bibliothek, 1987.

Orcibal, Jean. 'Leibniz et l'irénisme d'Antoine Arnauld'. In *Studia Leibnitiana: Supplementa 17 [and] 18—Leibniz à Paris (1672–1676)*, ed. Kurt Müller, Heinrich Schepers, and Wilhelm Totok, vol. 2, 15–20. Wiesbaden: Steiner, 1978.

Paasch, Kathrin. *Die Bibliothek des Johann Christian von Boineburg (1622–1672): Ein Beitrag zur Bibliotheksgeschichte des Polyhistorismus*. Berlin: Logos, 2003.

Palumbo, Margherita. 'Leibniz as Librarian'. In *The Oxford Handbook of Leibniz*, ed. Maria Rosa Antognazza, 609–21. Oxford: Oxford University Press, 2014.

Parkinson, G. H. R. 'Sufficient Reason and Human Freedom in the *Confessio Philosophi*'. In *The Young Leibniz and His Philosophy (1646–76)*, ed. Stuart Brown, 199–222. Dordrecht: Kluwer, 1999.

Paterson, William. *The Occasion of Scotland's Decay in Trade, with a Proper Expedient for Recovery Thereof, and the Increase of Our Wealth*. 1705.

Pelletier, Arnaud. 'Leibniz's Anti-Scepticism'. In *Scepticism in the Eighteenth Century. Enlightenment, Lumières, Aufklärung*, ed. Sébastien Charles and Plinio Junqueira Smith, 45–61. Dordrecht: Springer, 2012.

Perrault, Claude. 'Des sens exterieurs' In *Oeuvres diverses de physique et mechanique*, ed. Claude Perrault and Pierre Perrault, vol. 2, 517–95. Amsterdam: Jean Frédéric Bernard, 1727.

———. 'La mécanique des animaux'. In *Oeuvres diverses de physique et mechanique*, ed. Claude Perrault and Pierre Perrault, vol. 2, 329–409. Amsterdam: Jean Frédéric Bernard, 1727.

Perrault, Claude, and Pierre Perrault. *Oeuvres diverses de physique et mechanique*. 2 vols. Amsterdam: Jean Frédéric Bernard, 1727.

Peters, Hermann. 'Leibniz als Chemiker'. *Archiv für die Geschichte der Naturwissenschaften und der Technik* 7 (1916): 85–86.

Piro, Francesco. 'The Sellers of a Sweet Powder. Leibniz on Letters'. https://www.academia.edu/460136/The_sellers_of_a_sweet_powder_Leibniz_on_Letters. 1999.

Pocock, J. G. A. *The Machiavellian Moment: Florentine Political Thought and the Atlantic Republican Tradition*. Princeton, NJ: Princeton University Press, 1975.

Pococke, Edward. *Theological Works . . . to Which Is Prefixed, an Account of His Life and Writings by L. Twells*, ed. Leonard Twells. 2 vols. London: R. Gosling, 1740.

Pomian, Krzysztof. 'De la lettre au périodique: la circulation des informations dans les milieu des historiens au XVIIe siècle'. *Organon* 10 (1974): 25–43.

———. 'République des lettres: idée utopique et réalité vécue'. *Le Débat* 130 (2004): 154–70.

Popkin, R. H. 'Leibniz and the French Skeptics'. *Revue internationale de philosophie* 20 (1966): 228–48.

Preyat, Fabrice. *Le Petit concile de Bossuet et la Christianisation des Mœurs*. Berlin: LitVerlag, 2007.

Principe, Lawrence. *The Transmutations of Chymistry*. Chicago: University of Chicago Press, 2020.

Prinzler, Heinz. 'Aus der Geschichte der Entdeckung des Phosphors'. *Phosphorus, Sulfur, and Silicon and the Related Elements* 78 (1993): 1–13.

Quintan, Paul. 'La Raison, la Certitude, la Foi: quelques sur les préliminaires de l'acte de foi selon Huet'. In *Pierre-Daniel Huet (1630–1721): Actes du Colloque de Caen (12–13 novembre 1994)*, ed. Soûad Guellouz, 83–97. Paris: PSCL, 1994.

———. 'Le statut de l'apologétique chrétienne, le caractère raisonnable de la foi'. In *Pierre-Daniel Huet (1630–1721): Actes du Colloque de Caen (12–12 novembre 1994)*, ed. Soûad Guellouz, 197–210. Paris: PSCL, 1994.

Ramati, Ayval. 'Harmony at a Distance: Leibniz's Scientific Academies'. *Isis* 87 (1996): 430–52.

Ramus, Petrus. *Scholarum mathematicarum, libri unus et triginta*. Basel, 1569.

Ranum, Orest. *Artisans of Glory: Writers and Historical Thought in Seventeenth-Century France*. Chapel Hill: University of North Carolina Press, 1980.

Rateau, Paul. 'La "Drôle de pensée" (1675) ou quand Leibniz rêvait de faire de Paris une fête'. *Archives de Philosophie* 86, no. 2 (2023): 141–57.

Recous, Noémie. 'S'intégrer dans la République des Lettres. Le cas de Nicolas Fatio de Duillier (1681–1688)'. *Revue Historique* 677 (2016): 83–11.

Rescher, Nicholas. *On Leibniz: Expanded Edition*. Pittsburgh: University of Pittsburgh Press, 2013.

Richard, Francis. 'Le dictionnaire de d'Herbelot'. In *Istanbul et les langues orientales*, ed. Frédéric Hitzel, 79–88, Paris: L'Harmattan, 1997.

Richelet, Pierre. *Dictionnaire françois*. 2 vols. Geneva: Jean Herman Widerhold, 1680.

Riley, Patrick, ed. *Leibniz's Political Writings*. 2nd ed. Cambridge: Cambridge University Press, 1988.

Ritter, Paul. *Kritischer Katalog der Leibniz-Manuskripten*. Berlin: Akademie, 1908.

Robin, Jean-Luc. 'L'Académie des plaisirs de Leibniz, ou comment la Drôle de pensée entend promouvoir la science nouvelle'. *Seventeenth-century French Studies* 26 (2004): 17–80.

———. 'L'imaginaire scientifique au théâtre: le Tartuffe comme pratique et théorie de l'expérimentation'. *Seventeenth-century French Studies* 25 (2003): 145–56.

———. 'La théâtralisation de la conceptualité scientifique et ses enjeux'. *Biblio* 174 (2007): 331–41.

Robinet, André. *Leibniz: le meilleur des mondes par la balance de l'Europe*. Paris: PUF, 1994.
Roche, Daniel. *Les Républicains des lettres. Gens de culture et Lumière au XVIIIe siècle*. Paris: Fayard, 1988.
Rodis-Lewis, Geneviève, ed. *Lettres de Leibniz à Arnauld d'après un manuscrit inédit*. Paris: PUF, 1952.
Roger, Jacques. 'La politique intellectuelle de Colbert et l'installation de Christiaan Huygens à Paris'. In *Huygens et la France: Table ronde du Centre national de la recherche scientifique, Paris, 27-29 mars 1979*, ed. René Taton, 41–48. Paris: Vrin, 1981.
Roinila, Marku. 'G. W. Leibniz and Scientific Societies'. *Journal of Technology Management* 46 (2009): 165–79.
Rossi, Paolo. *Logic and the Art of Memory: The Quest for a Universal Language*, trans. S. Clucas. London: Continuum, 2006.
Rothkrug, Lionel. *Opposition to Louis XIV: The Political and Social Origins of the French Enlightenment*. Princeton, NJ: Princeton University Press, 1965.
Roscher, Wilhelm. 'Die österreichische Nationalökonomik unter Kaiser Leopold I'. In *Jahrbücher für Nationalökonomie und Statistik* 2, 1864.
Roux, Sophie. *L'Essai de logique de Mariotte. Archéologie des idées d'un savant ordinaire*. Paris: Classiques Garnier, 2011.
Rudolph, Hartmut. 'Scientific Organizations and Learned Societies'. In *The Oxford Handbook of Leibniz*, ed. Maria Rosa Antognazza, 543–62. Oxford: Oxford University Press, 2014.
Russell, Bertrand. *A Critical Exposition of the Philosophy of Leibniz*. Cambridge: Cambridge University Press, 1900.
Salomon-Bayet, Claire. 'Les académies scientifiques: Leibniz et l'Académie royale des Sciences, 1672-1676'. In *Studia Leibnitiana: Supplementa 17 [and] 18—Leibniz à Paris (1672-1676)*, ed. Kurt Müller, Heinrich Schepers, and Wilhelm Totok, 155–70. Wiesbaden: Steiner, 1978.
Saring, Hans. 'Crafft, Daniel'. In *Neue Deutsche Biographie*, vol. 3, 387. Berlin: Duncker & Humbolt, 1957. https://www.deutsche-biographie.de/pnd122899687.html.
Schaffer, Simon. 'The Show That Never Ends: Perpetual Motion in the Early Eighteenth Century'. *British Journal for the History of Science* 28, no. 2 (1995): 157–89.
Scheel, Günter. 'Hermann Conring als historisch-politischer Ratgeber'. In *Hermann Conring (1606-1681): Beiträge zu Leben und Werk*, ed. Michael Stolleis, 271–302. Berlin: Duncker & Humbolt, 1983.
Schepers, Heinrich. 'Demonstrationes Catholicae—Leibniz' großer Plan: Ein rationales Friedensprojekt für Europa'. In *Pluralität der Perspektiven und Einheit der Wahrheit im Werk von G. W. Leibniz: Beiträge zu seinem philosophischen, theologischen und politischen Denken*, ed. Friedrich Beiderbeck and Stephan Waldhoff, 3–15. Berlin: Akademie Verlag, 2011.
Schmaltz, Tad. *Radical Cartesianism: The French Reception of Descartes*. Cambridge: Cambridge University Press, 2002.
Schmidt, Georg. *Geschichte des Alten Reiches: Staat und Nation in der Frühen Neuzeit, 1495-1806*. Munich: Beck, 1999.
Schnath, Georg. *Geschichte Hannovers im Zeitalter der neunten Kur und der englischen Sukzession 1674-1714*. Hildesheim, Ger.: A. Lax, 1976.

Schneiders, Werner. 'Respublic optima: Zur metaphysischen und moralischen Fundierung der Politik bei Leibniz'. *Studia Leibnitiana* 9, no. 1 (1977): 1–26.

———. 'Sozietätspläne und Sozialutopie bei Leibniz'. *Studia Leibnitiana* 7, no. 1 (1975): 58–80.

Shapin, Steven. 'The House of Experiment in Seventeenth Century England'. *Isis* 79 (1988): 373–404.

———. 'The Mind Is Its Own Place: Science and Solitude in Seventeenth-Century England'. *Science in Context* 4 (1991): 191–218.

———. *Never Pure: Historical Studies of Science as If It Was Produced by People with Bodies, Situated in Time, Space, Culture, and Society, and Struggling for Credibility and Authority*. Baltimore: Johns Hopkins University Press, 2010.

———. 'Of Gods and Kings: Natural Philosophy and Politics in the Lebniz–Clarke Disputes'. *Isis* 72 (1981): 187–215.

———. 'A Scholar and a Gentleman: The Problematic Identity of the Scientific Practitioner in Early Modern England'. *History of Science* 29 (1991): 279–327.

Shapin, Steven, and Schaffer, Simon. *Leviathan and the Air-pump: Hobbes, Boyle and the Experimental Life*. Princeton, NJ: Princeton University Press, 1986.

Shelford, April. 'Of Sceptres and Censors: Biblical Interpretation and Censorship in Seventeenth-Century France'. *French History* 20 (2006): 161–81.

———. 'Thinking Geometrically in Pierre-Daniel Huet's "Demonstratio evangelica" (1679)'. *Journal of the History of Ideas* 63 (2002): 599–617.

———. *Transforming the Republic of Letters: Pierre-Daniel Huet and European Intellectual Life, 1650–1720*. Rochester, NY: University of Rochester Press, 2007.

Slack, Paul. *From Reformation to Improvement: Public Welfare in Early Modern England*. Oxford: Oxford University Press, 2014.

Sleigh, Robert. *Leibniz and Arnauld: A Commentary on Their Correspondence*. New Haven, CT: Yale University Press, 1990.

Smith, Pamela. *The Business of Alchemy: Science and Culture in the Holy Roman Empire*. Princeton, NJ: Princeton University Press, 1994.

Snyder, Jon. *Dissimulation and the Culture of Secrecy in Early Modern Europe*. Berkeley: University of California Press, 2009.

Soll, Jacob. 'The Antiquary and the Information State: Colbert's Archives, Secret Histories, and the Affair of the *Régale*, 1663–1682'. *French Historical Studies* 31 (2008): 3–28.

———. *The Information Master: Jean-Baptiste Colbert's Secret State Intelligence System*. Ann Arbor: University of Michigan Press, 2009.

———. *Publishing 'The Prince': History, Reading, and the Birth of Political Criticism*. Ann Arbor: University of Michigan Press, 2005.

Solomon, Howard. *Public Welfare, Science, and Propaganda in Seventeenth-Century France: The Innovations of Théophraste Renaudot*. Princeton, NJ: Princeton University Press, 1972.

Soninno, Paul. *Louis XIV and the Origins of the Dutch War*. Cambridge: Cambridge University Press, 1988.

Stanitzek, Georg. 'Projector' (2015). In *Encyclopedia of Early Modern History Online*, ed. Graeme Dunphy and Andrew Gow. Brill, 2023. https://doi.org/10.1163/2352-0272_emho_COM_025965.

Stegeman, Saskia. *Patronage and Services in the Republic of Letters: The Network of Theodorus Janssonius Van Almeloveen (1657-1712)*. Amsterdam: Apa-Holland Universiteits Pers, 2005.

Stewart, Matthew. *The Courtier and the Heretic: Leibniz, Spinoza and the Fate of God in the Modern World*. New Haven, CT: Yale University Press, 2006.

Stolleis, Michael. 'Die Einheit der Wissenschaften'. In *Hermann Conring (1606-1681): Beiträge zu Leben und Werk*, ed. Michael Stolleis, 11-34. Berlin: Duncker & Humbolt, 1983.

———, ed. *Hermann Conring (1606-1681): Beiträge zu Leben und Werk*. Berlin: Duncker & Humbolt, 1983.

Strickland, Lloyd. 'Leibniz's Egyptian Plan (1671-1672): From Holy War to Ecumenism'. *Intellectual History Review* 26, no. 4 (2016): 461-76.

Stroup, Alice. 'Nicolas Hartsoeker, savant hollandais associé de l'Académie des Sciences et espion de Louis XIV'. *Cahiers d'Histoire des Sciences et Techniques* 47 (1999): 201-23.

———. 'Science, politique et conscience aux débuts de l'Académie royale des sciences.' *Revue de Synthèse* 114, no. 3 (1993): 421-53.

Tantner, Anton. 'Intelligence Offices in the Habsburg Monarchy'. In *News Networks in Early Modern Europe*, ed. Joad Raymond and Noah Moxham, 443-64. Leiden: Brill, 2016.

Taton, René, ed. *Huygens et la France: Table ronde du Centre national de la recherche scientifique, Paris, 27-29 mars 1979*. Paris: Vrin, 1981.

———. 'Huygens et l'Académie royale des sciences'. In *Huygens et la France*, ed. René Taton, 57-68. Paris: Vrin, 1981.

———. *Les origines de l'Académie royale des sciences*. Paris: Palais de la découverte, 1966.

Teich, Mikulas. 'Interdisciplinarity in J. J. Becher's Thought'. *History of European Ideas* 9 (1988): 145-60.

Thompson, Richard. *Lothar Franz von Schönborn and the Diplomacy of the Electorate of Mainz*. Dordrecht: Springer, 1973.

Tönnies, Ferdinand. 'Leibniz und Hobbes'. *Philosophische Monatshefte* 23 (1887): 557-73.

Totaro, Pina. 'On the Recently Discovered Vatican Manuscript of Spinoza's *Ethics*'. *Journal of the History of Philosophy* 51 (2013): 465-76.

Totok, Wilhelm. 'Leibniz als Wissenschaftsorganisator'. In *Leibniz. Sein Leben, sein Wirken, seine Welt*, ed. Carl Haase and Wilhelm Totok, 293-320. Hanover: Verlag für Literatur, 1966.

Troitzsch, Ulrich. 'Erfinder, Forscher und Projektemacher. Der Aufstieg der praktischen Wissenschaften'. In *Macht des Wissens. Die Entstehung der modernen Wissensgesellschaft*, ed. Richard van Dülmen and Sina Rauschenberg, 439-64. Weimar: Böhlau, 2004.

Truchet, Jacques. 'Les arts du spectacle et le triomphe de la théâtralite'. In *XVIIe siècle: Diversité et cohérence*, 403-13. Paris: Berger-Levrault, 1992.

Ueberweg, Friedrich, Max Frischeisen-Köhler, and Willy Moog. *Die Philosophie der Neuzeit bis zum Ende des XVIII Jahrhunderts*. Berlin: Mittler, 1924.

Uffenbach, Zacharias von. *Herrn Zacharias Conrad von Uffenbach Merckwürdige Reise durch Niedersachsen Holland und Engelland.* 3 vols. Ulm und Memmingen, Ger.: Johann Friedrich Gaum, 1753-54.

Utermöhlen, Gerda. 'Der Briefwechsel des Gottfried Wilhelm Leibniz–die umfangreichste Korrespondenz des 17. Jahrhunderts und der république des lettres'. In *Probleme der Briefedition*, ed. Wolfgang Frühwald et al., 87-104. Bonn: DFG, 1977.

Van Damme, Stephane. '"The World Is Too Large": Philosophical Mobility and Urban Space in Seventeenth and Eighteenth Century Paris'. *French Historical Studies* 29 (2006): 379-406.

van der Lugt, Mara. *Bayle, Jurieu, and the Dictionnaire historique et critique.* Oxford: Oxford University Press, 2016.

van Houdt, Toon, Jan Papy, Gilbert Tournoy, and Constant Matheeussen, eds. *Self-Presentation and Social Identification: The Rhetoric and Pragmatics of Letter Writing in Early Modern Times.* Leuven: Leuven University Press, 2002.

Vernière, Paul. *Spinoza et la pensée française avant la Révoliution.* Paris: PUF, 1954.

Volphilhac-Auger, Catherine. *Ad Usum Delphini: l'Antiquité au miroir du Grand Siècle.* Grenoble: ELLUG, 2000.

Voltaire (François Arouet). *Siècle de Louis XIV*, ed. Charles Louandre. Paris, 1874.

Wahl, Charlotte. '"Im tunckeln ist ein blinder so guth als ein sehender". Zu Leibniz' Beschäftigung mit Leuchtstoffen'. In *Der Philosoph im U-Boot: praktische Wissenschaft und Technik im Kontext von Gottfried Wilhelm Leibniz*, ed. Michael Kempe, 225-59. Hanover: Gottfried-Wilhelm-Leibniz-Bibliothek, 2015.

———. 'Die Gier nach Ruhm unter dem Mantel der Bescheidenheit: Verbergen und Irreführen in der Mathematik um 1700'. In *G. W. Leibniz und der Gelehrtenhabitus: Anomymität, Pseudonymität, Camouflage*, ed. Wenchao Li and Simona Noreik, 101-26. Weimar: Böhlau Verlag, 2016.

———. 'Leibniz' Beziehung nach Hamburg'. *Mitteilungen der Mathematischen Gesellschaft in Hamburg* 37 (2017): 175-202.

———. 'Naturwissenschaft und Akademiegedanke in Leibniz' Mainzer Zeit'. In *Leibniz in Mainz. Europäische Dimensionen der Mainzer Wirkungsperiode*, ed. Irene Dingel, Michael Kempe, and Wenchao Li, 209-36. Göttingen: Vandenhoeck & Ruprecht, 2019.

Wakefield, André. *The Disordered Police State: German Cameralism as Science and Practice.* Chicago: University of Chicago Press, 2009.

———. 'Leibniz and the Wind Machines'. *Osiris* 25 (2010): 171-88.

Wallmann Johannes. 'Helmstedter Theologie in Conrings Zeit'. In *Hermann Conring (1606-1681): Beiträge zu Leben und Werk*, ed. Michael Stolleis, 35-54. Berlin: Duncker & Humbolt, 1983.

Waquet, Françoise. *Le Modèle francais et l'Italie savante: conscience de soi et perception de l'autre dans la République des Lettres (1660-1750).* Rome: École française, 1989.

Waquet, Françoise, and Hans Bots, eds. *La République des Lettres.* Paris: Belin, 1997.

Watson, Richard, and Marjorie Grene. *Malebranche's First and Last Critics: Simon Foucher and Dortous de Mairan.* Carbondale: Southern Illinoius University Press, 1995.

Whitmer, Kelly. *The Halle Orphanage.* Chicago: University of Chicago Press, 2015.

Wiedeburg, Paul. *Der Junge Leibniz, das Reich und Europa*, part 2, vol. 1: *Europäische Politik*. Wiesbaden: Steiner-Verlag, 1970.

Wiener, Philip. 'Leibniz's Project of a Public Exhibition of Scientific Inventions'. *Journal of the History of Ideas* 1 (1940): 232–40.

Wilson, Catherine. 'Leibniz on War and Peace and the Common Good'. In *Für unser Glück und das Glück anderer*, ed. Wenchao Li, Helena Iwasinski, and Simona Noreik, vol. 1, 33–62. Hildesheim, Ger.: Olms, 2016.

Wright, J. P. 'The Embodied Soul in Seventeenth-Century French Medicine'. *Canadian Bulletin of Medical History* 8 (1991): 21–42.

Youssef, Ahmed. *La fascination d'Egypte*. Paris: l'Harmattan, 1998.

Zedler, Johann Heinrich. *Grosses Vollständiges Universal-Lexikon*. Halle & Leipzig, 1737.

INDEX

Abulfeda: early work of, 105; *Geography*, 106; Leibniz's quest for manuscript, 107–8
Academia Naturae Curiosorum, 16
Academicism, 63
Académie de Montmor, 50, 109
Académie de Physique, 70
Académie des Sciences, 7, 12, 20, 33–34, 50, 85, 97, 102, 110, 112, 113; Leibniz's calculating machine, 116–18; Parisian, 141; smelting furnaces, 168
Académie Royale des Sciences, 95
'Academies of Games', 89, 92, 264
Academy of Sciences, Leibniz founding a German, 18–19
Account of the Present State of the Republic of Letters, translation of, 258–61
Ad Usum Delphini series, 200
Advis pour dresser une bibliothèque (Naudé), 131
Aerial Noctiluca or New Phenomena and a Process of a Factual Self-shining Substance, The (Boyle), 175
Age de Louis XIV, L' (Voltaire), 6
Albert the Great, 18
Alsted, Johann Heinrich, *Encyclopaedia*, 15
Ancient Academy, 65
Andreae, Johannes, utopian models of, 15
animism, Perrault's, 62
Anna Eleonore, Princess of Hessen-Darmstadt, 129
Antibarbarus (Nizolio), 24
anti-Cartesians, 61
Apollo, 261
Aquinas, Thomas, 257
Archimedes, 258, 266
Aristotle, 74, 82, 146, 245, 252, 256, 257, 261; history of philosophy, 101
Arnauld, Antoine: Claude and, 102; Ferrand advising Leibniz on, 103; Isaac Lemaître de Sacy and, 55; Leibniz and, 11, 35, 54–62; recommending Leibniz, 37

artisans and craftsmen, Parisian, 43–44
Art Poétique (Boileau), 2
atheism, 45, 201, 212
August, Ernst: accession of, 235; Friedrich's successor, 188, 189; Leibniz on library, 132
Augustine, 257

Bacon, Francis, 146; archetype, 15; history of philosophy, 100
Beauval, Henri Basnage de, Huet and, 71
Becher, Johann Joachim: *Foolish Wisdom and Wise Foolishness*, 174, 184; gold extraction from sand, 181–83; gold-making scheme, 138; Leibniz and, 11, 145, 152, 170, 175–85; topics of interest to, 176; on transmutation of silver to gold, 173
Bernard, Edward, Orientalist scholar, 105
Berthet, Père: Jesuit, 35; mathematician, 103, 214
Biagioli, Mario, self-fashioning of, 157–58
bibliothèque à sa phantasie. *See* library
Bibliothèque orientale (d'Herbelot), 105
Boineburg, Johann Christian von: convert to Catholicism, 55; patron of Leibniz, 2, 11, 18, 23–31, 119, 192, 237; recommending Leibniz, 34, 36–38
Boineburg, Philipp Wilhelm, Leibniz as tutor to, 38, 120
Bossuet, Jacques Bénigne: *Discours sur l'histoire universelle*, 72; Leibniz and, 5, 101, 152, 219; Leibniz and Huet, 100
Bouillau, Ismaël, mathematician, 48
Boyle, Robert: Leibniz and, 48; phosphorus sample, 175
Brand, Hennig: Leibniz's handling of, 188–89; Leibniz's impression of, 171; phosphorus and, 170–75, 220; physician, 151; project, 158; on transmutation of silver to gold, 173
Breger, Herbert: on Becher, 185; on Becher and Leibniz, 175

Brevis demonstratio erroris memorabilis Cartesii et aliorum circa legem naturae (Leibniz), 234–35
Brosseau, Christophe, 192; Douceur's claim and, 187; Leibniz with, 192–93
Brown, Stuart, on 'philosophical alliance' between Leibniz and Foucher, 64
Brunswick-Luneburg mining office, 165
Buot, Jacques, mathematician, 59
Burnett, Thomas, Huet and, 71
Bury, Emmanuel, analysing Huet, 73

Caesarinus Furstenerius (Leibniz), 134
calculating machine, Leibniz's, 108–18, 169
Campanella, Tommaso, utopian model of, 15
Capella, Martianus, 253; *De nuptiis Philologiae et Mercurii*, 75
Carcavy, Pierre de: background of, 108–9; Ferrand recommending, to Leibniz, 103–4; Leibniz and, 34, 59, 79, 162; Leibniz's calculating machine and, 108–18; translation of Leibniz to (1671), 248–51
Cardano, Girolamo, utopian model of, 15
Cartesian(s), 61: dogmatism, 63; dualism, 63; intuitionism, 85; Jansenists, 56; Leibniz refuting, 91; Leibniz rejecting, 81–82; methodology, 212–13; motion, 50; philosophy, 53–54; principles of knowledge, 64
Cartesianism, 11, 53, 54, 61, 63, 100, 212, 237; Huet countering, 73–74; Jansenist, 78; rise of, 101; threat to Christianity in, 56
Casaubon, Meric, 252
Cassini, Giovanni Domenico: Leibniz and, 34, 47; recruitment of, 96
Catholic Church, 39
Catholic doctrine of transubstantiation, 56, 57
Catholicism, 55, 239
Cavalieri, mathematical literature, 50
Chapelain, Jean: Conring and, 205; on direction of literary life in France, 98; Huet and, 70; letter to, 88, 98, 99
Characteristica geometrica (Leibniz), 213
Cherubim, Père, *La nature et presage des Cometes*, 220

Chevreuse, Charles Honoré d'Albert, duc de, 35, 117, 118; Leibniz on phosphorus, 172, 226, 229, 231–32
Christian doctrines, 100
Christianity, threat of Cartesianism to, 56
Christina (Queen) of Sweden, 71
Chrysippus, 256
Church of France, 55
citizenship of the mind, ideal of, 3
Claude, Jean: Arnauld and, 102; Calvinist minister, 56
Clerselier, Claude, Leibniz and, 35
Codex Hamburgensis, 71
Colbert, Jean-Baptiste, 42; Ferrand and, 101; grand design of commerce and policy, 43; as King's finance minister, 95; Leibniz and, 226–31, 233; Leibniz on, 111, 112, 113, 124, 223; Republic of Letters, 237, 241; strategy concerning a state Republic of Letters, 96–98; translation of letter from Leibniz to (1679), 266–68; translation of letter from Leibniz to (1682), 268–69; vision of learning, 94
Collège de Clermont, 47
Collins, John, Leibniz and, 48
Comenius, Johann Amos, design of 'house of wisdom', 15
Comiers, Claude: inventor, 216; philosopher, 195
common good, achieving, 93–94
conatus: concept of, 14; term, 57n24
Confessio Naturae (Leibniz), 57
Confessio philosophi (Leibniz), 46, 60
Conring, Hermann: Chapelain and, 98; Leibniz and, 192, 204–9, 238; letter to, 70, 71, 88; recruitment of, 96; Republic of Letters, 109
Consilium Aegyptiacum (Leibniz), 26n14, 71, 107
Consilium de maris mediterranei dominio et commerciis regi christianissimo vindicandis (Conring), 205
Consilium de Scribenda Historia Naturali, 142
Conspectus operum Aethiopicum (Vansleben), 107
consubstantiation, Lutheran doctrine of, 57
Consultatio de Naturae Cognitione, 142
Copernicus, 18

INDEX

Cordemoy, Géraud de, philosopher, 42
Corpus Juris Reconcinnatum, 14
Cotelier, Jean-Baptiste, Leibniz and, 35
Council of Trent, 42
Courtès, Huguette, 'impossible dialogue', 60
court librarian, Leibniz as, 131–34
Crafft, Johann Daniel: devotion to public good, 166–67; entrepreneurial drive, 164–65; financing of, 169–70; as informant on Becher to Leibniz, 179–81, 219; Leibniz and, 11, 114–15, 163, 238; Leibniz on, 110, 138–39; phosphorus and, 170–75
Criticism of the Bible (Simon), 219
Critique de la Recherche de la verité (Foucher), 62n53, 63, 65
Crusades and Saint Louis, Ferrand on, 101
curiosities, validity of, 238
Cusson, Jean, publisher of *Journal des Sçavans*, 41

Dacier, André, humanist, 72
Dacier, Anne, humanist, 72
Dauphin, Leibniz to (1675), translation, 254–58
De arte combinatoria (Leibniz), 114
de Bessy, Frénicle, geometer, 48
De corporum concursu (Leibniz), 212
Defoe, Daniel, on 'Projecting Age', 8
Deimerbroeck, Johann van, publication of de la Fayette's novel, 71
De jure suprematus ac legationis principium Germaniae (Leibniz), 135
de la Loubère, Simon, geometric reason, 214
De la Recherche de la Vérité (Malebranche), 2, 55, 61, 63
de la Roque, Jean Paul: *Journal des Sçavans*, 35, 191; journal editor, 222–24; Leibniz and, 194; Leibniz on phosphorus, 172
de Lauravy, Jean, on royal power in marriage, 39
Democritus, 256, 266
Demonstratio Evangelica (Huet), 72, 76–78
Demonstrationes Catholicae (Leibniz), 24, 45–46, 141
De nuptiis Philologiae et Mercurii (Capella), 75
De origini iuris Germanici (Conring), 204

de Sacy, Isaac Lemaître, Arnauld and, 55
des Billettes, Gilles Filleau, Leibniz and, 34, 115
Descartes, René: absolute intellectualism of, 78; Cartesianism and, 212; on existence of external world, 64, 65, 66; history of philosophy, 100; Leibniz and, 65; Leibniz on admiration for, 81; mathematical literature, 50, 54; physical teachings, 85; proof of God, 213
de Scudéry, Madeleine, Leibniz and, 35
Desgabets, Dom Robert, Malebranche and, 63
design, definition of, 7
de St Vincent, Grégoire, mathematical literature, 50
De summa rerum (Leibniz), 46
Dew, Nicholas, on Abulfeda's *Geography*, 106
d'Harouis, Nicolas, Ollivier and, 196
d'Herbelot, Barthélemy, *Bibliothèque orientale*, 105
Discourse on Metaphysics (Leibniz), 61
Discours sur l'histoire universelle (Bossuet), 72
Dissertatio de arte combinatoria (Leibniz), 13, 178
Douceur, Noel, on malleability of cast iron, 186–87
'Drôle de Pensée' (Leibniz). See 'Funny Thought'
ducal library, Leibniz and Brosseau building, 192–93
duc de Chevreuse
duc de Roannez, 34
du Fresne du Cange, Charles, Leibniz and, 35; Mainz's envoy in Paris, 38, 115, 119
du Fresne, Marc-Joseph Marion, 115; death of, 119
Durand, Guillaume, 257

Eckhard, Arnold, Leibniz and, 191, 213, 214
éclater, word, 29
Egyptian plan, Leibniz's, 25, 26, 27, 29, 37, 45, 94
Eisenhardt, Johann, Leibniz and, 192, 211
Elector of Mainz, Crafft and, 165–66
Elector of Saxony, Crafft and, 166
Elements (Euclid), 59
Eleonore, Anna, Hessen-Darmstadt, 129

Elers, Martin, mercantile project-maker, 161
Elizabeth (Princess), 212
Eloge (Fontenelle), 1
Elsevier, Daniel, bookseller, 133
Elsholz, Johann Sigismund, Leibniz and, 192
Eltz, Friedrich Casimir zu, Leibniz and, 130
employment, Leibniz's search for potential, 120–25
Encyclopaedia (Alsted), 15
Ens perfectissimum, Descartes's proof of, 213
entrepreneurs, as projectors, 7–9
Entretien de Philarète et d'Eugène (Leibniz), 135, 186
Ernst August, elector of Hanover: accession of, 235; Friedrich's successor, 188, 189; Leibniz on library of, 132
Ethica (Spinoza), 186, 219
Ethica and Opera Posthuma (Spinoza), 186
exaggeration, Leibniz and, 136, 145, 148
Exposition of the Catholic Doctrine (Bossuet), 152

Fabri, Honoré, mathematician, 214
Fermat, Pierre de, Carcavy and, 108–9
Ferrand, Louis: background of, 101–2; Leibniz and, 11, 45, 55, 102–4; Leibniz on calculating machine, 116–17; pursuit of Abulfeda's *Geography*, 101–8; Republic of Letters, 109, 219
Ferrarius Locupletatus (Parisino), 102
Filleau des Billettes, Gilles, Leibniz and, 34, 115
Fleischer, Tobias, as librarian, 131
Fogel, Martin, Leibniz acquiring collection of, 134
Fontenelle, Bernard de, on Leibniz, 1
Foolish Wisdom and Wise Foolishness (Becher), 174, 184
fortune, 27–28, 30
Foucher, Simon, Leibniz and, 35, 62–67, 191
François I, king of France, 74
Frederick of Güldenlow, Count Ulrik, employment offer to Leibniz, 120
Free Imperial Cities, 177
French Académie, 20. *See also* Académie des Sciences
French East India Company, 96

Friedrich, Johann (Duke): death of, 3, 139–40, 235; as employer to Leibniz, 2, 40, 121; Leibniz and, 11, 28, 45, 191; Leibniz and Colbert, 226; Leibniz and control of projects, 148–53; Leibniz and ideas for public good, 97n12; Leibniz as 'walking encyclopedia' at court of Hanover, 136–39; Leibniz confiding to, 58; Leibniz correspondence with, 122–24, 129–59; Leibniz' employment in court of, 2; Leibniz introduction to phosphorus, 171–72; Leibniz negotiating on cast iron process, 186–87; Leibniz on Arnauld, 60; Leibniz on Becher's process, 181–82; Leibniz on benefits, 238; Leibniz on Carcavy, 109; Leibniz on Crafft, 166; Leibniz reporting news to, 99; Leibniz settling in court of, 121; Leibniz's letter to, 154; parents of, 129; plans and public welfare, 141; steel production, 167; translation of Leibniz to (1679), 264–66
'Funny Thought' (Leibniz), 89–94; translation of, 261–64
Fürstenberg, Wilhelm Egon von, betrayal of empire, 40
Furstenerius, Caesarinus, Leibniz's pseudonym, 134–35

Galileo, 100, 146, 259; Carcavy and, 109; history of philosophy, 100; role in science, 158; on satellites of Jupiter, 157
Galland, Antoine, Thévenot hiring, 106
Gallois, Jean: Leibniz and, 51–52, 136, 191, 194, 199, 225–30, 232–34; mathematics, 214; proposing Leibniz to Académie, 117–18
Gamans, Père, Jesuit, 35
Gassendi, Pierre, 108; Huet and, 70
Geography (Abulfeda), Ferrand and pursuit of, 101–8
Geography (Ptolemy), 105
Georg, duke of Brunswick-Lüneburg-Calenberg, 129
Germany, Leibniz in, 125–26
Gervais, Père, Jesuit, 35
glory, Leibniz on, 85–86
God's plan, achieving common good, 93–94
gold extraction from sand, Becher's process, 181–83

INDEX [293]

Graevius, Johann Georg: Huet and, 71, 75, 75n26; Leibniz's letter to, 99; Leibniz on Becher to, 178
Grandamicus, Pater, on magnetism, 103
Gravel, Abbé de, 102n34; as intermediary for Leibniz, 104, 107, 108; peace conference invitation and, 124
Gresham College, 15
Grote, Otto, Leibniz and, 130
Gruber, Caspar, bookseller, 133
Gudius, Marquard, classical scholar, 79
Guericke, Otto von: Republic of Letters, 109; vacuum experiments, 115
Guiliemi Pacidii plus ultra, praefatio (Leibniz), 82
Guldin, mathematical literature, 50

Habbeus, Christian: Leibniz proposal to, 43, 119; Philipp and, 201; recommending Leibniz, 123; recommending Leibniz to Duke Friedrich, 130; translation of Leibniz to (1673), 253–54
Habermas, Jürgen, interpretations of public sphere, 96
Hansen, Adolf, Leibniz and, 192, 193–200, 217, 223
Hardouin, Jean, classicist, 72
Hardy, Claude, geometer and orientalist, 48
Hartzing, Peter: Leibniz collecting information on, 186; pumping plan for mining, 149
Harz mines, pumping system in, 142, 149–50
Harz mountains, 165
Harz project, 148, 151, 158
Hauerstein, Thomas Heinrich, bookseller, 133
Hertel, court librarian, 133
Hesenthaler, Magnus, Leibniz and, on Schickard manuscript, 107–8
Hessen Rheinfels, Ernst von, 145
Hevelius, Johannes, recruitment of, 96
Hippocrates, 258
Historia inventionis Phosphori (Leibniz), 170
Hobbes, Thomas: high-profile thinker, 57; Leibniz and, 14, 49; Leibniz's letter to, 99; moral teachings, 85
Hochstrasser, Tim, on Leibniz and success at court, 155

Hofmann, Joseph E., on Leibniz's mathematic work, 4n11
Holy Roman Emperor(s), 37, 204
Holy Roman Empire, 13, 135, 177; advancement of science, 16
holy war, 26
Hooke, Robert: calculating machine and, 151; Leibniz and, 48; mathematics, 214
Horb, J. H., Leibniz and, 45, 55
Huet, Pierre-Daniel: on Cartesianism, 73–74; *Demonstratio Evangelica*, 72; Leibniz and, 35, 39, 191, 194, 218, 238; Leibniz and, in defence of religion, 76–82; relation of Leibniz to, 70–75; translation of Leibniz to (1673), 251–53
Hugo, Ludolf, Leibniz and, 130
humanities, Leibniz and Friedrich, 146–47
Huygens, Christiaan: Leibniz and, 11, 47, 49–51, 191, 214, 226, 229–34, 238; Leibniz meeting, 34; Leibniz on Becher's process to, 182; proposing Leibniz to Académie, 117–18; recruitment of, 96
Hypothesis physica nova (Leibniz), 14, 48, 130, 213

inventors, as projectors, 7–9
iron processing, Crafft's, 169
Islamic theology, engagement with, 26n14
Israel, Menasseh Ben, Huet debating with, 76

Jansenists, 55, 56, 60
Jesuit order, 16, 17
Jesuits, 47, 56
Jesus Christ, 79
Johann Friedrich, duke of Hanover: death of, 3, 139–40, 235; as employer to Leibniz, 2, 40, 121; Leibniz and, 11, 28, 45, 191; Leibniz and Colbert, 226; Leibniz and control of projects, 148–53; Leibniz and ideas for public good, 97n12; Leibniz as 'walking encyclopedia' at court of Hanover, 136–39; Leibniz confiding to, 58; Leibniz correspondence with, 122–24, 129–59; Leibniz's employment in court of, 2; Leibniz introduction to phosphorus, 171–72; Leibniz negotiating on cast iron process, 186–87; Leibniz on Arnauld, 60; Leibniz on Becher's process, 181–82; Leibniz on benefits, 238; Leibniz on Carcavy, 109;

Johann Friedrich, duke of Hanover (*continued*)
Leibniz on Crafft, 166; Leibniz reporting news to, 99; Leibniz settling in court of, 121; Leibniz's letter to, 154; parents of, 129; plans and public welfare, 141; steel production, 167; translation of Leibniz to (1679), 264–66

Johann Georg II, elector of Saxony, Crafft and, 166

Johann Georg III, elector of Saxony, accession of, 203

Jones, Matthew L., on calculating machine, 4n11

Journal des Sçavans (journal), 6, 35, 41, 51, 191, 194, 222, 226

Jungius, Joachim: library of, 134; Societas Ereunetica, 16

jurisprudence, 49; philosophy and, 47

jurisprudentia rationalis, 13

Justel, Henri: Leibniz and, 34, 191, 194, 197, 214, 220; on practicalities of life, 216; salon, 71; war and progress of *Ad Usum Delphini* series, 200

Kahm, Johann Carl: Leibniz and, 130, 159; Leibniz's letters and, 130, 131; Leibniz's proposal to, 119

Kempe, Michael, on 'Funny Thought', 89

Kepler, Johannes, 18

Kochanski, Adam, Leibniz on Becher to, 180

Kornmann, medical physician, 222

Kozebue, Jakob, duke's doctor, 151

Kunckel, Johann, on phosphorus, 171

Kunst- und Werkhaus, Becher founding, 178

Laboratorium Chymicum (Becher), 171

Laerke, Mogens, on Leibniz's relationship with Spinoza, 4n11

Lambeck, Peter: as imperial librarian, 24, 179; and Leibniz on Carcavy, 109

La nature et presage des Cometes (Cherubim), 220

Lantin, Jean-Baptiste, Leibniz meeting Foucher through, 64

La Perpétuité de la Foi catholique touchant L'Eucharistie (Arnauld and Nicole), 55

Lasser, Hermann Andreas, Leibniz and, 13

lead production, projectors, 168

Lebrun, Charles-François, characterization of 'faiseurs de projets', 8–9

Leibniz, Gottfried Wilhelm: Becher and, 175–85; Brosseau with, building the ducal library, 192–93; calculating machine, 108–18, 169; calculating machine and Ollivier, 196–97; Cartesian methodology, 212–13; Cartesian philosophy, 53–54; confession of greatest faults, 36–37; Conring and, 204–9; on copy of Abulfeda manuscript, 106–8; correspondence with Duke Johann Friedrich, 129–59; correspondence with Christian Philipp, 200–204; as court librarian, 131–34; Crafft and, 164–75; defence of applied science and technology, 21–22; description of himself, 1; determination to remain in Paris, 124; devotion to public good, 166–67, 207; double standards of, 185–89; elevating the population, 17–20; in employ of Duke Friedrich, 121; existence as projector, 9; Ferrand and, 102–4; Foucher and, 62–67; Hansen and, 193–200; historical research and, 211–12; ideas for public good, 97n12, 97–98; instrumentalizing phosphorus, 225–35; inventions of, 7; learning from France for Germany, 125–26; learning in Paris, 45–52; on legal reform, 97–101; letters in German and French, 155; mathematical discoveries, 213–14; mathematics, 47–52; natural philosophy, 46–47; Oldenburg and, 201; Paris sojourn (1672–76), 2–3, 5–6, 10–12, 25–27, 40–44, 125–26; on political economy, 42–43; politics of scholarship, 12; projects and specialness of, in Germany, 139–48; as publicist and agent in Paris, 37–44; representation of, 10–11; Republic of Letters and, 3–7, 11–12; retracing philosophical progress, 100–101; rhetoric and rapport with Duke Friedrich, 153–59; role and activities in Republic of Letters, 69–75, 87, 132–34; roles and personae of, 3–4; scholarship of, 1–2, 3; scientific activity, 42; on search for potential employment, 120–25; secrecy and control of, 148–53;

INDEX [295]

self-representation of, 10; setting self up as master informer, 218–24; sharing scientific and technical curiosities, 214–17; on societies, 18–20; status in Republic of Letters, 155; universal philosophical language, 143; as 'walking encyclopedia' at court of Hanover, 136–39; widening his sphere of influence, 134–36
Leibniz to Carcavy (November 1671), translation of letter, 248–51
Leibniz to Christian Habbeus (1673), translation of letter, 253–54
Leibniz to Colbert (1679), translation of letter, 266–68
Leibniz to Colbert (1682), translation of letter, 268–69
Leibniz to Duke Johann Friedrich (1679), translation of letter, 264–66
Leibniz to Pierre-Daniel Huet (1673), translation of letter, 251–53
Leibniz to the Dauphin (1675), translation of letter, 254–58
Leidenfrost, Friedrich Wilhelm, Leibniz and, 130
Leopold (Emperor), 83, 99
Leopold I, Austrian emperor, 16, 40, 83, 99–100, 177
Lettres Provinciales (Pascal), 102
Leviathan and the Air-Pump (Shapin and Schaffer), 90
Leyser, Johann, theologian, 45
L'histoire critique du Vieux Testament (Simon), 78
library: ducal, Leibniz and Brosseau building, 192–93; as 'encyclopaedia' and 'inventory', 131–34; Leibniz's ideal, 132
Löffler, Simon, on Arnauld's indictment of Jesuit morality, 55
Logique (Arnauld), 55
Logique (Mariotte), 219
Louis XIII, (King of France), 42
Louis XIV (King of France), 22, 25, 60, 119, 135; Colbert as his finance minister, 95; Leibniz addressing, 85; Leibniz for, 234; Leibniz on patronage of, 86; Leibniz's proposal, 29, 30; Melchoir Friedrich von Schönborn and plan to, 38; model for patronage of learning, 224; Republic of Letters dedicated to, 84; rule of, 105

Louvre, 33
Lutheran doctrine of consubstantiation, 57

'Mala Franciae' (Leibniz), 44
Malebranche, Nicolas: Arnauld and, 55, 61; Desgabets and, 63; Leibniz and, 191; Oratorian, 35
Malebranche, Nicole, Oratorian, 35
Mariotte, Edme: Douceur's claim and, 187; Leibniz and, 48; philosopher, 191; physics and geometry, 215
Mars Christianissimus (Leibniz), 44
Mersenne, Marin, Carcavy and, 109
master informer, Leibniz setting self as, 218–24
mathematics, Leibniz and, 47–52, 146, 147, 213–14
Mazarin (Cardinal), 214
Ménage, Gilles: historian, 45; Huet and, 70
Mechanica sive de motu (Wallis), 48
Mecklenburg-Schwerin, Christian von (Duke), Leibniz's letters to, 121–22, 241
Medicean Stars, Galileo and, 157
Meditationes de Cognitione, Veritate, et Ideis (Leibniz), 234
Memorandum on Trade (Colbert), 95
Mercury, 84
Methodus Didactica (Becher), 178
Metternich, Lothar Friedrich von, denying Leibniz a position, 119–20
microscopes, discovery, 215
miracle, word, 30
Mohr, Georg, Leibniz and, 34
Montausier, duc de, 70; Leibniz's letter, 100, 101
More, Thomas, utopian models of, 15
Morell, André, Leibniz and, 34, 38
multiperspectivism, 86

Nachlass, 5
Naudé, Gabriel: *Advis pour dresser une bibliothèque*, 131; Huet and, 70
Neuburg, Philipp Wilhelm von, candidacy of, 25
New Discourse (Conring), 206
New Testament, Old Testament prophesy fulfillment in, 76
Nicole, Pierre: Arnauld and, 55, 59; Leibniz and, 35
Nijmegen peace conference, 135

Nizolio, Mario, *Antibarbarus*, 24
Nouveau système (Leibniz), 67
Nouveaux éléments de géométrie (Arnauld), 55, 59
Nouvelle façon d'hydromètres (Foucher), 63
Nouvelles de la République des Lettres, 233
Nova Methodus (Leibniz), 206
'Nova Methodus Discendae Docendaeque Jurisprudentiae' (Leibniz), 13
Nova Methodus pro Maximis et Minimis (Leibniz), 242
novelties, 28
Novus secundarum et ulterioris ordinis radicum in analyticis usus (Fermat), 109

Ockham, 257
Odyssey (Homer), 83
Oldenburg, Republic of Letters, 109
Oldenburg, Henry; Huet and, 71–72; Leibniz and, 34, 51, 201; phosphorus and, 172; promoting scientific activity, 48–49; Republic of Letters and, 109
Old Testament: authorship of, 77; prophesy fulfillment in New Testament, 76
Ollivier, Leibniz's arithmetic machine and, 196–97
Opera posthuma (Spinoza), 185–86
opportunities, 28–29
Ordo Caritatis, 142
Oriental Indian society, 17
Orientalism, 237
Oriental learning, Ferrand's proficiency in, 101–2
Ottoman Empire, 41, 106, 206
Ottoman threat, 105
Oudard, René, Leibniz and, 35
Ozanam, Jacques: on quadrature of circle and prime numbers, 215; self-taught geometer, 47

Pacidius Philalethi (Leibniz), 212
Paderborn, Ferdinand von, 192
Pallas Alexander, 261
Papin, Denis: Leibniz meeting, 34; pressure cooker invention, 216
Paracelsus, 18, 101, 257; Descartes and, 101
Pardies, Ignace Gaston, mathematician, 59
Paris: *honnête* sociability in, 90–91; Leibniz as publicist and agent in, 37–44; Leibniz encountering, 33–44; Leibniz's sojourn in (1672–76), 2–3, 5–6, 10–12, 25–27, 40–44, 125–26
Parisian honnête sociability, 90–91
Parisino, Michaele Baudrand, *Ferrarius Locupletatus*, 102
Paris sojourn (1672–76), Leibniz's, 2–3, 5–6, 10–12
Parmenides, 256
Pascal, mathematical literature, 50, 54
Pascal, Blaise: Leibniz on work of, 59; mathematical literature, 50, 54
Pascaline, Pascal's calculator, 109
Peace of Nijmegen, Philipp and, 203–4
Peace of the Church, 56, 60
Pell, John, mathematician, 48
Pentateuch, Simon on Mosaic authorship of, 78
Périer, Gilberte, Leibniz befriending, 34
Périer, Louis, Leibniz befriending, 34
perpetual light (*lumière perpétuelle*), phosphorus as, 172–73
Perrault, Claude: 'Des sens exterieurs', 62n50; Leibniz and, 34; rejection of Cartesian mechanism, 62
Petit Traité des Agréments (le Chevalier de Mère), 194
Phèdre (Racine), 195
Phaedo (Plato), 65
Philipp, Christian, 212; on Becher, 183–84; Leibniz and, 192, 219; Leibniz on Brand's claim, 188; Leibniz with, 200–204
philosopher, rise of figure of, 9n34
Philosopher's stone, 161, 249
philosophy, Leibniz, 46–47
phosphorus: discovery of, 226–27; instrumentalizing, 225–35; possible military applications, 230–31; production of, 10, 11
phosphorus production, Crafft and Leibniz, 170–75
Physical Treatise (Mariotte), 214
Placcius, Vincenz, Leibniz and, 192
plague outbreak, Austria, 168
Plato, 54, 256, 258; political thought, 17
Plato's Academy, philosophy of, 63
Pliny, 216, 253
political control, Leibniz's interest in, 92–93
Politischer Diskurs (Becher), 183

Pomponne, Arnauld de, minister, 31
Pomponne, Simon Arnauld de, Leibniz introduction to, 37
'Portrait of a Prince', Leibniz's, 100
Postel, Guillaume, Orientalist scholar, 105
Prestet, Jean, mathematics, 47
Prince Casimir of Hanau, 177
Prince de Condé, Leibniz and, 35
project (*projet*): Leibniz's, 139–48; term, 7
'Projecting Age', Defoe on, 8
project maker, problem of, 9n36
projector(s): ambiguity of, 9; idea of, 8–9; imaginary calculations of, 8–9n33; inventors or entrepreneurs as, 7–9; Leibniz's existence as, 9, 92; Leibniz's reputation as, 148; Leibniz's secrecy and control, 148–53; range of ventures, 161–64
projeter, verb, 9
Protestantism, 239
Protogaea sive de prima facie telluris (Leibniz), 235
Psychosophie (Becher), 178, 183
public good, 28, 28n23; Leibniz's devotion to, 166–67, 207
Pufendorf, Esaias, representative of Swedish crown, 201
Pufendorf, Samuel, Philipp and, 201
Pyrrhonism, 63
Pythagoras, 84, 256, 260

Querelle des anciens et des modernes, 11, 237

Rabel, Johann, Philipp and, 201
Ramus, Petrus, on practical mathematics, 15
Ratio corporis juris reconcinnandi (Leibniz and Lasser), 14, 206
Ratio Corporis Juris Reconcinnandi (Leibniz), 206
realism, 238
reality, speculating, 10
Real Presence, doctrine of, 56, 57
Regnaud, François, numerical species, 48
Regnauld, François, Leibniz and, 198
Reichshofsrat post, 168, 179, 180
Relatio Codicis Juris Gentium Diplomatici (Leibniz), 70
'Relation de l'état présent de la République des Lettres' (Leibniz): draft A, 84–85, 258–60; draft B, 84, 260; draft C, 84, 260; translation of, 258–61
Relations de Voyages (Thévenot), 106
religion, scholarship and history in defence of, 76–82
Republic of Letters, 33; challenge of 'old', 83; Colbert's information system, 96n6; Colbert's strategy concerning, 96–98; Conring and, 205;; Huygens and, 49; key players, 109; Leibniz and, 3–7, 11–12; Leibniz on, 218, 221; Leibniz's position in, 137; Leibniz's conception of, 87, 132–34; Leibniz's role and activities in, 69–75; Leibniz's status in, 155; translation of account of present state of, 258–61; as transnational community of scholars, 5. *See also* 'Relation de l'état présent de la République des Lettres'
respublica literaria, 23
Rheinburg, Wenzel von, gold-making scheme, 138
Roannez, Artus III Gouffier, duc de, Leibniz and, 34
Roberval, Gilles de: Académie and, 117, 118; Carcavy and, 109; death of, 117
Roehmer, Ole, Leibniz and, 34, 47
Rohault, Jacques, lectures on physics, 63
Rolamb, Anke, Leibniz and, 35
Roman Caesars, 204
Roman Church, 55
Royal Academy of Sciences, 34, 36
Royal Observatory, 116
Royal Society, London, 15, 19, 20, 34, 39, 48, 49, 72, 110, 114, 141

St Vincent, Grégoire de, mathematical literature, 50
Saring, Hans, describing Crafft, 166
Sauveur, Jacques, analysis and geometry, 48
Scaliger, Joseph, critic of Huet, 73
scepticism, 53, 67, 217; Academic, 63, 65; Justel's, 216; Leibniz's, of Becher, 182
Schaffer, Simon, on scientific discourse, 90
Schickard, Wilhelm: Thévenot and, 106; transcript of, 107, 108
Schönborn, Johann Philipp von: elector of Holy Roman Empire, 13; Leibniz's patron, 14, 18, 24, 33
Schönborn, Melchior Friedrich von, Leibniz's patron, 71, 110

Schuller, Georg Hermann: lead production, 168; Leibniz and, 191, 221; preparation of Spinoza's *Ethica* and *Opera Posthuma*, 185–86
science, evolution of, 3
scientific curiosities, 214–17
Scotus, 257
scriptures, scholarship and historical defence, 76–82
Seckendorff, Ludwig Viet von, Huet and, 71
Shapin, Steven, on scientific discourse, 90
silk production: Crafft, 169; Henry IV's introduction of, 184
Simon, Richard: on authorship of Pentateuch, 77–78; *Criticism of the Bible*, 219
sincerity, Leibniz's, 145
Siver, Heinrich, disciple of Jungius, 134
Sluse, mathematical literature, 50
smelting furnaces, Crafft's, 168
Smith, Pamela: on Becher and Leibniz, 175; on 'liminal individual', 9
social order, rational, 93
'Societät und Wirtschaft' (Leibniz), 21
Societas confessionum conciliatrix (Leibniz), 16
Societas Ereunetica, Jungius as founder of, 16
Societas Eruditorum Germaniae, 16
Societas Philadelphica (Leibniz), 16, 178
Societas Theophilorum, 142
'Society and Economy' (Societät und Wirtschaft) (1671), translation, 245–47
Socrates, 256
Spanish Netherlands, 205
Spannheim, Ezechiel, on manuscripts, 79
speculating, 10
Spinoza, Baruch: *Ethica*, 185–86; high-profile thinker, 57; *Opera Posthuma*, 185–86; posthumous works, 219; Schuller's confidant, 185–86; *Tractatus theologico-politicus*, 77, 78
Spitzel, Gottlieb, Orientalist, 77
steel production, Crafft and Leibniz, 167
Steno, Nicolas (Niels Stensen): Chapelain and, 98; Leibniz and, 191
Stensen, Neils, Leibniz and, 191
Subterranan Physica (Becher), 179
Suchay, Jean-Baptiste, philologist, 72

technical curiosities, 214–17
Théophraste Renaudot's Bureau d'Adresse, 140
Thévenot, Melchisédech: on copy of Abulfeda, 107; Leibniz and, 35; on Orientalist 'culture of curiosity', 105; *Relations de Voyages*, 106
Theaetetus (Plato), 65
'Theatre of Nature', Becher's, 178
Theoria cum Praxi (Leibniz), 165
Theoria motus abstracti (Leibniz), 14, 48, 51, 57, 98, 130
Thirty Years' War, 15, 26, 170, 176, 177, 205, 207
Thomasius, Jakob, Becher and, 178
Thoynard, Nicolas, Leibniz and, 35
Torricelli: Carcavy and, 109; mathematical literature, 50
Tractatus theologico-politicus (Spinoza), 77, 78
Traité de l'origine des romans (Huet), 71
Traité des plus belles bibliothèques (Jacob), 131
transubstantiation: Catholic doctrine of, 56, 57; Roman Church and, 55; Trentine doctrine of, 56
Treatise on Colours (Mariotte), 219
Treatise on Plants (Mariotte), 214–15
Treatise on the Perpetuity of the Faith (Arnauld), 58
Trentine doctrine of transubstantiation, 56
Trifolium Hollandicum (Becher), 182
Trinity, mystery of, 24
Triple Alliance, 25, 205 Holland, England, and Sweden, 25
Truchet, Jacques, on triumph of theatricality, 90
truth, concept of, 10
Tschirnhaus, Ehrenfried Walther von: admission to Académie, 234; Leibniz and, 34, 139, 212, 214, 219; Leibniz on Crafft, 166; mediation of, 185

Uffenbach, Zacharias von, visiting Leibniz, 239
universal philosophical language, Leibniz's plan, 143
University of Altdorf, 13
University of Helmstedt, 206, 211
University of Paris, 5

University of Rinteln, 191, 222
University of Tübingen, 107

Vagetius, Johann, disciple of Jungius, 134
Valens, Vettius, astrologer, 71
van Helmont, Descartes and, 101
Vansleben, Johann Michael, *Conspectus operum Aethiopicum*, 107
Verjus, Louis de: Brosseau and, 192; diplomat, 192
Vierort, Jakob, alchemist, 161
Virgil, 261
Voltaire, Age of Louis XIV, 6
von Hessen-Rheinfels, Landgrave Ernst, intermediary of, 58, 61
von Holten, Albert, Leibniz's letter to, 114
von Hornigk, Philipp Wilhelm, Crafft on, 168
von Koenigsmarck (Count), 193

von Rheita, Capuchin Anton, telescope, 220
Vossius, Isaac, Chapelain and, 98

Wagner, Gabriel, Huet and, 71
Wallis: mathematical literature, 50; Republic of Letters, 109
Wallis, John: Leibniz and, 48, 49; professor of geometry, 48, 71
Walter, Christian, Leibniz and, 221
Walter, Christian Albrecht, commenting on war, 200
Weigel, Erhard, Philipp and, 201
West India Company, 177
Wicquefort, Abraham de, library of, 133
Witzendorff, Hieronymus von, Leibniz and, 130

Zunner, Johann David, bookseller, 133

A NOTE ON THE TYPE

THIS BOOK has been composed in Miller, a Scotch Roman typeface designed by Matthew Carter and first released by Font Bureau in 1997. It resembles Monticello, the typeface developed for The Papers of Thomas Jefferson in the 1940s by C. H. Griffith and P. J. Conkwright and reinterpreted in digital form by Carter in 2003.

Pleasant Jefferson ("P. J.") Conkwright (1905–1986) was Typographer at Princeton University Press from 1939 to 1970. He was an acclaimed book designer and AIGA Medalist.

The ornament used throughout this book was designed by Pierre Simon Fournier (1712–1768) and was a favorite of Conkwright's, used in his design of the *Princeton University Library Chronicle*.